Intelligent Systems Reference Library

Volume 104

About this Series

The aim of this series is to publish a Reference Library, including novel advances and developments in all aspects of Intelligent Systems in an easily accessible and well structured form. The series includes reference works, handbooks, compendia, textbooks, well-structured monographs, dictionaries, and encyclopedias. It contains well integrated knowledge and current information in the field of Intelligent Systems. The series covers the theory, applications, and design methods of Intelligent Systems. Virtually all disciplines such as engineering, computer science, avionics, business, e-commerce, environment, healthcare, physics and life science are included.

More information about this series at http://www.springer.com/series/8578

Robert B. Fisher · Yun-Heh Chen-Burger
Daniela Giordano · Lynda Hardman
Fang-Pang Lin
Editors

Fish4Knowledge: Collecting and Analyzing Massive Coral Reef Fish Video Data

Editors
Robert B. Fisher
School of Informatics
University of Edinburgh
Edinburgh
UK

Yun-Heh Chen-Burger
Heriot-Watt University
Edinburgh
UK

Daniela Giordano
Universita Degli Studi Di Catania
Catania
Italy

Lynda Hardman
Centrum Wiskunde & Informatica (CWI)
Amsterdam
The Netherlands

Fang-Pang Lin
National Center of High Performance Computing
Hsinchu
Taiwan

ISSN 1868-4394 ISSN 1868-4408 (electronic)
Intelligent Systems Reference Library
ISBN 978-3-319-30206-5 ISBN 978-3-319-30208-9 (eBook)
DOI 10.1007/978-3-319-30208-9

Library of Congress Control Number: 2016932519

Printed on acid-free paper

This Springer imprint is published by Springer Nature
The registered company is Springer International Publishing AG Switzerland

To Mies, Hannah, Phoebe and Lars—with thanks for their patience

To Albert and Benjamin for their love and support

To Alberto, Robin, and Karin for their loving support and encouragement

To my parents Esin and Cengiz for their endless support

To Li-Chun and Matthew

Preface

This book was conceived near the end of the Fish4Knowledge project as a way of communicating the achievements of the project to the scientific community. Many scientific projects are very successful with journal and conference publications, but it is rare to find an extended document that gives a full overview of a project, describing not only the original contributions but also the necessary infrastructure aspects. This book belongs to the latter category—it gives a brief introduction to almost all aspects of the project in a series of 18 short chapters. This exposes a range of topics, and also a view of how all of the topics fit together into the full project. It is not a "popular science" account of the project, i.e., it does not go into the personalities, motivations, and challenges behind the project. Instead, it is a technical book describing the scientific and engineering of the project. But by keeping the chapters short, we have tried to make the content accessible to the broader scientific public, particularly for the biological and computer science communities. Each chapter cites more extensive descriptions of the content from the more than 50 technical publications arising from the project.

The idea for the Fish4Knowledge project started to germinate from discussions and visits between the Edinburgh and Taiwan partners, originally as an ecological monitoring project based on video data captured off the coast of Taiwan. Later, we saw the European Union Framework Seventh Programme call for proposals on the topic of Digital Libraries and Content. This call came at the start of the scientific community's widespread interest in "big data." As a consequence of this convergence, we conceived of a project that would combine computer vision, large datasets and databases, supercomputer processing, and intelligent information presentation methods. Clearly, the proposal was successful and the resulting project ran from October 1, 2010 through September 30, 2013. In the end, we recorded and analyzed about 90,000 hours (90 TB) of video from nine cameras off the coast of Taiwan, detecting and tracking over 1 billion fish, and recording their details in an SQL database approaching 500 GB in size. The results are analyzable and viewable by marine ecologists using the facet-based user interface that the project developed.

You might wonder if the authors are going to get rich from the royalties arising from this book? With more than ten authors to share the royalties, we thought that each author's share would hardly be more than enough to take their patient friends and family out for a nice dinner to celebrate the book. So, instead, we decided that the royalties should be donated to the FishBase project (FishBase.org) that provided us with much useful, and free, background knowledge about the species. We greatly appreciate this excellent and free resource.

I (Bob) would like to make a final, personal comment—although I was the coordinator of the Fish4Knowledge project, it was more a first amongst equals situation. There was a great deal of enthusiasm by all project members, both senior and early career researchers, and great cooperation and collaboration by all. It made it easy to coordinate the project (and this book). It was also a fun project, where the consortium meetings rotated around the different partners' locations. This included two meetings in Taiwan—great food, a bouncy boat trip to LanYu Island (one of the recording sites) and a bit of team snorkeling around the fish that we had seen so much of in the videos. It was a real pleasure working with everyone on this fun and scientifically interesting project!

Edinburgh, UK
October 2015

Robert B. Fisher
Yun-Heh Chen-Burger

Acknowledgments

First, we thank the funders of the research that is described in this book. Much of the Fish4Knowledge project was funded by the European Union Seventh Framework Programme [FP7/2007-2013] under grant agreement 257024, addressing Objective ICT-2009.4.3: Intelligent Information Management, Challenge 4: Digital Libraries and Content. Taiwan's NARL/NCHC was funded by the Taiwan National Science Council under grant agreement NSC 101-2923-I-492-002-MY2. We also thank the Taiwan Power Company, Taiwan Ocean Research Institute, and Kenting National Park for sharing the video datasets collected from their underwater observatories at Nanwan, Lanyu and Houbi Lake of Taiwan, which enabled this research. The underwater observation project was funded by Taiwan Power Company. We thank the Third Nuclear Power Plant of Taipower for logistical support. Also, we thank the marine ecologists, led by T.Y. Fan, from the National Museum of Marine Biology and Aquarium, Taiwan for camera deployments and maintenance at the research sites.

We thank National Center for High-performance Computing (NCHC), Taiwan and Academic Sinica, Taiwan, for their efforts in capturing those valuable underwater marine life videos and their tireless endeavors to combat regular typhoons and open sea conditions in maintaining the high quality of videos, to assist us achieve the best possible processing results. We extend our gratitude to the staff in the System Administration Division of NCHC, especially Yin-Yu Shig, for his great help and patience for our rule-breaking requests to use the supercomputing platform, ALPS.

The CWI team is grateful for the valuable contributions of Elvira Arslanova; Prof. Shao, and his colleagues from Academia Sinica; Prof. Fan, Dr. Hai Chiang and their colleagues from the National Museum of Marine Science & Technology of Taiwan; Dr.Ir. Nagelkerke, Dr. Tulp and their colleagues from the IMARES Research Institute of Wageningen University; Dr. Lavaleye, Dr. Duineveld and their colleagues from the Royal Institute for Sea Research (NIOZ); Tiziano Perrucci and Martin van Harmelen.

We thank the OpenVCE.net project for the building shell and the Virtual University of Edinburgh (Vue) for co-sponsoring the cost of the virtual land in Second Life and OpenSimulator which host copies of our gallery.

The University of Edinburgh team would like to thank EPCC and Iain Rae from the School of Informatics for assistance with Oracle (previously Sun) Grid Engine. Also, we greatly appreciate the help of Omer F. Rana (Cardiff University) and Rafael Tolosana-Calasanz (University of Zaragoza) who collaborated on the workflow performance evaluation.

The University of Edinburgh and University of Catania teams would like to thank the many known and unknown people who contributed to the ground truth data generation for fish detection, fish tracking, and fish species classification. Bob thanks the Institute for Advanced Studies at the University of Western Australia for the support during a sabbatical visit, which gave him space to finish the assembly of the book, and especially Mohammed Bennamoun for hosting him and all the helpful discussions.

We acknowledge the great support from the external people involved with the project, including the project's Scientific Advisory Board:

Kwang-Tsao Shao (Biodiversity Research Center, Academica Sinica)
Steffen Staab (University of Koblenz and Landau)
Konstantinos Stergiou (Aristotle University of Thessaloniki)
Monique Thonnat (INRIA)

We thank the EU Project Officer Stefano Bertolo, who was great—good suggestions, good encouragement, minimizing bureaucracy, and the EU scientific reviewers: Jenny Benois-Pineau, Anna Bosch Rue, and Rafael Garcia for their support and advice.

Finally, we thank the Springer team, and especially Prof. Lakhmi Jain, University of South Australia, for his support in developing the opportunity to publish this book with Springer.

Contents

Chapter 1
Overview of the Fish4Knowledge Project

Robert B. Fisher, Kwang-Tsao Shao and Yun-Heh Chen-Burger

Abstract This chapter introduces and gives an overview of the EC funded research project called Fish4Knowledge, which investigated 'big-data' issues arising from processing 87 thousand hours of video to detect and analyze 1.4 billion tropical coral reef fish. The chapter starts with a brief tour of the project, and then gives some of the background to the project and researchers.

1.1 Introduction

The study of marine ecosystems is vital for understanding environmental effects, such as climate change and the effects of pollution, but is extremely difficult because of the practical difficulties with obtaining large amounts of data. Undersea video data is usable but is tedious to analyze (for both raw video analysis and abstraction over massive sets of observations), and is mainly done by hand or with hand-crafted computational tools. Fish4Knowledge developed methods that allow a major increase in the ability to use video data for investigating marine ecology questions. This is achieved by: (1) Video analysis that automatically extracts information about the

R.B. Fisher (✉)
School of Informatics, University of Edinburgh,
10 Crichton St, Edinburgh EH8 9AB, UK
e-mail: rbf@inf.ed.ac.uk

K.-T. Shao
Biodiversity Research Center, Academia Sinica, No. 128, Sec. 2,
Academia Rd. Nankang, Taipei, Taiwan, ROC
e-mail: zoskt@gate.sinica.edu.tw

Y.-H. Chen-Burger
School of Mathematical & Computer Sciences, Heriot-Watt University,
Edinburgh EH14 4AS, UK
e-mail: y.j.chenburger@hw.ac.uk

R.B. Fisher et al. (eds.), *Fish4Knowledge: Collecting and Analyzing Massive Coral Reef Fish Video Data*, Intelligent Systems Reference Library 104,
DOI 10.1007/978-3-319-30208-9_1

observed marine animals, which is recorded in an observation database. (2) User interfaces that allow researchers to formulate and answer higher level questions over that database without needing specialist programming skills.

The project concept was to acquire undersea video data from 9 cameras off the coast of Taiwan in coral reef areas and to detect and track fish observed in the videos, which are then recognized according to their species. A database recording the data extracted from all processed videos was created. A user interface was developed that allows marine ecologists to assess the distribution of fish by time, date, species, and location. A typical frame from one of the videos can be seen here:

As an indication of achievement, the project recorded 524 thousand unique videos, each 10 min long, resulting in 87 thousand hours of video (91 Tb). From these, 1.4 billion individual fish instances were detected, which were tracked, resulting in 145 million trajectories. The fish in these trajectories were then analyzed by the species recognition algorithms. These algorithms were also capable of eliminating non-recognizable detections (which are mainly detections of non-fish image artifacts, such as sunlight refractions on the coral or ocean floor). After this analysis, 81 million trajectories of recognized fish remained. All detection, tracking and species recognition data was stored in an SQL database, containing about 400 Gb of processed results. The project required about 400 core-years of processing to compute all of the detections and recognitions. Improved computational efficiency (e.g., 10–50 times faster) could have been achieved by recoding the algorithms from Matlab into C/C++, but this was a research project and it was felt that the extra human effort was better spent on the research issues, rather than the efficiency issues.

This chapter gives a quick overview of the project, but interested readers can find more details in the body of the book, and even more details at the project web site: http://groups.inf.ed.ac.uk/f4k/ which includes links to project publications, datasets, and code. The fish detection, tracking, recognition and unusual behavior ground truth data are publicly available. A subset of the raw videos and the full processed results are also publicly available. The user interface is publicly available at: http://gleoncentral.nchc.org.tw/.

1.2 A Quick Tour of the Project

As described above the project was proposed as a "big data" project, where the big data comes from a set of video cameras instead of a (more common) text, image or sensor database. The project was aimed at exploring the sorts of support that big data could provide to marine ecologist researchers, as described in Chap. 2. Because one of the research partners was from Taiwan, the marine environment source for the data was also from Taiwan, in particular from coral reefs from three different locations around the coastline. The marine biological background to the project is described in Sect. 1.3.

The project acquired video data from 9 cameras at various places off the coast of Taiwan. Further information about the video capture, recording and storage can be found in Chap. 4. In the end, the project recorded 524 K unique videos, each 10 min long, resulting in 87 thousand hours of video (91 Tb). Offshore capture, local collection and upload to the central supercomputer required a substantial network of components, illustrated here:

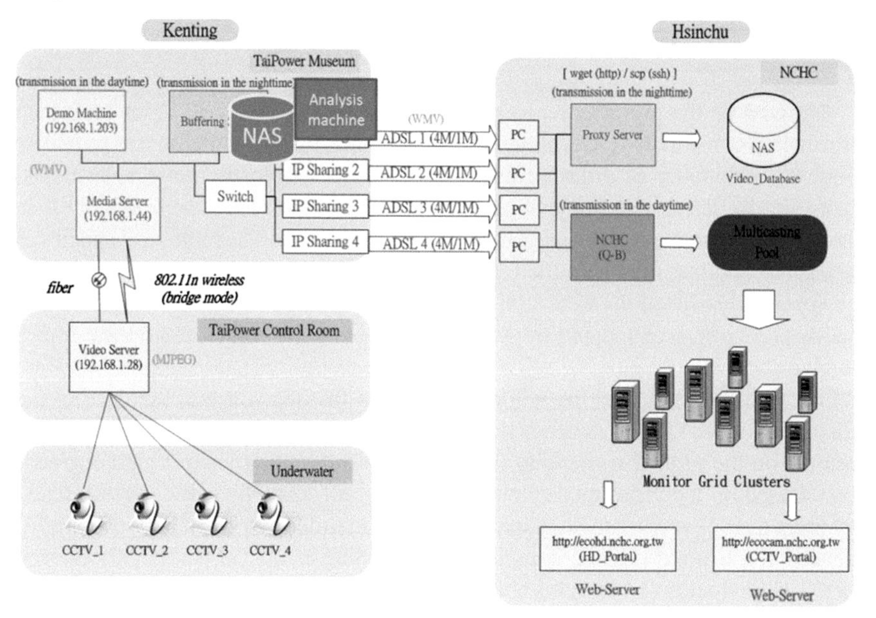

Sample images from 9 of the cameras can be seen here:

One goal of the project was to analyze the acquired video, in order to extract information of use to marine ecologists. In this project, that information was primarily about the abundance of different fish species. Accordingly, the project developed algorithms to detect novel objects appearing in the video, which were mainly fish. These fish were tracked for as long as they remained visible. The biological species of the tracked fish was then identified. Finally, all the information about the detections and species classifications were stored in a database, which is used when generating user displays of fish abundance or other information. More details about these topics are given below.

Fish detection and tracking worked best when the detector was matched to the video quality; hence, we developed a video quality classification algorithm (93 % accurate on the ground-truth data) that identified different types of video degradations. Based on the video quality classification of all 524086 videos, we estimated that there are not very many high quality videos (in order of top left to bottom right) Algae (9 %), Blurred (35 %), Complex (7 %), Coding errors (21 %), Highly blurred (12 %), High quality (14 %), plus Unknown (1 %—not shown). Sample images from the 6 main video degradations are:

Given the video quality estimate, the fish detection algorithm was then applied, resulting in 1.44 billion fish detections grouped into 145 million tracks (e.g., the same fish seen in multiple frames). The underwater scene is a difficult environment for detection because of the algae on lens, moving plants, changing lighting, caustics. Further, because of the difficulty of data capture remotely in the field, much of the image data is of size 320×240 with a low frame rate (5–8 fps). Our estimated detection rate is 81 % (combining both the true and false detections in the commonly F_1 score) and the frame-to-frame Correct Tracking Decision rate is 82 %. More information about water quality assessment, fish detection and tracking can be seen in Chaps. 9 and 10.

Examples of two detections and trackings overlaid against the background are shown here:

The detected fish were then classified into different species using a machine learning based algorithm. To train the classifier, we needed samples of fish from the different species. A ground-truthing algorithm was developed based on clustering samples in advance, and then using humans to clean up the groups and label the species. More details of this and other ground-truthing algorithms are given in Chap. 14. Although we did find 35 distinct species in the ground truthing process, many of the species only had a few examples (too few to train the classifier properly), so we ultimately recognized only the top 23 species. The top 15 species accounted for more than 97 % of the observed fish in a very unbalanced dataset. (The top species, *Dascyllus reticulatus*, accounted for more than 40 % of the observations.) After training the hierarchical classifier, we achieved 97 % classification accuracy (on the ground truth dataset of 25+ thousand fish) when considering all fish, and an average of 75 % accuracy when averaged over all 15 species. More information about the recognition algorithms is given in Chap. 11. An image of the 35 most common species and the numbers contained in the ground truth are shown below (number of detections, with number of trajectories in parentheses). Altogether, the images shown here were from 27470 fish detected from 8780 trajectories.

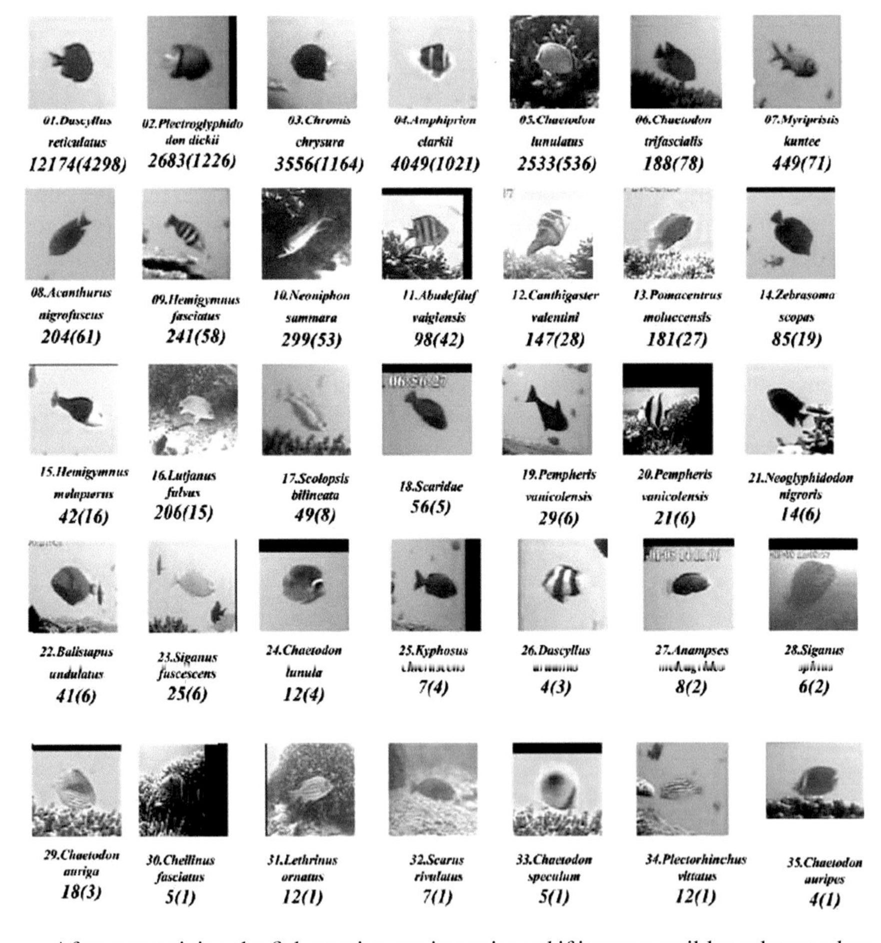

After recognizing the fish species, we investigated if it was possible to detect when fish were performing rare activities. We developed a clustering-based hierarchical classification algorithm that used the fish trajectories. Applied to *D. reticulatus*, we estimated that the algorithm was able to recognize rare behaviors with an estimated F_1 detection rate of 0.91 (a trade-off between true, false and missed detections. 1.0

is perfect performance). More details can be seen in Chap. 12. Examples of rare trajectories are here:

Most of the processing was done on a dedicated 96 core machine (left), but more was also done using up to 1000 cores on the Windrider machine (right), both shown here:

More details of the system hardware can be found in Chap. 3. An intelligent queueing process and monitor was developed to keep track of the processing of the 524 thousand videos, where each video needed to be classified, fish detected and tracked, and then recognized. One aspect of the complexity of this process was monitoring processing failures. As the system had many cores and disks, and much communication, as well as used experimental image processing software, processing tasks would fail occasionally. The queue manager was responsible for detecting this and retrying, using an SQL database to manage the tasks. More details of the intelligent workflow management process are given in Chap. 8.

A substantial software architecture was needed to support both the scale of the processing and the constant development of the research software. The system software architecture was oriented around a shared video and SQL database, where processes would extract and store data into and out of the databases. This allowed each stage of the processing to be run independently, provided only that the previous stage had been run on that video. Moreover, as the project developed, new versions of the software would be developed. It was decided to keep all processed results, so all results were tagged with the version numbers of the algorithms that produced them. In order to speed up the most common types of queries, summary tables aggregating results over hour, day, week, month, camera, video quality type and species were created. More details of the result SQL data storage configuration can be found in Chap. 5, more of the overall software architecture can be found in Chap. 6, and logical table structure in Chap. 7. Altogether, the raw SQL recording all of the detected and recognized fish totaled about 400 Gb.

Interrogating such a large and complicated result database is not easy. Although writing specialized queries in SQL is possible, we believed that the typical marine biologist, at least in the early stages of the data analysis, would need a more user-friendly tool. Accordingly, we developed a faceted user-interface that allowed a user to select different views of the full database, selecting by date, time, site/camera, species, and certainty score. The interface and examples of its use are described in Chap. 13 and a sample screenshot can be seen below, which shows, at the top, the average number of fish detections per 10 min video clip, as a function of the time of day. The different colors show the cumulative counts for the different species. The lower section of the display shows some of the selection settings that the user can make, in this case 'All' species, but only video from site NPP-3 in 2011.

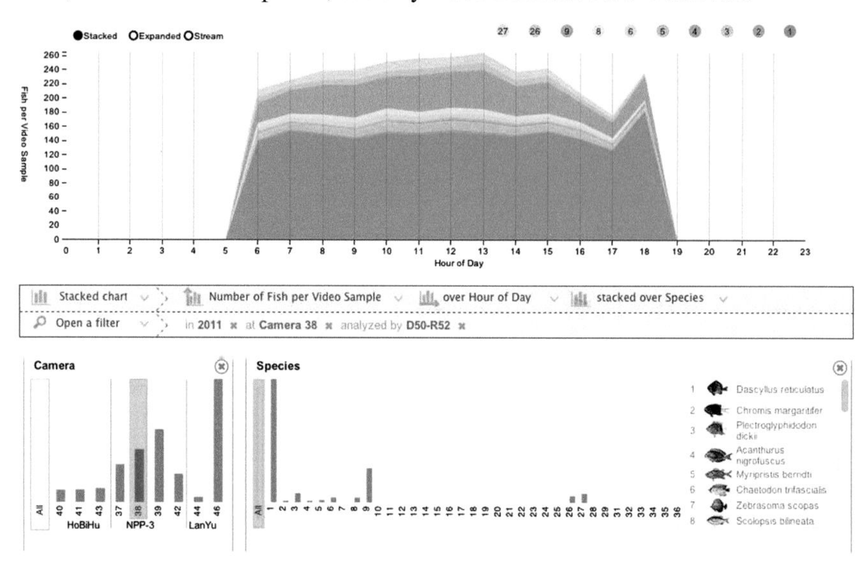

Chapter 16 explores some of the questions that can be investigated using the database, such as "Does the number of fish per video vary according to the time of day?". The plot below is aggregated over all days, cameras, qualities and species and is the median number of fish per hour, with a robust estimation of the ± 1 standard deviation error bars. The black curve is over all species, cyan: *D. reticulatus*, red: *S. bilineata*, green: *P. dickii*, and blue: *A. clarkii*. The plot shows a slight increase in the median value of the total count at dawn and dusk, but the variances are quite large. It's hard to make any conclusions for the individual species.

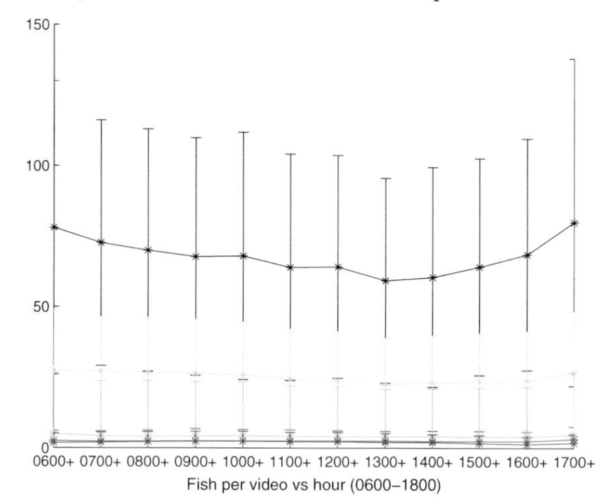

Given that the database results are a function of a long chain of processes, physical, electronic and computational, errors accumulate. As a start to addressing this 'big-data' issue, Chap. 15 looks at some approaches to uncertainty management.

To finish the book, we have a fun little chapter (Chap. 17) describing a Second Life pavilion developed for the Fish4Knowledge project, with an example screenshot below. This site was intended for an innovative approach to public-engagement.

Finally, we give some critical assessments of the project and possible future directions in Chap. 18, including a list of project publications and dissemination activities.

1.3 Background Information About the Studied Marine Environments

One of the underwater video capture systems was installed inside the intake bay of the Third Nuclear Power Plant (NPP3) (21°57′N; 120°45′E) of Taiwan Power Company (Taipower). NPP3 is located on the western side of Nanwan Bay at the southern tip of Taiwan (Fig. 1.1) and it began to operate its first generator in 1984. Because the bay is protected by a pocket-like dyke (Fig. 1.2 left), fragile coral species such as *Acropora* can grow very well and are less damaged by typhoons except for the strong ones. For example, in 1984, a severe tropical storm Wynne hit Nanwan directly which caused a change of the coral reef fish assemblage afterward (Jan et al. 2001).

More importantly, the intake area is fully protected as a no-take area where nobody can get in; only researchers can apply for permission to enter and dive on weekdays. Thus, this small (about 30,000 m^2 in area) and shallow (10–15 m in depth) bay has become the most effective Marine Protection Area and coral reef fish paradise in Taiwan. This is the other reason why we chose this safe place to install the real-time wireless video camera monitoring system. In 2003, Taipower granted a joint research project to National Museum of Marine Biology & Aquarium (NMMBA), Biodiversity Research Center of Academia Sinica (BRCAS) and National Center for High-Performance Computing (NCHC) to develop this underwater video monitoring and exhibition system. All video image data were stored at NCHC and images were accessible online via Internet broadcasting. Four real-time video frames taken from four camera heads could be watched from computer or mobile devices (Jan et al. 2007) (Fig. 1.2 right).

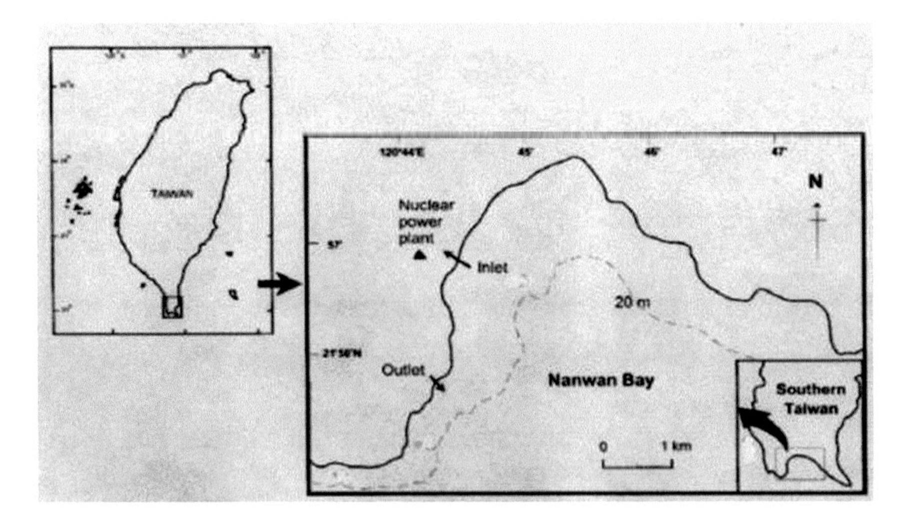

Fig. 1.1 Sketch showing the inlet of Nanwan Bay and the study site where the video camera was installed. *Inlet* site of the water intake constructed by the nuclear power plant. *Outlet* outlet of the water discharge canal

Fig. 1.2 *Left* aerial view showing the inlet bay of the 3rd Nuclear Power Plant. (Photograph credit: Chen Ming-Ming) *Right* underwater video capture system installed in front of one coral patch. (photograph credit: Dr. Tung-Yung Fan)

Fig. 1.3 Damselfishes, particularly *Chromis viridis* hiding in the branches of *Acropora* coral. (Photograph credit: Dr. Tung-Yung Fan)

Taipower also made a 3D film called "The Eden of Fishes" in 2014 to introduce the effect of marine area protection and the film is now screening at both the Southern Exhibition Hall of NPP3 at Houbihu and Northern Exhibition Hall of NPP3 at Wanli simultaneously. A total number of 230 species of reef fishes have been recorded after conducting the joint research project mentioned above. A guide book to introduce these coral reef fishes inside the intake bay was also published (Shao et al. 2015). Due to the shelter provided by prosperous *Acropora* and the continuous supply of zooplankton brought in by the cooling water, the fish assemblage here is dominated by zooplankton feeders with large aggregations of *Dascyllus* spp. and *Chromis* spp. (Fig. 1.3). The feeding composition is rather different from other fish assemblages

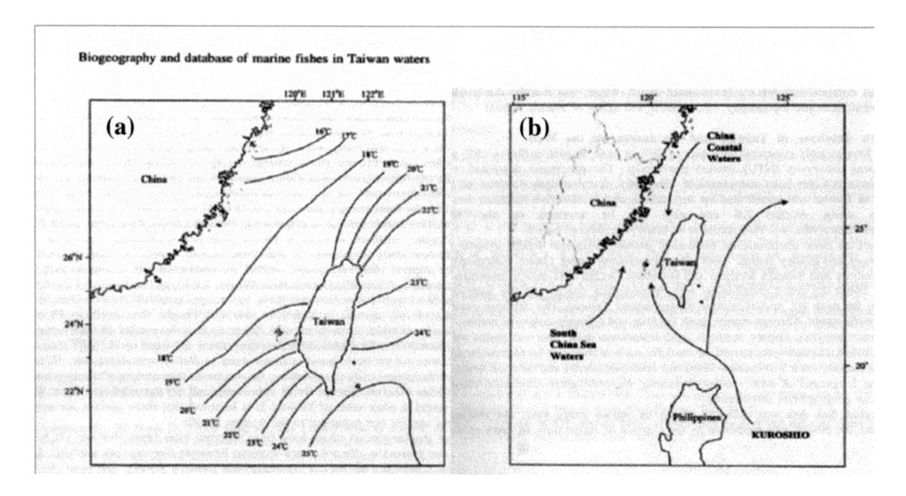

Fig. 1.4 Isothermal line of sea water temperature in winter season (**a**) and the three ocean currents around Taiwan (**b**)

at the outlet area and other places in Kenting National Park where carnivores and omnivores are dominant.

Nanwan Bay is located in Taiwan's first marine park, Kenting National Park, in southern Taiwan. The Park was established in 1982 to protect the most beautiful coral reef system in Taiwan. Kenting is located at the top of The Coral Triangle which is the hot spot of marine biodiversity in the world. The total numbers of coral species and reef fish species recorded so far have reached 300 and 1154 species, respectively (Dai 2011; Chen et al. 2010). The water temperature here is between 20–30 °C, much warmer than the water temperature in northern Taiwan (15–30 °C). The difference between northern and southern parts of Taiwan can reach 5 °C in the winter which creates different marine species composition in the northern and southern fish fauna. The reason for the temperature difference is that Taiwan is surrounded by three ocean currents: Kuroshio along the southern and eastern coasts all year round, South China Sea current on the western coast in summer and cold China coastal current along the west in the winter. As a result, the isothermal line of sea water temperature surrounding Taiwan is an oblique bisection, not horizontal like the Tropic of Cancer (Fig. 1.4) (Shao et al. 1999).

Additionally, Taiwan is located at the intersection of three Large Marine Ecosystems (or Ecogeographic Region) of the East China sea, the South China Sea and the Philippine Sea which has the "ecotone" effect and brings different kinds of marine life from the three regions into Taiwanese waters. Furthermore, the topography, substratum and water depth of Taiwanese territorial waters are quite diversified which give rise to various marine habitats and ecosystems. Although the total length of coastline in Taiwan is about 1100 Km it is 1600 Km if the island of Penghu (Pescadores), Hsiao-Liu-Chiu, Green Island and Orchid Island are included. This is not very long compared to other marine countries, but Taiwan has almost all of the different kinds of marine habitats. There are coral reef, soft bottom, hard bottom, estuary, mangrove, sea grass bed, algal reef, sandy barrier lagoon, open ocean and deep sea, and even cold seep and hydrothermal vents. The high habitat diversity creates high marine biodiversity in Taiwan. According to the "Fish Database of Taiwan" (http://fishdb.sinica.edu.tw), Taiwan possesses more than 3,100 species of fishes, about 1/10 of the world's total, even though its total land area is only 36,000 km^2 which is about 0.025 % of the world's total.

Owing to the large quantities of intake cooling water discharged from the reactors, several physical parameters (e.g., water temperature, nearshore ocean currents), chemical parameters (e.g., chlorine, radionuclide, trace elements), marine organisms (e.g., phytoplankton, zooplankton, benthic invertebrates, corals and fishes) and fishery resources in the waters around NPP3 were monitored during the plant's operation period. The National Science Committee on Problems of the Environment, Academia Sinica (SCOPE/AS) started this long-term project in 1979 (Su et al. 1989). Hung et al. (1998) concluded that the operator of the two units of NPP3 had not produced detectable effects on the marine ecosystem except for coral bleaching caused by the thermal discharge along the outlets of NPP3. Coral bleaching was serious in 1998 and 2007 when El Niño and La Niña affected the Central Western Pacific. However, in 2007, Taiwan's coral bleaching was not as serious compared to other neighborhood countries since there was an upwelling of cold water intruding into Nanwan Bay from outside (Lee et al. 1997). Also, the typhoons passing by southern Taiwan occasionally in the summer season can cause air and water temperature to go down and give corals a break (Fan T.Y., pers. comm.). Recent study on the pocilloporid corals from regions characterized by unstable temperatures, such as those exposed to periodic upwelling, display a remarkable degree of phenotypic plasticity (Mayfield et al. 2013). On the contrary, the cold water intrusion which suddenly lowers water temperature by up to 7–10 °C can kill some fishes if they cannot escape. Events like this happened in November 1989 and July 2008, and were reported in the newspaper.

On top of water temperature changes at Kenting, anthropogenic factors also affected the marine ecological environment after the National Park was established and started to attract tourists. The negative impacts come from farmland or habitat destruction along the Kenting coast and slope land to accommodate constructions such as hotels, restaurants and recreational facilities. These developments in turn led to an increase of raw sewage discharges as well as heavy sediments and suspension particles discharged directly into Nanwan Bay after rainfall. Overfishing, including on coral reef fishes, to satisfy seafood consumption of tourists is another serious problem. Both the first marine long-term ecological research project supported by National Science Council during 2001–2005 and a human impact monitoring project granted by National Park Authority since 2001 demonstrated this impact. The monthly nitrogen loading can be explained by the rainfall and numbers of tourists (Meng et al. 2008). Enriched water not only could cause seaweed to outgrow corals, but it could also trigger sea anemone outbreaks; this is exactly what happened during 2008–2010 in Kenting areas. Liu et al. (2009), using ECOPATH/ECOSIM, found that the interaction between overfishing and eutrophication was the main reason why the coral ecosystem degraded in the past 10 years.

1.4 Project Context, Objectives and Achievements

The project was designed as a next generation big data experiment, in which the data feed was live video observing undersea coral reef formations and fish (as contrasted with most previous video analysis that observes people). The justification for this project concept was that it would push the research boundaries in the ability to: (1) remotely record and store video data, (2) detect, track and recognize objects in a difficult visual environment (water and illumination disturbances, uncontrolled targets, unbalanced species composition), (3) present large amounts of extracted noisy information in a manner usable to marine ecologists, but without requiring them to be computer programmers, and (4) process and store the data acquired in a flexible and efficient manner. The project was aimed at 'big data', whereby the project would acquire an image database of about 2 billion video frames, from which we extracted 1+ billion fish (images recoverable but not explicitly stored) and their corresponding descriptors (explicitly stored). The resulting database is of the same order as the world's largest image databases (Google had 10+ billion images in 2010, Flickr had an estimated 7 billion images in 2013).

The main project objectives were technological, to develop improved methods for:

1. Capturing, storing and accessing massive amounts of video and RDF data in a timely manner.
2. Detecting moving targets (mainly fish) in the noisy undersea environment.
3. Recognizing fish species by integrating multiple 2D perspectively distorted views over time.
4. Characterizing interactions between the fish.
5. Exploiting fish, system function and system capability ontologies to convert queries into workflow sequences.
6. Helping marine ecologists explore the fish database, to explore and answer questions about the fish population.

Many of the research directions were motivated by interviews with marine ecologists, to help understand their research questions and needs.

The key achievements/discoveries/innovations of the project were:

1. **Advancing the image analysis technology for video for moving object detection**: through development of new methods for background modeling usable in both underwater and standard video, a new covariance particle filter able to handle multi-object occlusions and to track effectively objects with complex and unpredictable 3D trajectories, and a novel approach for discriminating objects of interest from the background, by integrating both objectness and motion properties.
2. **Novel methods for acquiring ground truth**: using clustering to group fish with similar appearances, to make ground truth cleaning and labeling more efficient.
3. **Novel methods for recognizing deforming similar shapes (fish)**: through development of a hierarchical classifier with a post-classification rejection filter,

which worked under variable lighting conditions, took advantage of temporal consistency, and overcame a large imbalance in the class sizes.
4. **A novel facet-based User Interface approach**: which allowed marine ecologists to select what and how to view the data, how to present the potential biases in the data, relevant ground-truth, and its impact on user trust.
5. **A novel Workflow management process**: that allowed tracking and controlling computation progress in a complex multi-processor/multi-resource computing platform where components occasionally fail.
6. **A novel interface between the datastores and the heterogeneous compute machines**: allowed large-scale task-parallel execution with considerable dataflow.
7. **Unique publicly available datasets**: A massive amount of ecological video was recorded—about 500 thousand 10 min clips at 5–8 frames per second, with about 40 % at 320 × 240 and the remainder at 640 × 480. This resulted in about 91 Tb of data, of which a 1 Tb subset is publicly available. A database of 1.4+ billion detected, tracked and recognized fish covering 23 species, which represent 99+ % of the observed fish (about 500 Gb) was collected. Finally, clips of all detected fish were compiled into a 1 Tb dataset. All datasets available from the project web site.

1.5 Project Team

The Fish4Knowledge project was undertaken by a substantial team, whose leaders were:

Yun-Heh Jessica Chen-Burger, Heriot-Watt University (United Kingdom)
Robert Fisher (coordinator), Univ. of Edinburgh (United Kingdom)
Daniela Giordano, Università di Catania (Italy)
Lynda Hardman, Centrum Wiskunde & Informatica (Netherlands)
Fang-Pang Lin, National Applied Research Laboratories (Taiwan)

The researchers and other team members were:

Elya Arslanova, Centrum Wiskunde & Informatica (Netherlands)
Emmanuelle Beauxis-Aussalet, Centrum Wiskunde & Informatica (Netherlands)
Bas Boom, Univ. of Edinburgh (UK)
Karen Chang, National Applied Research Laboratories (Taiwan)
Yi-Hsuan Chen, National Applied Research Laboratories (Taiwan)
Jia-Shin Cheng, National Applied Research Laboratories (Taiwan)
Fiona Clark, Univ. of Edinburgh (UK)
Jiyin He, Centrum Wiskunde & Informatica (Netherlands)
Isaak Kavasidis, Universita [accent ‘ on last a] di Catania (Italy)
Xuan (Phoenix) Huang, Univ. of Edinburgh (UK)
Sun-In Lin, National Applied Research Laboratories (Taiwan)
Shi-Wei Lo, National Applied Research Laboratories (Taiwan)

Gaya Nadarajan, Univ. of Edinburgh (UK)
Jacco van Ossenbruggen, Centrum Wiskunde & Informatica (Netherlands)
Kwang-Tsao Shao, Academia Sinica (Taiwan)
Roberto Di Salvo, Università di Catania (Italy)
Simone Palazzo, Università di Catania (Italy)
Yi-Haur Shiau, National Applied Research Laboratories (Taiwan)
Concetto Spampinato, Università di Catania (Italy)
Austin Tate, Univ. of Edinburgh (UK)
Kuo-Tai Tseng, National Applied Research Laboratories (Taiwan)
Cheng-Lin Yang, Informatics, University of Edinburgh (UK)

1.6 Conclusions

The Fish4Knowledge project was an exciting big-data project, novel in part because the source of the data was from video, rather than the more commonly explored text datasets from e.g., Google, Twitter, Facebook, etc. This novel data source enabled research advances in image analysis, system architecture and control, and user interfaces. It was also a fun project, involving a large cohesive team and the chance for some exciting coral reef field work.

References

Chen, J., K. Shao, R. Jan, J. Kuo, and J. Chen. 2010. *Marine Fishes in Kenting National Park*. Kenting National Park Administration. first revised edition.

Dai, C. 2011. *Eco-Tourism Map of Coral Reefs in Taiwan*, vol. 1. BookZone Publishing Company.

Hung, T., C. Huang, and K. Shao. 1998. Ecological survey of coastal water adjacent to nuclear power plants in Taiwan. *Chemistry and Ecology* 15: 129–142.

Jan, R., J. Chen, C. Lin, and K. Shao. 2001. Long-term monitoring of the coral reef fishes communities around a nuclear power plant. *Aquatic Ecology* 35: 233–243.

Jan, R.Q., Y.T. Shao, F.P. Lin, T.Y. Fan, Y.Y. Tu, H.S. Tsai, and K.T. Shao. 2007. An underwater camera system for real-time coral reef fish monitoring. *Raffles Bulletin of Zoology Supplement* 14: 273–279.

Lee, H., S. Chao, K. Fan, Y. Wang, and N. Ling. 1997. Tidallly induced upwelling in a semi-enclosed basin: Nanwan Bay. *Continental Shelf Research* 19: 671–690.

Liu, P.J., K.T. Shao, R.Q. Jan, T.Y. Fan, S.L. Wong, J.S. Hwang, J.P. Chen, C.C. Chen, and H.J. Lin. 2009. A trophic model of fringing coral reefs in Nanwan Bay, Southern Taiwan suggests overfishing. *Marine Environmental Research* 68: 106–117.

Mayfield, A.B., T. Fan, and C. Chen. 2013. Physiological acclimation to elevated temperature in a reef-building coral from an upwelling environment. *Coral Reef* 32: 909–921.

Meng, P., H. Lee, J. Wang, C. Chen, H. Lin, K. Tew, and W. Hsieh. 2008. A long-term survey on anthropogenic impacts to the water quality of coral reefs, Southern Taiwan. *Environmental Pollution* 156(1): 67–75.

Shao, K., J. Chen, and S. Wang. 1999. Biogeography and database of marine fishes in Taiwan waters. *Society For Ichthyol* 673–680.

Shao, K., J. Tsai and C. Chen. 2015. The Eden of fishes. Marine protected areas at the intake bay of the 3rd nuclear power plant in Taiwan. Taipower Company Press, Nov 2015, ISBN 978-986-04-6453-5.

Su, J., T. Hung, Y. Chiang, T. Tan, K. Chang, C. Huang, C. Huang, K. Shao, P. Huang, K. Lee, K. Fan, and S. Yeh. 1989. *An ecological and environmental survey on the waters adjacent to the southern nuclear power plant*. Taipei: SCOPE/ROC, Academia Sinica. 70.

Chapter 2
User Information Needs

Emma Beauxis-Aussalet and Lynda Hardman

Abstract Computer vision technology has been considered in marine ecology research as a innovative, promising data collection method. It contrasts with traditional practices in the information that is collected, and its inherent errors and biases. Ecology research is based on the analysis of biological characteristics (e.g., species, size, age, distribution, density, behaviors), while computer vision focuses on visual characteristics that are not necessarily related to biological concepts (e.g., contours, contrasts, color histograms, background model). It is challenging for ecologists to assess the scientific validity of surveys performed on the basis of image analysis. User information needs may not be fully addressed by image features, or may not be reliable enough. We gathered user requirements for supporting ecology research based on computer vision technologies, and identified those we can address within the Fish4Knowledge project. We particularly investigated the uncertainty inherent to computer vision technology, and the means to support users in considering uncertainty when interpreting information on fish populations. We introduce potential biases and uncertainty factors that can impact the scientific validity of interpretations drawn from computer vision results. We conclude by introducing potential approaches for providing users with evaluations of the uncertainties introduced at each information processing step.

2.1 Introduction

Requirements for the scientific study of fish population concern both (i) the kind of *measures* that need to be performed for specific studies (Table 2.1), and (ii) the *sampling method* i.e., the conditions under which measurements need to be performed (e.g., repeating measurements at timeframes, locations, or other environmental

E. Beauxis-Aussalet (✉) · L. Hardman
Centrum Wiskunde & Informatica, Science Park 123, 1098 Amsterdam, The Netherlands
e-mail: Emmanuelle.Beauxis-Aussalet@cwi.nl

L. Hardman
e-mail: Lynda.Hardman@cwi.nl

R.B. Fisher et al. (eds.), *Fish4Knowledge: Collecting and Analyzing Massive Coral Reef Fish Video Data*, Intelligent Systems Reference Library 104,
DOI 10.1007/978-3-319-30208-9_2

conditions of interest). The Fish4Knowledge project developed technologies providing *measurements* of fish populations. Provided with such technology, ecologists can study fish populations at the locations or periods of interest, applying the *sampling method* appropriate for their study.

Measurements are never perfect, whether they are performed with novel computer vision technology, or with more traditional data collection techniques. They contain errors such as misidentified species or undetected fish. The *sampling method* can be an additional source of uncertainty. For instance, too few measurements may be performed on benthic zones (i.e., ecosystems on the sea floor). The information needs and uncertainty issues related to *sampling methods* were not in the scope of the Fish4Knowledge project, and are only briefly discussed in this chapter. We refer to Cochran (1977) for further information on sampling methods.

In this chapter, we discuss the kind of *measurements* that can be performed through computer vision. We first introduce the essential measures for ecology research on fish populations (Sect. 2.2), and the data collection methods that can provide such measurements (Sect. 2.3). We detail the biases at stake with computer vision compared to other data collection methods in Sect. 2.4. Finally, Sect. 2.5 discusses the uncertainty factors involved when applying our computer vision technology. It considers uncertainty issues arising both with the computer vision algorithms, and with the in-situ application conditions (e.g., the impact of fields of view and image quality on computer vision uncertainty). It introduces the information needs for controlling the uncertainty in computer vision results.

2.2 Information Needs for Ecology Research on Fish Populations

A large variety of ecology studies rely on monitoring fish populations. For instance, monitoring fish populations takes part in studies that aim at describing ecosystems' typology (e.g., types of habitats, distributions of animal and plant species, and feeding habits i.e., *trophic* chains), evaluating differences between ecosystems under different conditions (e.g., before and after environmental events such as typhoons, or human disturbances such as construction works), or investigating specific characteristics of species (e.g., daily routines, reproduction seasons, and maturity phases). Across this variety of topics, most studies rely on similar measurements performed on fish populations, and on similar sampling methods to decide on when and where to perform the measurements.

Measuring fish populations—The most basic measures of fish population are fish counts and species identification (Gibson et al. 2001; Magurran 2004). With this information, ecologists investigate questions such as how many fish occurred in specific time periods and locations, what were their species, what is the proportion of each species in the overall population (i.e., the *species composition*), what is their distribution and density over areas, or what is the total number of species

Table 2.1 Information required for studying aspects of population dynamics, and ability of data collection methods to extract the necessary information

	Fish counts	Species identification	Behavior identification	Fish body size
Research topic				
Population dynamics	Mandatory	Mandatory	Optional	Important
Trophic systems	Mandatory	Mandatory	Important	Important
Reproduction	Mandatory	Mandatory	Important	Important
Migration	Mandatory	Mandatory	Optional	Optional
Data collection method				
Experimental fishery	+	+/ ++[a]	−	++
Commercial fishery	+	+	−	+
Diving observation	+	+	++	+
Manual image analysis	+	+	+	−/+[b]
Computer vision	+	+	−/+[c]	−/+[b]

The signs indicate whether data collection methods: − cannot supply the information, + can supply the information, ++ can supply the most precise information
[a]Fish dissection, sometimes performed after experimental fishing, is the most accurate technique for recognizing coral reef species that are visually similar
[b]Information supplied if stereoscopic vision, or calibrated distance camera-background
[c] The state-of-the-art does not fully address the wide scope of fish behavior variety

(i.e., the *species richness*). Other widely-spread information needs are fish body size and behavior identification. From fish body size, ecologists derive fish age and maturity, as well as reproductive cycles (e.g., presence of offspring). From fish behavior (e.g., mating, feeding, nursing, aggressiveness), ecologists derive fish maturity and reproductive cycles too, but also seasonal cycles and food chains (i.e., *trophic systems* describing which species feed on which species, and how often). User information needs concern the study of population dynamics in general, i.e., how species abundances evolve over time, locations or environmental conditions. They also concern the study of three main phenomena influencing population dynamics: trophic systems, reproduction and migration. Each topic of study requires specific information, as summarized in Table 2.1. These user information needs are illustrated in Table 2.2 with typical questions ecologists seek to answer with our video monitoring system.

Sampling method—All studies require a correct sampling of fish counts for the species, time periods and locations of interest. For some studies of reproduction and migration, an extensive sampling of large areas and time periods covering one to several years is necessary. Sampling methods are well-developed in the ecology domain (Cochran 1977). Requirements for appropriate sampling basically consist of collecting information for subsets of locations and time periods that are representative

Table 2.2 Typical questions ecologists seek to answer (Deliverable 2.1 (Beauxis-Aussalet et al. 2012))

Q1	How many species appear and their abundance and body size in day and night including sunrise and sunset period
Q2	How many species appear and their abundance and body size in certain period of time (day, week, month, season or year). Species composition *[set of species and relative population sizes]* change within one period
Q6	Feeding, predator-prey, territorial, reproduction (mating, spawning or nursing) or other social or interaction behavior of various species
Q7	Growth rate of certain species for a certain colony or group of observed fishes
Q8	Population size change for certain species within a single period of time
Q10	Immigration or emigration rate of one group of fish inside one monitoring station or one coral head
Q11	Solitary, pairing or schooling behavior of fishes

of the overall ecosystem. Ecology research typically considers the different components of ecosystems, e.g., the types of habitats and their proportional land coverage. Samples are often collected in each part of the ecosystems, proportionally to their geographical coverage (i.e., *stratified random sampling* in Cochran 1977). Measurements are repeated to account for their variance. Measurements' variance contributes to the interpretation of the patterns observed in the collected data. Well-founded statistical methods, based on measurements' variance, allow to compute the probability that patterns observed in the data occurred by chance, and are not representative of the actual fish populations. These statistical methods are essential for ecology research, since they support the scientific validity of conclusions drawn on fish populations.

2.3 Data Collection Techniques

Computer vision is a relatively new technique for marine ecology. Marine ecologists traditionally rely on 3 main data collection techniques: experimental fishery, commercial fishery data, and diving observations. Additionally, the use of cameras has been rapidly developing as a promising technique.

Experimental and commercial fisheries—For experimental fishery, scientific vessels are used to catch fish at specific sampling locations and time periods, with calibrated nets or fish traps. Ecologists then perform measurements which sometimes include fish dissection. For collecting data from commercial fishery, two methods exist: data can be collected by ecologists onboard commercial vessels, or by non-scientific personnel of the fishery company. The latter involves trust issues and potential biases due to the experience of the person in charge of collecting the data (Kraan et al. 2013). Commercial fishery data have the advantage of offering large coverage of marine areas, but at the disadvantage of targeting only commercial species.

Diving observations—Divers can collect further information complementing fish counts and species identification. A variety of fish behaviors can be observed, whereas fishery data can only provide information of feeding and reproductive behaviors (e.g., through fish dissection revealing the content of fish stomach or the presence of offspring and eggs). Further, cryptic and benthic species (i.e., camouflaged or living on the seabed) are better sampled since they are unlikely to be caught in fishing nets. However, diving observations cannot provide perfect data as human observers can make mistakes, e.g., depending on their diving experience, or difficulties inherent to fish species or ecosystems.

Video technologies—Images are also widely used as a means of observation. Cameras are used at fixed locations, with or without baits attracting fish. They can be oriented toward the open sea, or toward the sea floor for sampling benthic ecosystems. For the latter, calibrating a fixed distance between cameras and sea floor allows the measurement of fish body size. Stereoscopic vision, i.e., the use of pairs of cameras, is a more precise technique for estimating fish body size. Divers also use handheld cameras, sometimes moved along transects (i.e., predetermined path on the sea floor covering a representative part of the ecosystem). Recent innovations in ecology practices particularly developed on Stereo Baited Remote Underwater Video systems (stereo-BRUV), where stereoscopic vision allow the measurement of fish body size (Langlois et al. 2006). Figure 2.1 shows examples of handheld and stereo-BRUV cameras.

Ecologists visually identify the fish and their species, and interpret their behavior. Computer vision has valuable potential as a replacement of tedious, time-consuming manual image analysis. The development of this technology can aim at extracting the same scope of information as for manual image analysis. To address user information needs, the primary computer vision task is the detection of fish and their species (see Chaps. 9–11). For behavior identification, the Fish4Knowledge project is supported by recent research addressing the detection of rare and abnormal behaviors (see Chap. 12). The project also benefit from experimentation with a behavior

Fig. 2.1 Example of handheld (*left*) and stereo-BRUV cameras (*right*). Photography by Peter Southwood, licensed under Creative Commons Attribution, "Diver swimming a transect for Reef Life Survey PB164684" (*left*), "Stereo BRUVS in action at Rheeders Reef P2277038" (*right*)

identification technique based on user-defined rules, and potentially applicable for collecting ground-truth sets of fish behaviors (Spampinato et al. 2013). But further technical challenges need to be addressed since the scope of fish behaviors is very diverse. For instance, the visual features representative of fish behaviors are difficult to specify. They vary depending on species for the same behavioral functions (e.g., each species feeds differently), and they often need to be analyzed overtime in several video frames, since some behaviors are not recognizable in a single image.

Impact of video technologies on sampling methods—Estimating the area covered by the cameras' field of view is essential to the design of sampling methods, and to the analysis of the collected data (e.g., to study fish density). But estimating the area covered by a camera is a difficult task. For instance, it requires controlling the distance within which information collection is possible, or is reliable enough (e.g., for detecting small fish). Such depth of field of view varies depending on camera lens, image quality, water turbidity, and the reliability computer vision software. Estimating the area covered by cameras is more subtle when baits are used. The strength and direction of currents modify the area in which animals can sense the bait, and thus the coverage of the sampled area (Taylor et al. 2013).

The use of fixed cameras, with continuous collection of measurements on fish population, is an important paradigm shift regarding the temporal coverage of the samples. It contrasts with common data collection methods that perform measurements during limited timeframes. Their temporal coverage is limited to the selected timeframes, and the measurements performed within a timeframe are intended to represent all the species living in the environment. With the Fish4Knowledge system, the temporal coverage is very large, with fish counts continuously measured over time. More precisely, since video streams are sequenced and stored and 10-min video samples, fish counts are repeatedly measured in small units of time, i.e., every 10 min. Ecologists can not assume that measurements performed on a 10-min video sample are representative of all the species living in the ecosystem. But they can assume that species occur in videos samples at their natural frequency.

Scope of the Fish4Knowledge project—Each data collection method has its own advantages and disadvantages, and no single method fits all types of ecology research. The requirements for selecting a data collection method comprise constraints on the types of ecosystem to access, the timeframes for performing the study, the human and material resources available, the funding for acquiring and maintaining equipments, the types of information that need to be collected, the measurements' potential errors and biases, and on the uncertainties that can be tolerated. The most important information needs, as summarized in Table 2.1, are addressed by a choice of data collection techniques. Computer vision potentially provide measurements of fish body size. But the Fish4Knowledge project was not provided with equipments for measuring it (e.g., stereoscopic vision). Detecting fish behavior is supported by advances such as those presented in Chap. 12 and Spampinato et al. (2013). But the large variety of fish behavior is seldom addressed. For instance, it is challenging to detect all the diverse feeding behaviors of a small set of species. Hence the Fish4Knowledge user interface focused on addressing two main user information needs: fish counts and species identification. With this information, ecologists can study population dynamics, i.e., the evolution of fish counts over time, locations, or other environmental conditions. Migrations and reproduction cycles are possible to study, on the condition of implementing an extensive sampling of the ecosystem. The next sections detail the potential errors and biases inherent to computer vision, and the related information needs for controlling the uncertainty issues.

2.4 Potential Biases

All data collection methods are imperfect and can yield errors and biases in measurements of fish populations. Some errors can be systematic and yield biased information, e.g., some species are potentially over- and under-represented. For example, cryptic species camouflaged amongst corals are typically under-estimated in fish counts because they are more difficult to detected. Data collection methods are thus always *selective*, i.e., specific parts of ecosystems and specific species are not consistently measured and their measurements are biased. From comparative studies of data collection methods (Cappo et al. 2004; Harvey et al. 2001; Lowry et al. 2012; Trevor et al. 2000) and from interviews with ecologists, we identified nine main forms of *selectivity* at stake with the common data collection methods discussed in Sect. 2.3. Data collection methods potentially bias the counts of nine types of species: *benthic* species (i.e., living on the sea floor), *sedentary* species (i.e., living in and around the cavities of coral heads), *schooling* species (i.e., living in dense groups), small species and young fish, *cryptic* species (i.e., camouflaged in the ecosystem), *shy* species (i.e., fleeing humans and boats), *look-alike* species (i.e., visually similar species), *rare* species (i.e., occurring at low frequency), and herbivorous or carnivorous species. Table 2.3 summarizes the potential biases implied by the main data collection methods. The Fish4Knowledge project uses cameras without bait, at fixed positions and not held by divers, and that can be positioned to observe benthic zones and coral heads. These settings limit potential biases in the counts of benthic, sedentary, shy, herbivorous and carnivorous species. Yet, biases are still at stake with sedentary, schooling, cryptic, look-alike and rare species, as well as small fish.

Sedentary and schooling species—Computer vision potentially over-estimates sedentary and schooling species because they are likely to repeatedly swim in and out of the camera field of view. Hence single individuals may be repeatedly counted. For instance, with our system, we observed potential over-estimation of a sedentary species called *Dascyllus reticulatus*. Schooling species may as well be under-estimated because fish in the group occlude each other and may remain undetected.

A method to overcome such biases with sedentary and schooling species consists of counting fish appearing in only one frame of the video footage. But this method is likely to further under-estimate rare species, since the chances they appear on one single frame are very low. Further, this method disables the analysis of visual features over several frames (e.g., fish trajectories) which is necessary for recognizing fish behavior, and identifying some species (i.e., if their swimming behavior is more discriminative than their visual appearance).

Table 2.3 Main biases with species that are potentially under- or over-estimated by data collection methods

	Experimental fishery	Commercial fishery	Diving observation	Manual image analysis	Computer vision
Benthic species	–[a]	–[a]	=	=	=
Sedentary species	–	–	=	=	= /+[b]
Schooling species	=	=	–/+	–/+	–/+[b]
Small fish	–/ =[c]	–/ =[c]	–/ =[d]	–/ =[d]	–/ =[d]
Shy species	–	–	–/ =[e]	–/ =[f]	–/ =[f]
Cryptic species	–	–	=	–	–
Look-alike species	=	=	–/+	–/+	–/+
Rare species	=	–	=	=	–/ =[g]
Herbivorous and carnivorous species	–/ =[h]	=	=	–/ =[h]	–/ =[h]

The signs indicate whether parts of ecosystems are likely to be + over-represented, = neither under- nor over-represented, – under-represented.
[a]Considering that the destructive use of trawl nets is not an option
[b]Some species often swim in and out of the camera field of view, yielding over-estimated fish counts
[c]Large granularity of nets' and fish traps' mesh can let small fish slip through
[d]Small fish may not be visually detectable from a large distance
[e]Cloaking procedures can allow the observation of shy fish
[f]With handheld cameras, some species flee from divers
[g]The recognition of all rare species may not be possible due to lack of ground-truth images
[h]Baits, if used, can attract either herbivorous or carnivorous species

Small fish—Detecting small fish is difficult for all data collection methods in Table 2.1. In the case of diving observation, manual image analysis and computer vision, this type of bias is limited if observations are performed within small depths of field of view. With large depths of field of view (e.g., observing the open sea), ecologists need to consider that small fish are sampled only in a limited range around cameras or divers.

Look-alike and cryptic species—Look-alike and cryptic species are difficult to detect for computer vision software and human observers. Look-alike species can be either over- or under-estimated, and cryptic species are very likely to be under-estimated. Ecologists need to apply specific methods for studying cryptic species. These involve either divers carefully scrutinizing sea floors or coral heads, or the use of toxicants forcing the fish to leave their camouflaged position. Data collection based on imagery is not suitable for their study.

Rare species—Under-estimations of rare species is due to the inability of computer vision software to recognize species for which there are insufficient image samples to train the recognition algorithm. This can be overcome by implementing the missing species recognition features, at the cost of collecting ground-truth for these species. More information on ground-truth collection requirements are discussed in Chap. 14.

2.5 Uncertainty Factors Impacting the Potential Biases

Ecologists are concerned with the reliability of information extracted using computer vision technologies. User needs for information on uncertainty issues are illustrated in Table 2.4 with typical questions ecologists seek to answer. Considering the entire population monitoring system, potential errors and biases are not only due to computer vision software. Uncertainty is also introduced throughout the in-situ deployment of the system. For example, some cameras may receive lower lighting, and yield poor image quality and more computer vision errors. For the Fish4Knowledge system, its in-situ deployment (see Chaps. 3–8) and its computer vision software (see Chaps. 9–11), we identified the 10 uncertainty factors summarized in Table 2.5.

Uncertainty factors due to computer vision software—The computer vision algorithms developed within the Fish4Knowledge project use sets of fish examples to learn how to detect fish and species, called ground-truth. They are manually annotated by experts, and often crowd-sourced (see Chap. 14). *Ground-Truth Quality* is essential to control the errors in computer vision results. Scarcity, *Image Quality* or annotation errors in ground-truth images potentially yield error-prone computer vision software. The Fish4Knowledge system processes images in two steps, fish detection and species recognition. *Fish Detection Errors* concern undetected fish (i.e., False Negatives) and non-fish objects identified as fish (i.e., False Positives). *Species Recognition Errors* concern species misidentifications, i.e., fish recognized as a species they do not actually belong to. *Fish Detection Errors* can impact *Species Recognition Errors*, i.e., species can be attributed to non-fish objects.

Table 2.4 Typical questions ecologists seek to answer w.r.t. uncertainty issues (Deliverable 2.1 (Beauxis-Aussalet et al. 2012))

Q13	In certain area or geographical region, how many species could be identified or recognized easily and how many species are difficult. The most important diagnostic character to distinguish some similar or sibling species
Q16	Comparison of the different study results between using diving observation or underwater real time video monitoring techniques. Or the advantage and disadvantage of using this new technique
Q17	The difference of using different camera lens and different angle width
Q20	Hardware and information technique problem and the possible improvement based on current technology development and how much cost they are

Table 2.5 Uncertainty factors introduced by computer vision software or in-situ system deployment

Factor	Description
Uncertainty due to computer vision algorithms	
Ground-truth quality	Ground-truth items may be scarce, represent the wrong objects, or odd fish appearances unlikely to yield representative fish model
Fish detection errors	Some fish may be undetected, and non-fish objects may be detected as fish
Species recognition errors	Some species may not be recognized, or confused with another
Uncertainty due to in-situ system deployment	
Field of view	Cameras may observe heterogeneous, incomparable ecosystems. Fixed cameras may shift overtime (e.g., with typhoons, maintenance)
Duplicated individuals	Fish swimming back and forth are repeatedly recorded. Rates of duplication vary among *Fields of view* (e.g., open sea or coral head) and species swimming behavior (e.g., sheltering in coral head), thus producing biases
Sampling coverage	The numbers of video samples collected for each condition of interest (e.g., areas, time periods) may not be sufficient for the statistical validity of conclusions derived from software outputs
Fragmentary processing	Some videos may be yet unprocessed, missing, or unusable (e.g., encoding errors)
Uncertainty due to both computer vision algorithms and deployment conditions	
Image quality	Recording conditions may impair [the] collected information, e.g., lighting conditions, turbidity, lens fouling, resolution, frame rate and compression
Biases emerging from noise	Data processing errors may be random (noise) or systematic (bias). Biases may emerge from the combined features of data collection (*Image Quality, Field of View*) and processing (*Fish Detection* and *Species Recognition Errors*)
Uncertainty in specific output	Errors in specific computer vision results may be extrapolated from errors measured in test conditions, compared to the conditions specific to subsets of computer vision results (*Image Quality, Field of View*)

Uncertainty factors due to in-situ deployment conditions—This source of uncertainty is usually not in the scope of computer vision software evaluations. Evaluations performed in computer vision research are intended to be valid for most applications of the algorithms, and are abstracted from case-specific application conditions. However, errors and biases in computer vision outputs can be significantly influenced by environmental conditions (e.g., water turbidity lowers *Image Quality* and may increase *Fish Detection Errors*), by the placement of cameras (e.g., some *Fields of View* may over-represent sedentary species), and by computational issues during video processing (e.g., missing videos yield *Fragmentary Processing*).

The uncertainty factors introduced when deploying the system interact with each other, and with the uncertainty factors inherent to computer vision algorithms.

The *Field of View* impacts the kind of ecosystems observed by each camera, as well as the size of areas within field of view depth. Hence it influences the *Sampling Coverage*. *Field of View* also impacts the chances of *Duplicated Individuals*, e.g., observing coral heads is more likely to yield overestimation of sedentary species than observing the open sea. The *Image Quality* of recordings is impacted by both camera features (e.g., lens), and time-varying environmental conditions (e.g., lighting, turbidity, biofouling). Different *Image Quality* can yield different levels of *Fish Detection* and *Species Recognition Errors*, and thus potential *Biases Emerging from Noise*. Finally, the initial *Sampling Coverage* allowed by the camera deployment over the ecosystem can be reduced by *Fragmentary Processing* of the videos, i.e., due to unprocessed or missing videos.

2.6 Conclusion

Computer vision technology has a great potential for ecology research. It can address essential information needs, while reducing the material cost and human effort involved with common data collection techniques. However, information extracted from video is not perfect, and for scientific usage, evaluations of uncertainty must be delivered to ecologists. The Fish4Knowledge project needs to addresses the challenge of providing both information about fish populations (Table 2.1), and about the uncertainty inherent to the computer vision system. The project needs to deliver fish detection and species recognition algorithms, to provide essential information for studying fish population dynamics, and potentially, for studying fish migration, reproduction and trophic systems (i.e., food chains). The project also needs to provide evaluations of the errors in fish detection and species recognition. It supports ecologists in estimating potential biases in computer vision end-results. Ecologists need to consider other uncertainty factors, such as image quality or missing videos. Means to assess and communicate uncertainty issues to ecologists are discussed further in Chaps. 13 and 15. Integrating information about these uncertainty issues is necessary to enable the scientific usage of Fish4Knowledge technologies.

References

Cappo, M., P. Speare, and G. De'ath. 2004. Comparison of baited remote underwater video stations (bruvs) and prawn (shrimp) trawls for assessments of fish biodiversity in inter-reefal areas of the great barrier reef marine park. *Journal of Experimental Marine Biology and Ecology* 302(2): 123–152.

Cochran, W.G. 1977. *Sampling techniques*, 3rd ed. New York: Wiley.

Beauxis-Aussalet, E., L. Hardman, J. van Ossenbruggen, J. He, C. Spampinato, and B. Boom. 2012. *D2.1 user information needs; d2.2 user scenarios and implementation plan; d2.3 component-based prototypes and evaluation criteria; d6.6 public query interface*. Fish4Knowledge Project: Technical Report Del 2.1/2.2/2.3/6.6. Accessed 4 Nov 2014.

Gibson, R., M. Barnes, and R. Atkinson. 2001. Practical measures of marine biodiversity based on relatedness of species. *Oceanography and Marine Biology* 39: 207–231.
Harvey, E., D. Fletcher, and M. Shortis. 2001. A comparison of the precision and accuracy of estimates of reef-fish lengths determined visually by divers with estimates produced by a stereo-video system. *Fisheries Bulletin* 99: 63–71.
Kraan, M., S. Uhlmann, J. Steenbergen, A. Van Helmond, and L. Van Hoof. 2013. The optimal process of self-sampling in fisheries: Lessons learned in The Netherlands. *Journal of Fish Biology* 83(4): 963–973.
Langlois, T., P. Chabanet, D. Pelletier, and E. Harvey. 2006. Baited underwater video for assessing reef fish populations in marine reserves. *Fisheries Newsletter - South Pacific Commission* 118: 53.
Lowry, M., H. Folpp, M. Gregson, and I. Suthers. 2012. Comparison of baited remote underwater video (bruv) and underwater visual census (uvc) for assessment of artificial reefs in estuaries. *Journal of Experimental Marine Biology and Ecology* 416–417: 243–253.
Magurran, A. 2004. *Measuring biological diversity*. Malden: Taylor & Francis.
Spampinato, C., E. Beauxis-Aussalet, S. Palazzo, C. Beyan, J. Van Ossenbruggen, J. He, B. Boom, and X. Huang. 2013. A rule-based event detection system for real-life underwater domain. *Machine Vision and Applications* 25(1): 99–117.
Taylor, M., J. Baker, and I. Suthers. 2013. Tidal currents, sampling effort and baited remote underwater video (bruv) surveys: Are we drawing the right conclusions? *Fisheries Research* 140: 96–104.
Trevor, J., B. Russell, and C. Russell. 2000. Detection of spatial variability in relative density of fishes: Comparison of visual census, angling, and baited underwater video. *Marine Ecology Progress Series* 198: 249–260.

Chapter 3
Supercomputing Resources

Jih-Sheng Chang, Sun-In Lin, Fang-Pang Lin and Hsiu-Mei Chou

Abstract The data analysis software of the Fish4Knowledge (F4K) system is composed of several image processing modules and the execution of the processes is conducted by a workflow engine. As image processing is a data-intensive computing task, it often involves many computational processes and frequent data transfer. Traditional high performance computing platforms usually require special applications to move data from storage facilities to the computing nodes, and data transfer can become the main bottleneck of the whole process workflow as the applications scale up. In this study, we adopted the concept of cloud computing and implemented a computing platform which combined virtualization and distributed computing technologies, to support image data analysis on a large scale. A uniform job execution interface was also developed to hide the complexity of resource assembly from the users and to simplify computing resource allocation and process execution.

3.1 Introduction

The Fish4Knowledge project is designed to answer questions from marine scientists by using analysis of video data collected from long term underwater monitoring of coral reef system (Chou et al. 2009; Lin et al. 2009). The Video/Image Processing

J.-S. Chang · S.-I. Lin
National Applied Research Laboratories #22, Keyuan Road,
Central Taiwan Science Park, Taichung 407, Taiwan
e-mail: jihsheng.chang@gmail.com

S.-I. Lin
e-mail: lsi@nchc.narl.org.tw

F.-P. Lin (✉) · H.-M. Chou
National Applied Research Laboratories #7, 6th RD Road,
Science Park, Hsinchu 300, Taiwan
e-mail: c00fpl00@nchc.narl.org.tw

H.-M. Chou
e-mail: hmchou@nchc.narl.org.tw

R.B. Fisher et al. (eds.), *Fish4Knowledge: Collecting and Analyzing Massive Coral Reef Fish Video Data*, Intelligent Systems Reference Library 104,
DOI 10.1007/978-3-319-30208-9_3

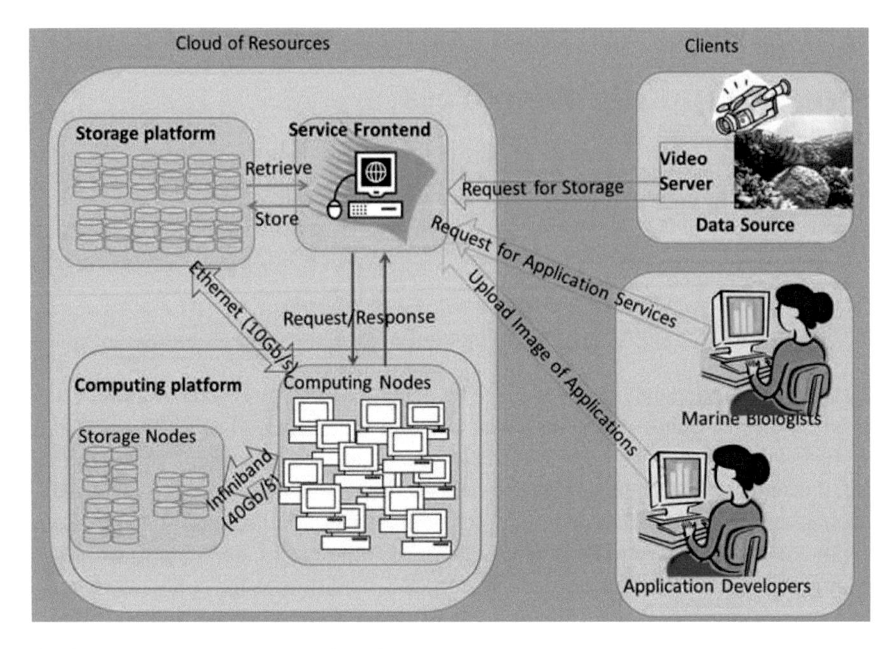

Fig. 3.1 Integration architecture of platforms

software is composed of several pipelined image processing tasks, and from there fish objects were extracted and recognized from the video data. Given the massive amount of video data to be analyzed within the scope of this project, the computational infrastructure that is required to process these large-scale data sets poses challenges to the system development.

One of the challenges to the infrastructure building for the F4K project was the diverse requirements of different components. For example, the video processing components (detection, tracking, and recognition) need fast computing, the database component needs a fast Input/Output interface, and workflow and User Interface components need stable networks which can transmit data flow seamlessly. To address this challenge we adopted an Infrastructure-as-a-Service (IaaS) model of cloud computing, where storage and computing resources are consolidated in one single access framework. Figure 3.1 shows the conceptual architecture of the infrastructure service framework. The three major components of the framework are: storage platform, computing platform, and service frontend. The integration of platforms is based on a 3-tier architecture shown in Fig. 3.2. The strength of this architecture is that it has presentation, processing, and data management logically separated which provides great flexibility and reusability (Tsai et al. 2008; Chung et al. 2010; Wu et al. 2009).

In this chapter, we focus on the building of the computing platform. For the storage platform and service front end please refer to Chaps. 4 and 5. The content of this chapter is organized as the following: description of the computing platform in Sect. 3.2, the process execution interface in Sect. 3.3, followed by conclusions in Sect. 3.4.

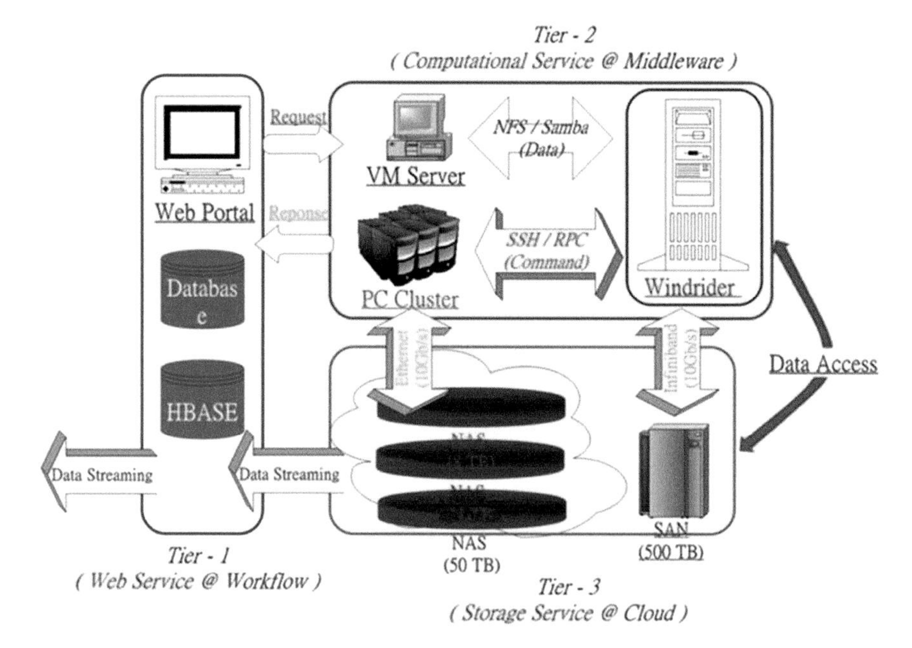

Fig. 3.2 Conceptual architecture of the infrastructure service framework

3.2 Computational Platform

In order to provide a flexible high-performance computing environment in support of F4K project, we created a heterogeneous computing platform composed of a supercomputer system and a Virtual Machine cluster. With the heterogeneous platform we can experiment with a variety of process execution strategies. The rest of this section will elaborate on the setting up of these two systems.

3.2.1 Supercomputing Platform

The supercomputing platform we used is a multi-core Symmetric MultiProcessing (SMP) cluster, Advanced Large-scale Parallel Supercluster (ALPS, also known as WindRider). The system uses AMDR Opteron 6100 processors, and has a total of 8 compute clusters, 1 large memory cluster, and over 25,600 compute cores. It offers an aggregate performance of over 177 TFLOPS (ALPS_team 2014). ALPS uses Platform LSF (Load Sharing Facility) for resource management.

3.2.2 The Virtual Machine Cluster Platform

In the F4K consortium, the research partners have different research cultures, such as different coding languages and environments, and data-handling schemes. Using laboratory-hosted servers to carry research individually will make integration extremely difficult given the diverse software and hardware, and maintenance costs are very high. To fulfill the diverse requirements from the consortium members we employed virtual machines to emulate different computing environments for the members.

We experimented with two different hosting models for the virtual machine platform, one used a single SMP multicore server to host multiple virtual machines, and the other used a 4+1 node PC cluster hosting a single virtual machine on each cluster node. The details of these two models will be described in the remainder of this section.

Virtual Machines on SMP Server

The server (known as gad246) is an off-the-shelf SMP machine. The system is equipped with a pool of homogeneous processors running independently, each processor executing different programs and working on different data and with the capability of sharing common resources. With these features, we can employ virtualization technology to simulate the system into a virtual cluster. The virtual cluster acts as a test-bed for integration of software components, and supports incremental migration of the software system from the development phase into production.

Virtual Machines on the Computer Cluster

A computer cluster consists of a set of connected computers that work together so that, in many respects, they can be viewed as a single system. Unlike a SMP machine, computer clusters have each node set to perform the same task, controlled and scheduled by software. With this feature a virtual cluster on a computer cluster is good for simulating a parallel computing environment. In this project, we used a 4+1 node computer cluster and had single virtual machine run on each node to form a 4+1 virtual cluster. The virtual cluster acted as auxiliary computing resources for the tasks that are less compute intensive.

3.3 Process Execution Interface

The Video/Image Processing software of the F4K project is comprised of several pipelined image processing tasks, and from there fish objects were extracted and recognized from the video data. The execution of the analysis pipeline is coordinated by a workflow engine which requests the system for the required resources, and then submits the execution to the assigned destination of the computing platform.

As mentioned above, the platform is a combination of two heterogenous computing systems—a supercomputer and a virtual cluster. These two platforms have different resource schedulers, an interface to bridge the two schedulers is required

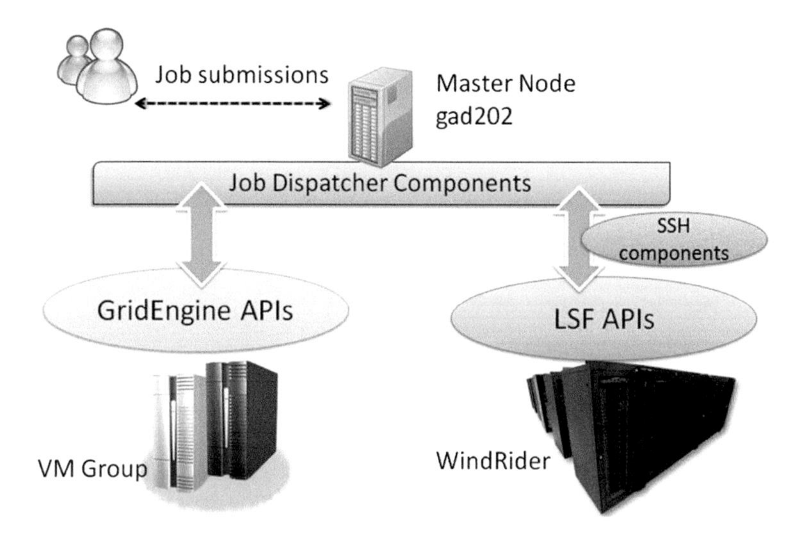

Fig. 3.3 Implementation model of job dispatch interface

so users can easily submit jobs without knowing the details of the schedulers. A middleware, Job Dispatcher, was designed for F4K developer during experimental stage as a test-bed to submit jobs to the proper system and to track job status. A conceptual diagram of the Job Dispatcher interface is shown in Fig. 3.3. The rest of this section will describe the design of the interface in detail.

3.3.1 Distributed Resource Management System

A distributed resource management system is a set of software products that enable distributed computing by providing end-to-end access to computing resources through load sharing and dynamic job scheduling within heterogeneous computing environments. We implemented the heterogenous computing platform by using two different resource management systems: Grid Engine for the virtual cluster platform and Platform LSF for the supercomputer platform. A brief introduction to these two systems is given in the following paragraph.

Grid Engine

There are 9 nodes in the virtual cluster platform, illustrated in Fig. 3.4. As shown in the illustration, Gad202 is the master node which is responsible for coordination of resources and coordinating the execution of the computing jobs. Data are shared through commonly accessed shared folders.

Figure 3.5 shows the job running model of the Grid Engine system (Grid_Engine_team 2014). There is a QMaster daemon running on a master node which is the main coordinator for computing resource management and job

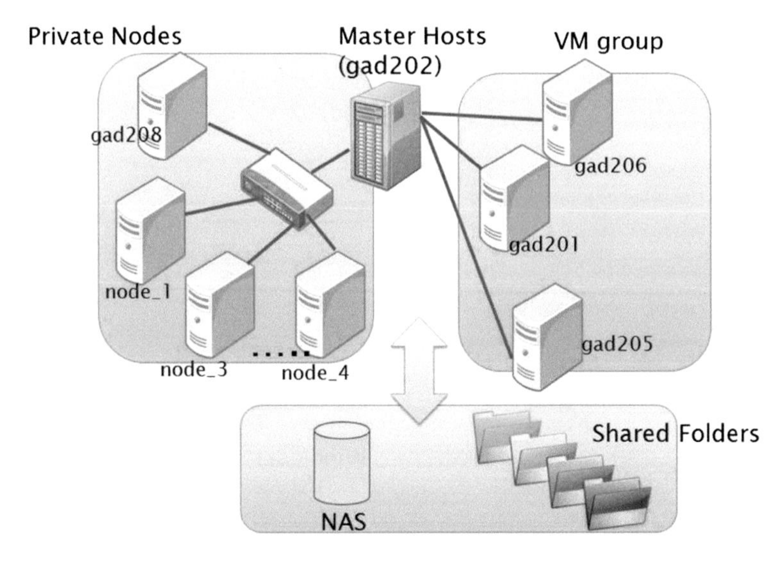

Fig. 3.4 The architecture of VM group

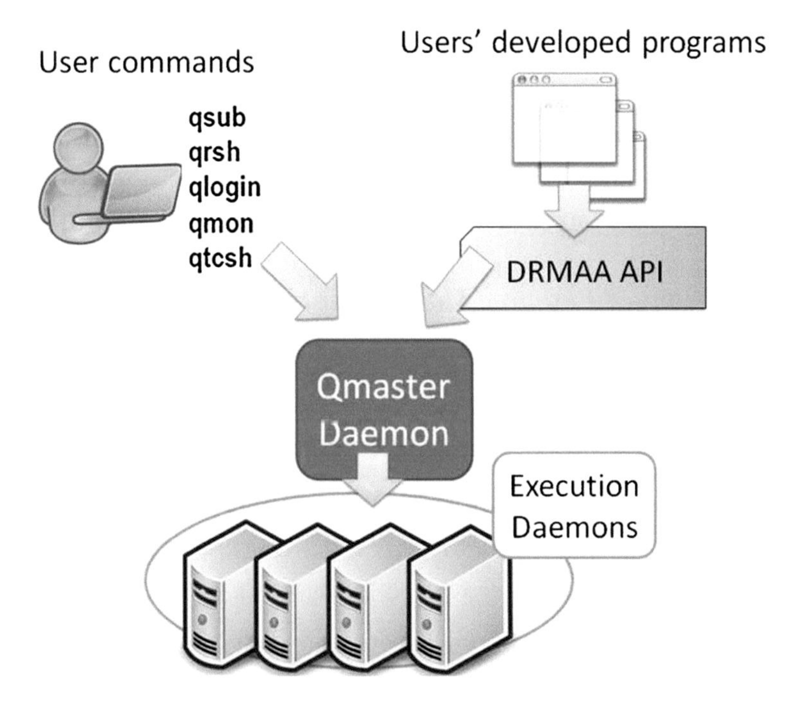

Fig. 3.5 The Qmaster and execution daemons are responsible for the communication between the master and salve nodes

scheduling. There are a few execution daemons deployed on the slave nodes. Communications between the master and slave nodes rely on the daemons. The master will deal with requests or commands from users by communicating with the execution daemons on slave nodes.

Users interact with Grid Engine using Grid Engine commands. Execution results and error messages will be recorded in the user's home directory, named following the convention of YourJob.sh.o7771 / YourJobName.sh.e7771 where 7771 is the job id of the specific job. Users can check these two files to know whether job execution is a success or not.

Platform LSF
Similar to Grid Engine, the Platform Load Sharing Facility (or simply LSF) is a workload management platform, job scheduler, for distributed HPC environments (IBM_team 2014). It can be used to execute batch jobs on networked computing systems on many different architectures. NCHC ALPS supercomputing platform uses Platform LSF to schedule workloads submitted to the platform. Users specify one of the preconfigured submit classes to queue jobs, and upon submission the job is routed to the appropriate LSF class according to the job criteria. All batch jobs must be submitted to a valid submit class. If the class doesn't exist, the job will be terminated and an error message will be issued. There is one preconfigured submit class without maximum time restriction, named as monos01 (2 computing nodes with 96 cores), dedicated for the F4K project.

Job Dispatcher
The DRMAA (Distributed Resource Management Application) API is a high-level Open Grid Forum API specification for portable programmatic access to cluster, grid, and cloud systems (DRMMA_Working_Group 2014). The scope of the API covers all the high level functionality required for applications to submit, control, and monitor jobs on execution resources in the DRM system. The DRMAA api is implemented for both Grid Engine and Platform LSF, and that advantage allowed us to develop a uniform job execution interface, Job Dispatcher, to bridge two computing platforms.

The Job Dispatcher is programmed in the C language and compiled as several standalone components, which is only used as a test-bed for F4K developer during development stage without inclusion in the final production run.

A Secure Shell protocol is adopted in the interface between the Grid Engine and Platform LSF systems, as illustrated in Fig. 3.6. APIs to control the system for job submission and controls from Grid Engine to Platform LSF are provided including job submission, job monitoring, job control, and job submission with dependency. The corresponding parts at Platform LSF to receive and control jobs are implemented on WindRider. In Platform LSF, each job possesses a current status code. Users can do a *job rget* to determine the job running status by checking the returned code.

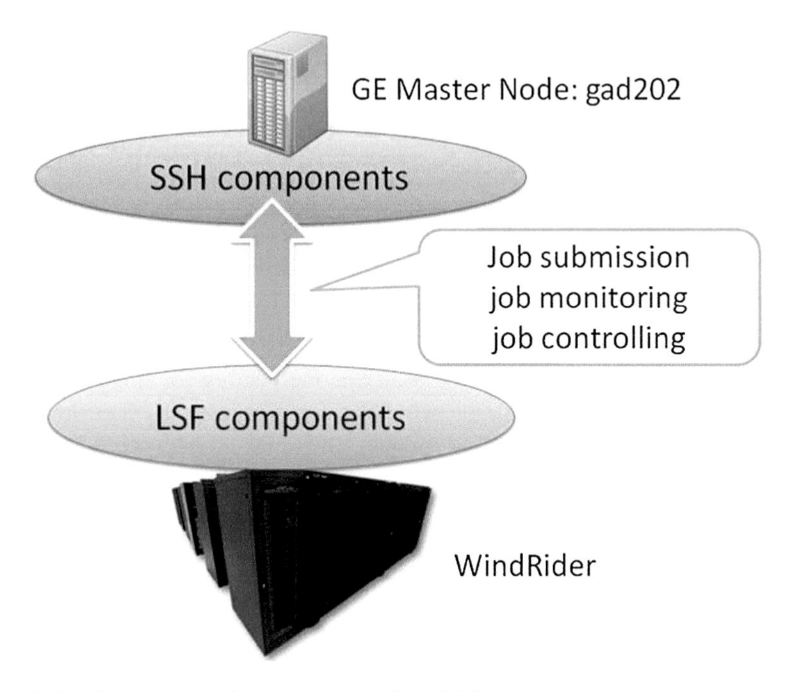

Fig. 3.6 Interface between grid engine and platform LSF

3.4 Summary

In this chapter, we described briefly two heterogeneous platforms for F4K components to execute compute intensive image processing tasks. A multicore supercomputing platform, Windrider, acted as the main execution platform with 96 cores, several open queues with varying capacities and the LSF resource scheduler. The VM cluster was the alternative platform which utilizes Oracle's Grid Engine and has the developer's full capacity in terms of processor usage. The following chapters will demonstrate that both platforms were used effectively for the compute intensive tasks of F4K. Furthermore, we experimented with a Job Dispatcher facility which acted as a bridge between the workflow and the two computing platforms. The Dispatcher, although not used in the final production run, gave valuable insight on how the workflow should connect to these two platforms. Virtualization technology is nowadays mature enough to support high performance computing on a large scale. In the future, we will expand our virtual machine cluster into a larger scale that is capable to support big data analysis in real time by using technologies like Hadoop Yarn (Apache 2014a), Storm (Apache 2014b), etc.

References

ALPS_team. 2014. Alps. http://www.nchc.org.tw/en/services/supercomputing/supercomputing_1/ar585f1.php. Accessed 16 Nov 2014.

Apache. 2014a. Hadoop. http://hadoop.apache.org/. Accessed 16 Nov 2014.

Apache. 2014b. Storm. https://storm.apache.org/. Accessed 16 Nov 2014.

Chou, H., Y. Shiau, S. Lo, S. Lin, and F. Lin. 2009. *A real-time ecological observation video streaming system based on grid architecture*. In *Proceeding HPC Asia (KaoShiung, Taiwan)*

Chung, T.-L., W.-Y. Chang, W.-F. Tsai, F.-P. Lin, E. Strandell, L.-C. Ku, J.-G. Lee, J.-Y. Chang, T.-H. Lee, J.-H. Wu, S.-C. Lin, M. Chen, Y.-H. Lee, K.-C. Chang, and Y.-F. Wang. 2010. Cyberinfrastructure for flood mitigation in taiwan. *Proceedings of the ICE—Water Management* 163(1): 3–11.

DRMMA_Working_Group. 2014. Open grid forum drmma api. http://www.drmaa.org/index.php. Accessed 16 Nov 2014.

Grid_Engine_team. 2014. Grid engine. http://gridscheduler.sourceforge.net/. Accessed 16 Nov 2014.

IBM_team. 2014. Ibm platform lsf. http://www2.nchc.org.tw/~a00yys00/lsf7/7.0.6/lsf_hpc_using/intro.html. Accessed 16 Nov 2014.

Lin, N.-C., T.-Y. Fan, F.-P. Lin, K.-T. Shao, and T.-H. Sheen. 2009. Monitoring of coral reefs at the intake inlet and outlet bay of the 3rd nuclear power plant in southern Taiwan. In *Proceeding of annual meeting of the fisheries society of Taiwan*.

Tsai, W.-F., W. Huang, F.-P. Lin, B. Hung, Y.-T. Wang, S. Shiau, S.-C. Lin, C.-H. Hsieh, H.-E. Yu, L.-L. Pan, and C.-L. Huang. 2008. The human-centered cyberinfratructure for scientific and engineering grid applications. *Journal of the Chinese Institute of Engineers* 31(7): 1127–1139.

Wu, J.-H., F.-P. Lin, and W.-F. Tsai. 2009. A scalable middleware for multimedia streaming. In *Proceeding of HPC Asia 2009, Kaohsiung, Taiwan*.

Chapter 4
Marine Video Data Capture and Storage

Sun-In Lin, Fang-Pang Lin and Hsiu-Mei Chou

Abstract We designed the architecture of the observation system to support capturing of high resolution videos. An intuitive design is to transmit videos to the storage site directly. There is, however, a risk of data lost caused by network instability and possibly electrical blackout. A set of hardware components is devised as the first tier processing and buffering device which stores videos temporarily while the video transmission process is waiting for a network channel. Higher bitrates, e.g., 5 mbps, are also tested in order to provide a clearer and more reliable data source for further video analysis. With the buffering space we are able to overcome network bandwidth limitations and transmit higher resolution videos in an effective way.

4.1 Introduction

The infrastructure to support the Fish4Knowledge project is composed of a number of networking components: video cameras in distributed locations continuously capturing and sending data streams, a massive storage system to store data, and high performance computing facilities to host data analysis tasks. The core issues to be addressed are how to do fast data query and retrieval with Tera-scale coupled repositories for video data and metadata, and how to accelerate the workflow process execution via compute parallelization. This chapter consolidates studies on data capturing, storing, and retrieving: the first is to maintain a sustainable data capturing system

S.-I. Lin
National Applied Research Laboratories, # 22, Keyuan Road,
Central Taiwan Science Park, Taichung 407, Taiwan
e-mail: lsi@nchc.narl.org.tw

F.-P. Lin (✉) · H.-M. Chou
National Applied Research Laboratories, # 7, 6th RD Road,
Science Park, Hsinchu 300, Taiwan
e-mail: c00fpl00@nchc.narl.org.tw

H.-M. Chou
e-mail: hmchou@nchc.narl.org.tw

R.B. Fisher et al. (eds.), *Fish4Knowledge: Collecting and Analyzing Massive Coral Reef Fish Video Data*, Intelligent Systems Reference Library 104,
DOI 10.1007/978-3-319-30208-9_4

which enables recording and transmission of high quality video data continuously, the second is the construction of a massive storage system which has 100 tera-scale capacity to store long term video data, and the third is the implementation of a database caching mechanism to boost data retrieval. The rest of this chapter is organized as follows: enhanced video capturing system in Sect. 4.2, massive data storage system in Sect. 4.3, the design of the database caching mechanism to improve data retrieval efficiency in Sect. 4.4, followed by a brief summary in Sect. 4.5. See also Chap. 3 for a detailed description of computing resources.

4.2 Enhanced Video Capturing System

The video capturing system was adapted from our previous work on real time video streaming systems based on a grid architecture (Chou et al. 2009; Shiau et al. 2010). The streaming system supports a variety of capture devices and video encoding formats that are commonly used. The original design of the streaming system was focused on supporting multiple types of capture device and compressing the video to a variety of bitrates for different network bandwidth requirements. In certain circumstances, for better viewing experiences, the videos were smoothed with interpolated dummy frames. However, for scientific data analysis, information might be lost during the compression, and interpolation can cause the wrong interpretation of the results.

To fulfill the requirement of precision for scientific discovery, we adjusted the capturing and compression methods to achieve higher video quality. However, higher video quality requires higher hardware capacity—for both computing and storage. Thus, an enhanced video capturing system was designed (Chou 2013). Figure 4.1 shows the architecture diagram of the enhanced video capturing system, which we elaborate on in the remainder of this section.

4.2.1 Better Video Server Management

To operate a system in an uncontrolled environment is a challenge given that environmental factors, such as temperature, humidity, and salinity, won't remain constant. In order to avoid environmental influences and maintain a stable system, the server that receives the video signal from the undersea cameras was located indoors with adequate air conditioning. Remote Power Management (RPM) was also installed to provide feasible system and service recovery after server crashes. This required the installation of a costly fiber channel network, however, it is well worthwhile as it reduced maintenance overhead. Moreover, to enhance video quality to the level that is needed for analysis, the resolution was increased to 640×480 and frame rate was raised to 24 fps without interpolated dummy frames. Table 4.1 lists the variations in format and capturing methods. Video is also de-interlaced (linear) to reduce interline

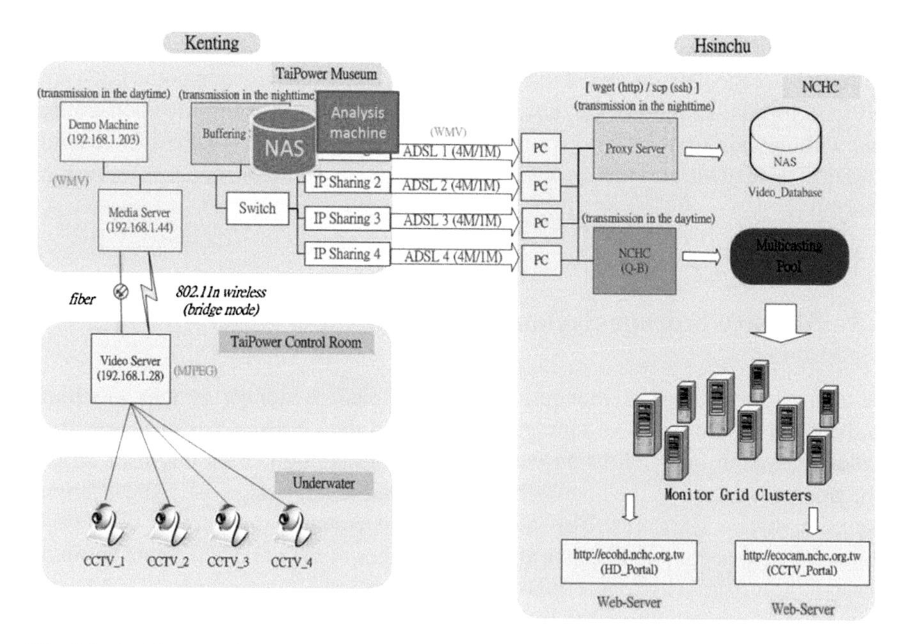

Fig. 4.1 Architecture diagram of enhanced video capturing system in the F4K project

Table 4.1 Combination of format and method for capturing and storing videos

Device	CCTV	CCTV	HD
Format	FLV	MPEG4	FLV
Resolution	640 × 480	640 × 480	1280 × 760
De-Interlace mode	Linear	Linear	None
FPS	24	24	30
Bit-Rate	1 M	5 M	20 M
Capturing method	Stream dump	Stream dump	Stream dump
Site	NPP3, Lan Yu, Hobihu	NPP3	NMMBA

jitter in the video, and the capturing method is changed to using stream dump which directly dumps video data from source streaming. The stream dump method can chop the continuous video stream into 10-minute clips more precisely.

4.2.2 Local Buffer Space

Due to network bandwidth constraints between observation sites and NCHC, it was not feasible to transmit and store real-time high bitrate video at the central storage of NCHC. Therefore, we implemented a local process solution by installing a new server (Local Processing Server) and 7.7 TB local storage (NAS Server) on site as buffering storage. We compared the time required to encode video in both H.264 and MPEG4 format, and found that encoding in H.264 format was too time consuming and with no gain in quality, so we implemented the system with MPEG4 encoding format. The high bitrate video provided clear and reliable image sequences for further analysis.

4.3 Massive Storage System

To address the storage challenges, a scale-out massive storage system was built to meet the requirements of storage capacity and data transfer performance. In a scale-out system, new hardware can be added and configured as the need arises. The main advantage of the approach is cost containment, along with more efficient use of hardware resources. The system we built contains three levels of storage components to accommodate data at different stages of the lifecycle: a data buffer at observatories to hold live data temporarily before they are transferred back to NCHC, a midterm storage pool to store data in the analysis stage, and long term storage to preserve critical data. The system allows data sharing by many different machines and operating systems through different network protocols, such as Network File System (NFS), HTTP, and Samba. The hierarchy of the storage system is illustrated in Fig. 4.2.

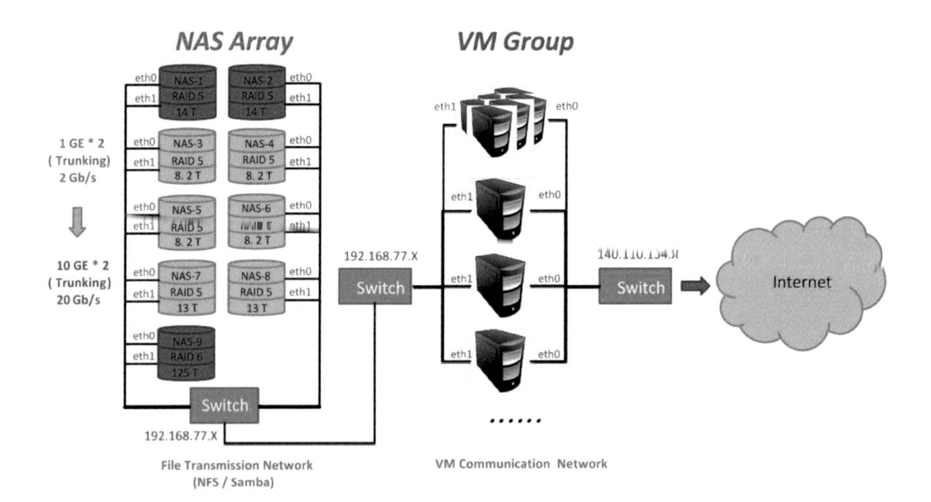

Fig. 4.2 The hierarchy of the massive storage system

Table 4.2 Summary of individual network-attached storage (NAS) units

NAS Storage	Size(TB)	Used(%)	Commment
NAS 1	14.0	60	Historical video storage
NAS 2	14.0	44	
NAS 3	8.2	1	
NAS 4	8.2	25	
NAS 5	8.2	96	
NAS 6	8.2	42	
NAS 7	13.0	8	VM NFS shared storage
NAS 8	13.0	0	Temp video storage
NAS 9	125.0	0	F4K data storage
NPP3 NAS	7.7	100	Storage in NPP3 site

4.3.1 Assembly of Storage Drives

Redundant Array of Independent Disks (RAID) is a commonly used data storage virtualization technology that combines multiple disk drive components into a logical unit for the purposes of data redundancy or performance improvement (Patterson et al. 1988). Different RAID modes provide different levels of security, performance overhead and storage capability. In our system, RAID 5 was adopted for video NAS to balance storage capability and security level. It also allows the storage array to not be destroyed by a single disk failure. The newly built storage (NAS 9) comes with 125 TB storage capability in RAID 6 mode using hardware RAID, which provides hardware acceleration to read and write data in RAID. In total, up to 220TB of storage capacity is built for data storing (see Table 4.2). The storage system was isolated from the public network to avoid direct network attacks. The access model to the storage system is shown in Fig. 4.3.

4.4 Improvement of Data Retrieval Efficiency

Locating a specific file from a file system can be time consuming when the number of files is large. One way to accelerate the process is by using a data catalog to organize the file list. In our system, a video database is designed to store information about video clips effectively. The video database is a catalog of video data files, which stores the information about where, when, and how the videos were captured. A universal unique identifier (UUID) was designed to identify every single file and is used as an index for referencing. A web service was developed to allow project components to retrieve specific videos for processing. The database schema is shown in Fig. 4.4. The full SQL description is in Appendix C.

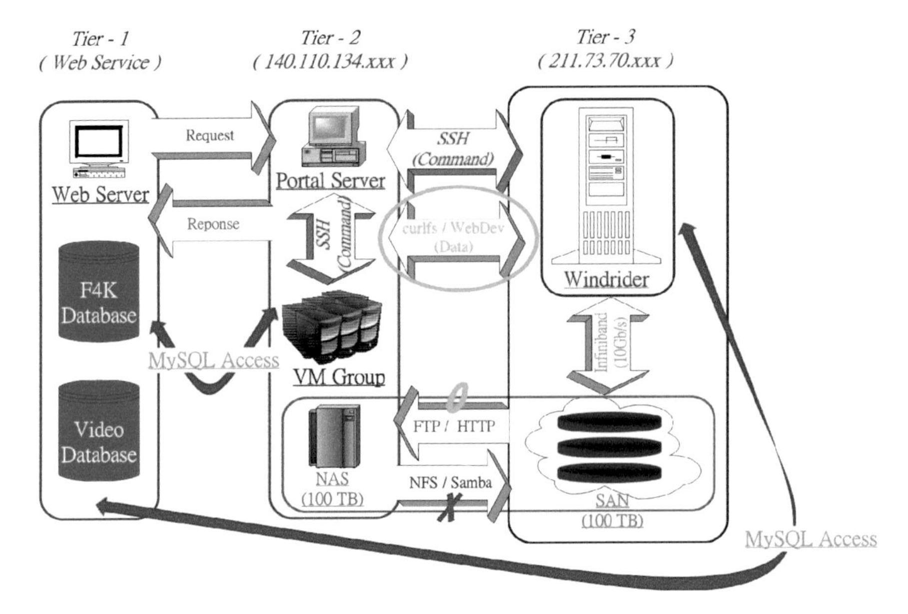

Fig. 4.3 Access to storage at NCHC

As there are more than 600 K video records in the database, database queries can be time consuming when the join of multiple tables is required to aggregate information. In our system we implemented two mechanisms to improve data retrieval efficiency: one with a unique key to identify the record, and the other is a five-tier database caching to reduce the I/O hit to the database by caching constantly accessed data in a temporary table.

4.4.1 Universally Unique Identifier

One of the methods to improve the performance of a database query is to use the Universally Unique Identifier (UUID) generated during the query in accordance with the key data as the only identified information or index in the database query (Warford 2009). Traditionally, crossing multiple tables and complicated fields increases the load on the database and I/O usage, greatly affecting the performance of the database query. Using an index can help quickly locate the information in the database, and the binary search tree is used in the traditional database to speed up the search of indexing. In this study, the most commonly used field is encrypted via MD5 to produce a fixed-length UUID, which is saved in the data table as an index to reduce the query on multiple table joining and field matching, and thus the query performance is improved and users can get the query results rapidly. Figure 4.5 describes the scheme to generate the UUID.

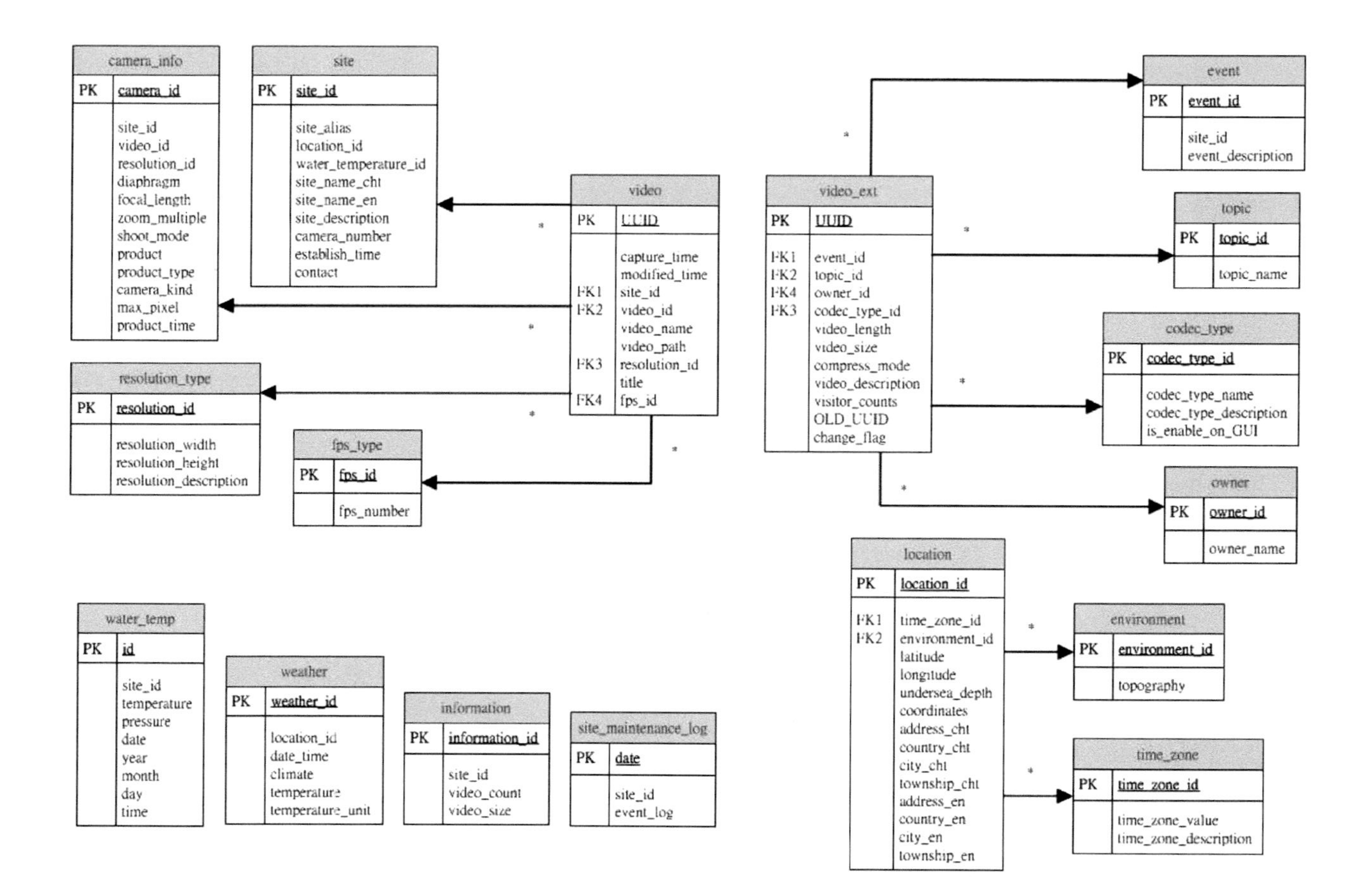

Fig. 4.4 Video database schema diagram

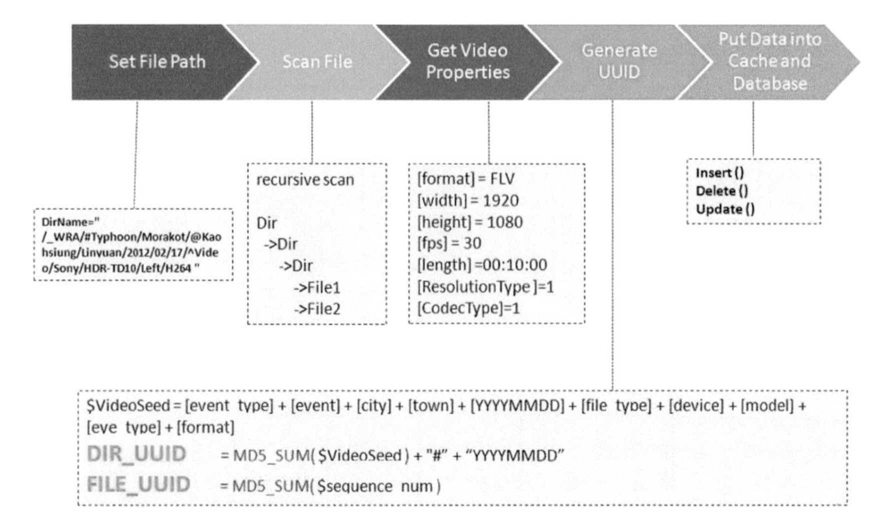

Fig. 4.5 UUID generation scheme

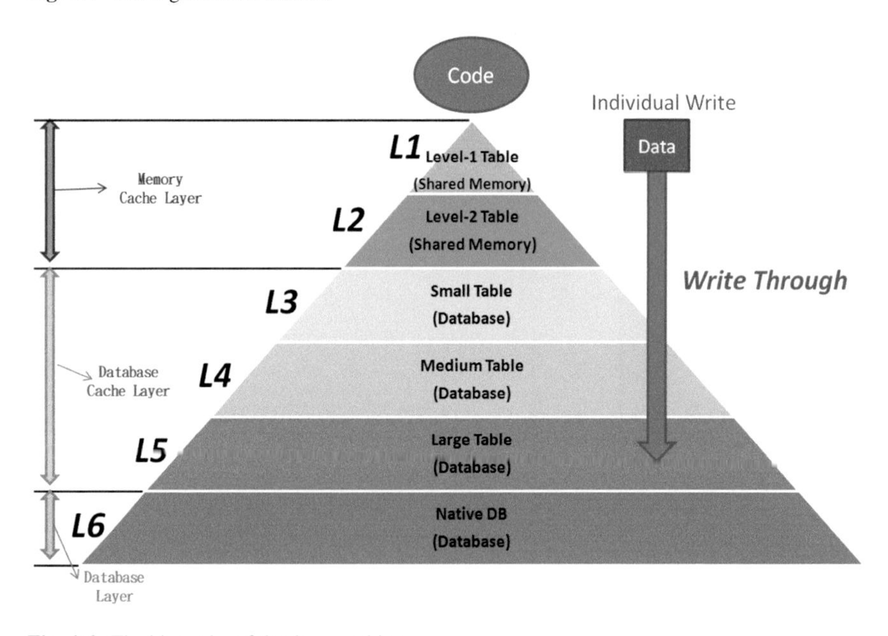

Fig. 4.6 The hierarchy of database caching

4.4.2 Database Caching

In addition to using the UUIDs as the indices, we also applied the concept of hierarchical computer memory to the database query, and adopted a five-layer cached architecture, shown in Fig. 4.6. When the data are committed into the database, in addition to writing data into the main database tables, cached tables are generated automatically. On the other hand, the cache tables are updated sequentially when data are retrieved in responding to users' queries. In this study, we implemented a hybrid cache, i.e., memory is used as the storage media for Layer-1 and Layer-2 cache, and database tables are used for Layer-3, Layer-4, and Layer-5 cache. The size of each cache layer is increased gradually from Layer-1 to Layer-5. The five-layer cache is a typical model for hybrid cache, applying different storage media and design principles and using different sizes of cache so as to reduce the search times and increase the cache hit rate, and thus improve the database access performance.

As the cache capacity of each layer is fixed, an alternative mechanism for data has to be used if there are new data to be added while the cache is full. For instance, as Fig. 4.7 shows, a ***round-robin queue*** is a basic and popular substitution mechanism to replace data based on the arrival order of the data, a.k.a. FIFO. Yet, this method does not work well for the cache efficiency: the data frequently used will be replaced by newly arrived data without any conditioning and thus it is not easy to keep important data in the cache sufficiently long.

We employed a cache substitution principle that exploits both the temporal and spatial characteristics of caching to achieve an optimal replacement rule. For example, LRU (Least Recently Used) and LFU (Least Frequently Used) cache replacement algorithms (Englander 2009; Wilson 2012) can be incorporated simultaneously. The frequency of use of data has the priority to replace the data. If multiple data have the same frequency of use, the earliest (oldest) datum will be replaced instead. Different

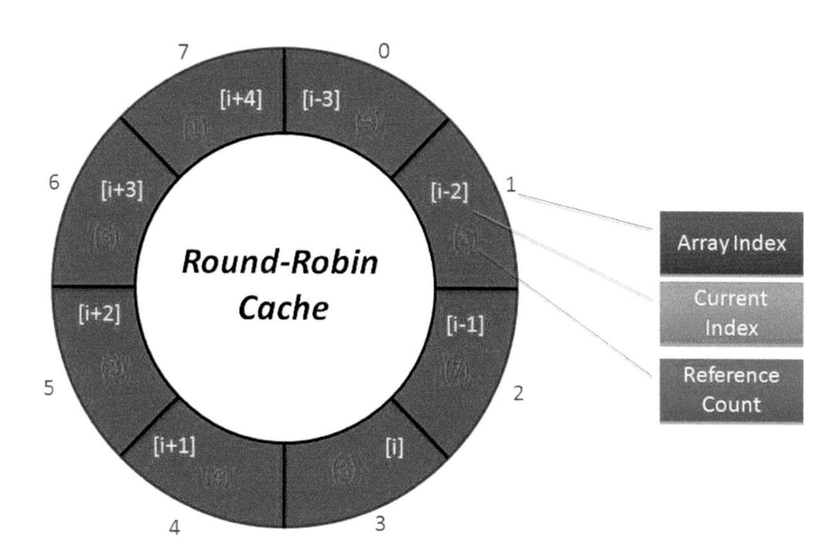

Fig. 4.7 Using the Round-Robin cache to assign data to cache tables

cache replacement schemes can be employed in different layers of cache to optimize the efficiency of overall cache usage, leading to a better performance as a result.

A hierarchical cache will speed up the database search, but when shared memory is used as the cache, we need to pay attention to possible race conditions that can occur during simultaneous data updates. We apply a 'Lock' mechanism to the shared memory cache to ensure cache coherence and prevent data errors when there are multiple concurrent writes.

4.5 Summary

To fit the requirement of best signal-to-noise ratio for better image analysis video resolution is pushed to the limit of 640×480 at 24 fps. A hierarchical storage system with 100 tera-byte scale was built to accommodate video collection. The best configuration of network linkage, which is the major factor of transferring speed between storage and computation facilities, was implemented. A relational table was created to record metadata about the video collection, and a database caching scheme is implemented to make data looking up effectively.

The 5-tier database caching scheme is a unique design among other similar caching implementations. In the scheme, shared memory is used as cache to boost query performance to the video database. The implementation of Universal Unique Identifier (UUID) hashed constant query information in a single code and avoided multiple table joins in the queries. With these two implementations we reduced the video retrieval time by many folds. In the future we plan to expand these two implementations to other scientific applications that require instant response of data retrieval, such as a bridge health monitoring and analysis system.

References

Chou, H. (2013). Video and rdf store. Technical report, Fish4Knowledge Deliverable D4.1.

Chou, H., Y. Shiau, S. Lo, S. Lin, and F. Lin. 2009. A real-time ecological observation video streaming system based on grid architecture. In *Proceedings HPC Asia (KaoShiung, Taiwan)*

Englander, I. (2009). The architecture of computer hardware, systems software, and networking: An information technology approach. Hoboken: Wiley.

Patterson, D., G.A. Gibson, and R. Katz. 1988. A case for redundant arrays of inexpensive disks (raid). In *Proceedings SIGMOD*, 109–116.

Shiau, Y.H., Y.H. Chen, K.T. Tseng, and J.S. Cheng. 2010. A real-time high-resolution underwater ecological observation streaming system. International Archives of the Photogrammetry, Remote Sensing and Spatial Information Science, vol. XXXVIII. Part 8.

Warford, J.S. (2009). *Computer systems*. Sudbury: Jones and Bartlett.

Wilson, J.R. (2012). Seven databases in seven weeks: A guide to modern database and the NoSQL movement. Pragmatic Bookshelf.

Chapter 5
Logical Data Resource Storage

Hsiu-Mei Chou

Abstract The Fish4Knowledge project involves several data-intensive computing tasks. These tasks need to access a large volume of video data that have been acquired at several remote seaside locations in Taiwan, and which are stored in a data center in central Taiwan. It is important that such data can be processed efficiently and effectively. The outcome of the project depends in part on how well the data are managed. Managing data can be challenging particularly when studies involve several researchers and/or when studies are conducted from multiple locations. The strategy of data management depends on the types of data involved, how data are collected and stored, and how they are used—throughout the research lifecycle. This chapter describes the implementation and optimization of data resource management for the Fish4Knowledge project.

5.1 Introduction

Like many scientific research projects (Gray et al. 2005), the data analyses of the Fish4Knowledge project are performed in hierarchical steps, as shown in the schematic architecture diagram (Fisher 2013) in Fig. 5.1. During the first phase (the frame analysis and sequence analysis boxes in Fig. 5.1), initial data are generated, based on a subset of properties as stored in the videos (e.g. via deploying fish detection algorithms). In the next phase (the Recognise/Catalogue box in the diagram of Fig. 5.1), data are transformed or aggregated in some way to support fish identification (e.g. via fish species recognition algorithms). Such analyses are often accompanied by complex joins among multiple datasets.

At the center of the Fish4Knowledge process flow are three data repositories: Videos, Metadata, and Live data, and these repositories contain data in different formats. The Videos repository stores historical video clips in flash video format

H.-M. Chou (✉)
National Applied Research Laboratories,
7, 6th RD Road, Science Park, 300 Hsinchu, Taiwan
e-mail: hmchou@nchc.narl.org.tw

R.B. Fisher et al. (eds.), *Fish4Knowledge: Collecting and Analyzing Massive Coral Reef Fish Video Data*, Intelligent Systems Reference Library 104,
DOI 10.1007/978-3-319-30208-9_5

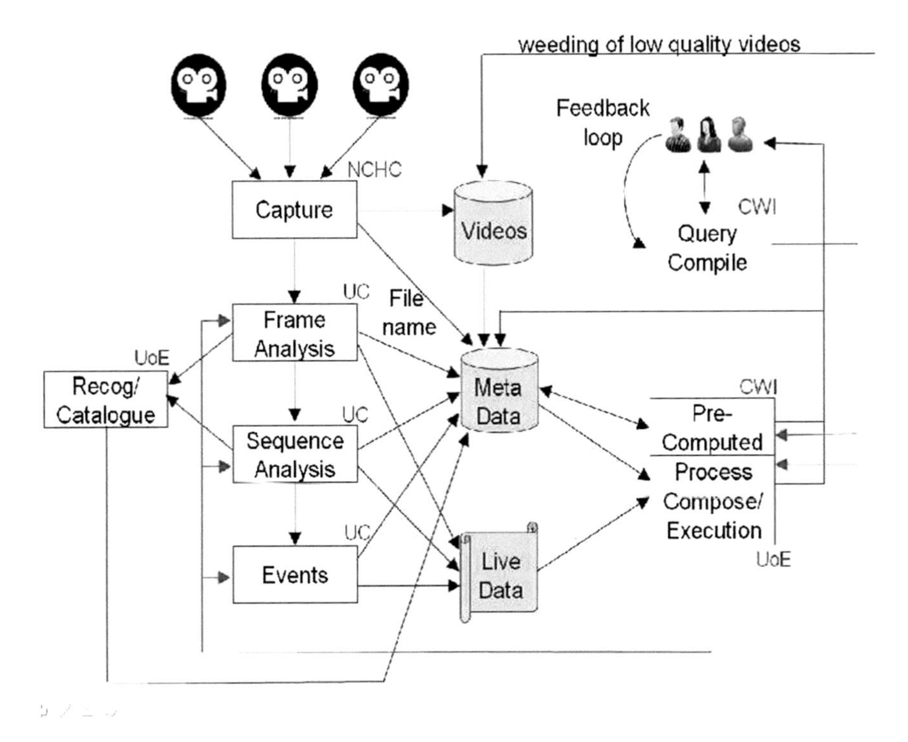

Fig. 5.1 Conceptual architecture diagram of Fish4Knowledge process flow

(Chou et al. 2009). In total, 524 K unique clips, each 10 min long, are stored with an accumulated total volume of 33 TB. The Metadata repository contains three types of information: (1) abstraction information, such as classes of video quality, bounding box of fish objects, tracking trajectories, and identified species, for each frame in each video; (2) dynamic information about process execution, such as parameters, and execution status, which are required by the workflow module; and (3) summaries of analytics results. The Live data repository contains a cluster of video servers sending real time video streams of data from the monitor sites and users can view them from any place with a standard web browser. See also Chap. 4 for details of Videos and Live data. For the diverse types of data being used and generated in the process cycle, a good data management scheme has to be in place to make the data flow between the processes smoothly.

The design of a proper data management scheme needs to consider the process of data generation and storage, the type of data involved, and the usage of data throughout the project lifecycle. It is a challenging task given that the project involves several researchers working from multiple locations. This chapter describes the implementation of our data management framework, which enables sharing and reusing of data, and reduces redundant data query processes.

5.2 Data Management

We give implementation details of the data management scheme for this project. The scheme includes a logical storage system and a data transfer method. The logical storage system organizes data collection in such a way that computer programs can quickly select items from within a collection. Most research projects use a database management system (DBMS) as the logical storage system. In this project the data handled have a clear relational structure with little heterogeneity so we chose a relational database to manage the data. Structured Query Language (SQL) is the standard query language used to perform tasks such as updating or retrieving data from a relational database. For the data transfer, incorporated with the system security restrictions, we adopted the http protocol to read/write data from/to storage directly. In the remainder of this section we will elaborate more details of the scheme. See also Chap. 7 for the Fish4Knowledge Database Structure.

5.2.1 Design of Database

A database schema is the layout of the data contained in the database including: the types of data, their properties, the relationships between them, and how they are organized. It is often described using the Entity/Relationship Model (ERM). The formalism used to draw the ERM diagram is illustrated as Fig. 5.2.

The F4K project includes two major databases: the Videos database and the Metadata database. Their schemas are described below.

We chose the database structure presented in Figs. 4.4, 5.3, 5.4 and 5.5 based partly on the major entities in the project (cameras/sites, videos, fish, and processing jobs), and partly around the amount of data and access expected for each of the tables (see Table 7.1, which shows that the majority of the SQL storage was used for recording detected fish data).

The cameras/sites tables recorded the camera location, type, cleaning logs, etc. The video tables recorded where and when the video was recorded, and with what parameters. There would be about 500 thousand records in this table, so encoding needed to be compact and constant data kept in other tables (e.g., the camera tables).

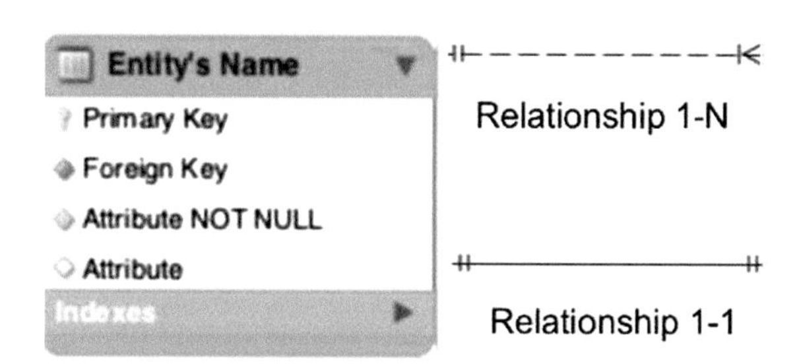

Fig. 5.2 Entity/Relationship formalism

The fish tables are the largest of the tables, with over 1.4 billion detection records, and about 140 million tracked fish records. The detected and tracked fish records were structured to avoid duplicate and constant values in the records, because of the amount of space they would require. Hence, a tracking linked to a number of detection records rather than duplicating all of the detection detail. The processing job tables were used to keep track of the processing stages for each video, and the associated details linked which versions of the algorithms had been used for the processing.

5.2.2 *Videos Database*

The Videos Database (VidDB) is a catalog of video clips where it stores the information about where, when, and how the videos were captured. The entities used for describing the VidDB are cameras and videos. The ERM diagram is shown in Fig. 5.3. The full SQL description is in Appendix C.

5.2.3 *MetaData Database*

The Metadata database is at the heart of the processes flow, as shown in the Fig. 5.1 architecture diagram. The entities stored in this database can be roughly divided into two groups: one for image processing results and the other for process flow control.

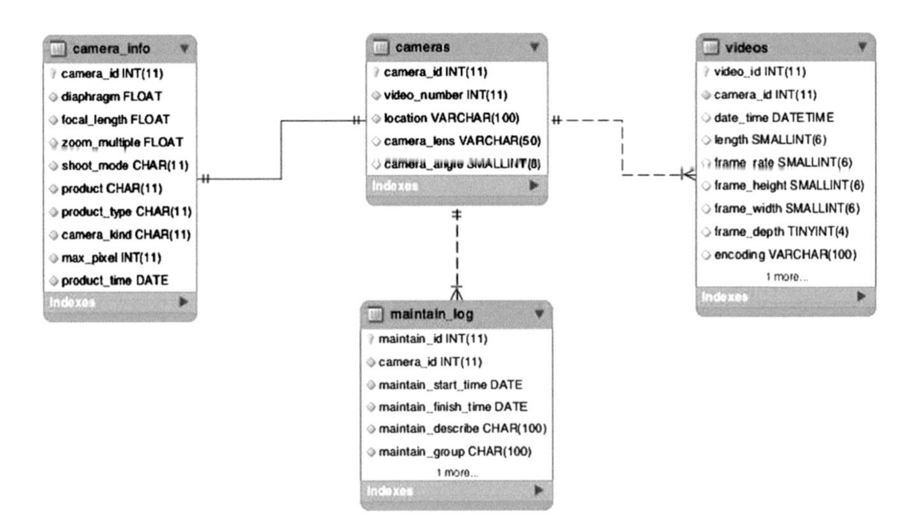

Fig. 5.3 ER model of videos database

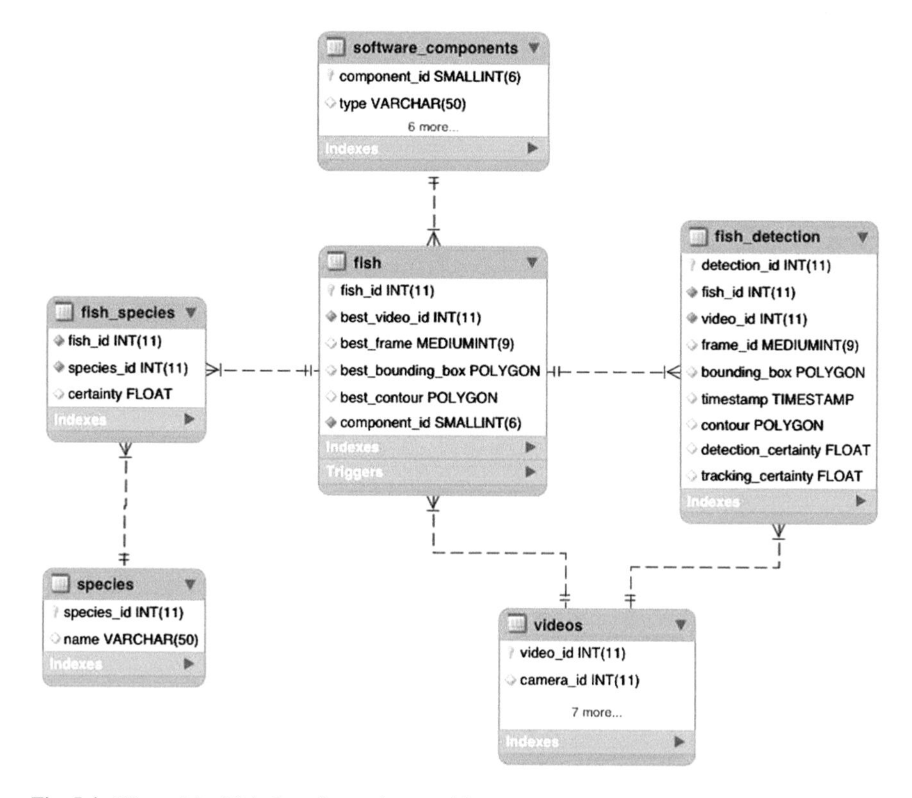

Fig. 5.4 ER model of fish detection and recognition

Image Processing Results
This group revolves around fish detection. The entities are the results of video processing in terms of detection of fish and recognition of species (Palazzo 2011). The schema is shown in Fig. 5.4.

Process Flow Control Group
This group of entities facilitates process status information sharing among all running processes. The design consideration is to enable fast information exchange, and immediate data storage among processes running distributively on computing resources. This group of entities centers on ***processed_video***, the schema is shown in Fig. 5.5.

5.3 Implementation

Because of budget limitations, we could not afford the high subscription fee required for a commercial product like Oracle, or MS SQL server. We went for an open source solution under this constraint. After conducting serial tests on two popular database

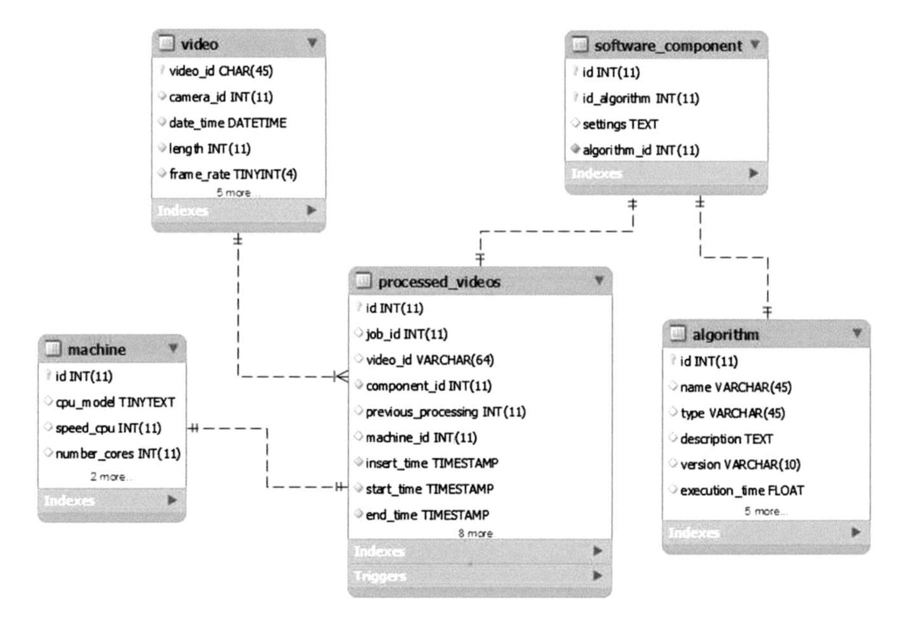

Fig. 5.5 ER model of process flow control

systems—PostgreSQL and MySQL (Postgre 2013; Oracle 2014), we found both systems have pros and cons. We choose to implement the Fish4Knowledge database using MySQL. This was mainly because the simplicity of implementation and the availability of connectors for any programming language.

Hardware

The performance of the DBMS system relies heavily on the hardware configuration. In this project we use a commodity off-the-shelf hardware server equipped with dual Intel Xeon 5560 CPU, 144MB DDR-3 memory, and a fiber channel interface to link SAN storage system. To assure the availability of the database server we implement master–slave replication to create a backup server, and the slave server can be brought up instantly if the master goes down (nixCraft 2007).

Performance

During the intensive processing phase of the project thousands of VIP processes were running at same time and these processes needed to communicate with the database instantly to retrieve parameters and store results. Thousands of job instances interacting simultaneously with the database can create a heavy workload on the database server which eventually became a bottleneck in the overall workflow. We identified disk I/O latency is the major bottleneck for data communication. To resolve this issue we developed a memory cache management mechanism, by caching often accessed data and objects in memory to reduce the number of times the disk database must be read.

5.4 Future Work

In our environment, the database is shared among all computing processes, and these processes can read/write to the same table at the same time. A deadlock can happen if the system doesn't schedule these read/write workloads properly. In general, the database server is smart enough to schedule workloads to prevent deadlock and return results instantly. However, in some cases, an internal locking mechanism is required to prioritize read/write sequences. In this case, internal table locking can cause deadlocks in the system. We encountered several such kinds of deadlocks and that degraded the system's performance seriously. Thus, we are working on a new design of an intelligent process monitoring agent that will manage the query workloads and pre-fetch information beforehand so the processes can run more effectively.

References

Chou, H., Y. Shiau, S. Lo, S. Lin, and F. Lin. 2009. A real-time ecological observation video streaming system based on grid architecture. In Proceedings of the HPC Asia. KaoShiung, Taiwan.

Fisher, R. B. et al. 2013. Fish4knowledge website. http://www.fish4knowledge.eu. Accessed 4 Nov 2014.

Gray, J., D.T. Liu, M.A. Nieto-Santisteban, A.S. Szalay, G. Heber, and D. DeWitt. 2005. Scientific data management in the coming decade. *ACM SIGMOD Record* 34(4): 35–41.

nixCraft. 2007. Mysql proxy load balancing and failover. http://www.cyberciti.biz/tips/mysql-proxy-howto.html. Accessed 4 Nov 2014.

Oracle. 2014. Mysql database system. http://www.mysql.com/. Accessed 4 Nov 2014.

Palazzo, S., C. Spampinato, and J. van Ossenbruggen. 2011. Fish4knowledge Deliverable D5.2—rdf/rdms datastore definition. Technical Report Del 5.2, Fish4Knowledge Project.

Postgre SQL_Global_Development_Group. 2013. Postgresql database system. http://www.postgresql.org. Accessed 4 Nov 2014.

Chapter 6
Software Architecture with Flexibility for the Data-Intensive Fish4Knowledge Project

Bastiaan J. Boom

Abstract The software architecture used in the Fish4Knowledge project allowed us to produce a system able to analyze 3 years of video footage and represent this data with an interface suitable for end-users (marine biologists/ecologists). To achieve this, the architecture design focused on four problems: (1) Data-Intensiveness, which allowed the system to process and store 528624 video clips of 10 min, resulting in 1445.41M fish observations. (2) Flexibility, which gave the developer freedom to design their own solutions within the larger system. (3) Dependency, which provided clear definitions of the output of the subsystems allowing researchers to work with each other's outputs. (4) Trust, which was important for the end-user to understand the uncertainties in the system and how to deal with them. The overall design of the final system used the database as means of communication for the different software components in order to deal with these challenges In this case, developers only had to make a database connection to obtain data they depended on. The main lesson learned in this project is that stable database definitions and visualizations are very important in Data-Intensive projects.

6.1 Introduction

The goal of the Fish4Knowledge project was to analyze 3 years of video footage from multiple cameras using a supercomputer (1000 processors) and present this kind of data with a user interface to marine biologists. In this chapter, we will focus on the Software Architecture developed to achieve such a system within a limited amount of time (3 years). In this research project, computer scientists from multiple disciplines (Computer Vision, High Performance Computing, Information Retrieval, Human Computer Interaction, Workflow Management) came together to develop this system. The Fish4Knowledge system (Boom et al. 2014) is unique in respect to: the amount of data it processed, the kind of data it analyzed, the sort of data that collected,

B.J. Boom (✉)
Cyclomedia, Van Voordenpark 1b, 5301 KP Zaltbommel, Holland, The Netherlands
e-mail: bas.boom12@gmail.com

R.B. Fisher et al. (eds.), *Fish4Knowledge: Collecting and Analyzing Massive Coral Reef Fish Video Data*, Intelligent Systems Reference Library 104,
DOI 10.1007/978-3-319-30208-9_6

the kind of analysis it presents. In this system, there were many open challenges, which requires some flexibility for the developers of the subsystems in order to tackle these challenges. But next to flexibility, developers are often dependent on results of others, so good definitions are necessary to connect the different subsystems. Finally, it is important that the system can be used where one of the biggest issues for users given the automatic analysis of data is if the data can be trusted. In the remainder of this section, we will focus on the four biggest software architect problems in the Fish4Knowledge system which are: Data-Intensiveness, Flexibility, Dependencies, Trust.

Data-Intensiveness: The Fish4Knowledge system had to process 528,624 video clips of 10 min (which is equal to 3671 days (24 h) or 10 years of video). Given these videos, we extracted both the locations of the fish in the videos over time and recognize the species of fish in the videos, resulting in 124.28M fish which are detected in multiple frames giving us 1445.41M observations. To process this kind of information within the 3 years time period the analysis software for fish detection and recognition is run on the supercomputer(s) in NCHC (Taiwan), (see also Chap. 3), using up to 1000 processors. All this information was stored in a database which could be accessed by our web-interface. Although the exact numbers were not known at the beginning of the project, the estimates computed before the project indicated that this would be a data-intensive project. Dealing with this kind of data was an important aspect of the design since it was not known if normal database software could deal with these numbers.

Flexibility: The Fish4Knowledge is a research project, which was faced with new challenges both in recognizing millions of fish images in videos, processing these videos and showing this kind of information to marine biologists. The design of the individual subsystems was not always clear, especially because researchers had to create novel solutions to tackle a diversity of problems. Freedom in the design of the research software is important, because both the requirements and expected output can change also partially because of the user needs and feedback on the system.

Dependencies: Although flexibility is important, it is also important that you can work with the output of partners, especially because researchers are often dependent on the data of other researchers. Given a data-intensive project like this, being able to understand the data of other researchers is important. Clear definitions need to be developed that allow researchers to communicate with each other.

Trust: A final very important issue especially in this system was trust. This system uses automatic video analysis software, which will probably never work perfectly. Detecting and recognizing fish is for humans often a difficult job, where annotations by marine biologists already shows disagreement in the results. Computers are not able to outperform marine biologists, partly because it is difficult to obtain enough information to learn the appearances of all possible fish species that are recorded by our cameras. Given that not all information is perfect in the system, the big challenge is how to make user aware of this and give them the tools to verify if the information in the system they would like to use is good enough.

In the remainder of this chapter, we discussed the system architect are developed within the Fish4Knowledge project that is able to tackle the important problems we identified above.

6.2 Software Design

6.2.1 Grand Design of Interaction

The main idea of the software architecture is to communicated by means of the storage facility (database). This means that the data that is processed by the individual software components (subsystems) is available to all other software components in the project, but more importantly to the end-users (marine biologists), this is first described in Boom and Fisher (2011). All software components will save their output into a database (see also Chap. 7). In this case, the software components only need an interface to the database to integrate themselves into the system. The database allows other software to query the information and process the data further or visualize the information to the end-user (see also Chap. 2). Because it is important to keep track of which software produced the information, components have to store a unique version identifier with the information which allows the system to keep track of the software that is responsible for writing the information. The main components are the fish detection and fish recognition software where there are up to 10 default versions and multiple other experimental versions not used in production.

6.2.2 Problem Verification of the Grand Design

The four biggest challenges discussed in the introduction of the system are: Data-Intensiveness, Flexibility, Dependencies, Trust. The grand design is centered around the database, where software components can work as agents that contribute information given that the necessary data is available. Given a *data-intensive* problem like this, using the database as means of communication between software scales well, because databases are often designed to deal with large amounts of data in a structured manner. The database gives *flexibility*, because new input and output can be defined if necessary by creating new tables, while the same interface to the database can be used. The database will also not place restrictions on the functionality that is offered by the software. On the other hand, it works well if there are *dependencies* between components, because the database definitions allow developers to define expected input and output. In this case, database definitions are not allowed to change except if other developers are consulted. It allows people even to develop software if the software component on which they depends does not work, because they can already manually import some of the expected information into the database allowing for quick prototyping even in cases of dependencies. Finally the *trust* in the system is managed by keeping track of the software which is responsible for the information. Because we are working with automatic analysis software, evaluation of this software allows us to identify how good this kind of software performs. This evaluation allows the users to check to what degree they can trust the software, but it also allows them to improve the evaluation if that is not good enough for them.

6.2.3 Practical Issues in Design Concerning the Database

During the design, we discovered that there will be a lot of data, so database solution had to deal with this amount of data. This data also comes in different formats, like video, image, ontologies, which a database together with other storage facilities have to deal with. However, by using the database component, we intend to give the other components a simple interface to the storage facilities without having to worry for instance about storing information in a distributed manner or different interfaces to retrieve different kind of information.

6.2.4 Software Components Within the Fish4Knowledge Project

In this section, we will give the overview of the system and how the software components interact with each other by means of the storage facilities:
First the videos from the underwater webcams are stored in the storage facilities (Chap. 7), the Fish Detection component (Chap. 9) will get the videos out of the storage facilities and will find the fish and label their location in the frames (fish location). The Fish tracking (Chap. 10) allows us to follow fish in multiple frames and also can contribute in behavior studies of fish. The Fish Detection/Tracking components will again store the obtained information (for example the fish locations) in the database. The Fish Description component will add certain descriptions to the stored fish (like the kind of tail the fish has, or the color of the fish). The Fish Recognition (Chap. 11) component will try to determine the exact species label (or family label) and will store this. Fish Clustering (Chap. 14) allows us to determine if fish are very similar to each other, which is used to annotate fish species allowing us to learn the appearance of different species. The Query Engine is able to retrieve all the information previously stored in the storage facilities, and for instance count the number of species X during the month December. The User Interface (Chap. 13) represents the information to the users, but also gives the user an interface to search through all the information in the storage facilities. The Workflow (Chap. 8) component will check which system resources are available to perform new jobs. For instance, it will keep track of the videos which have not been processed yet by the Fish Detection/Tracking component

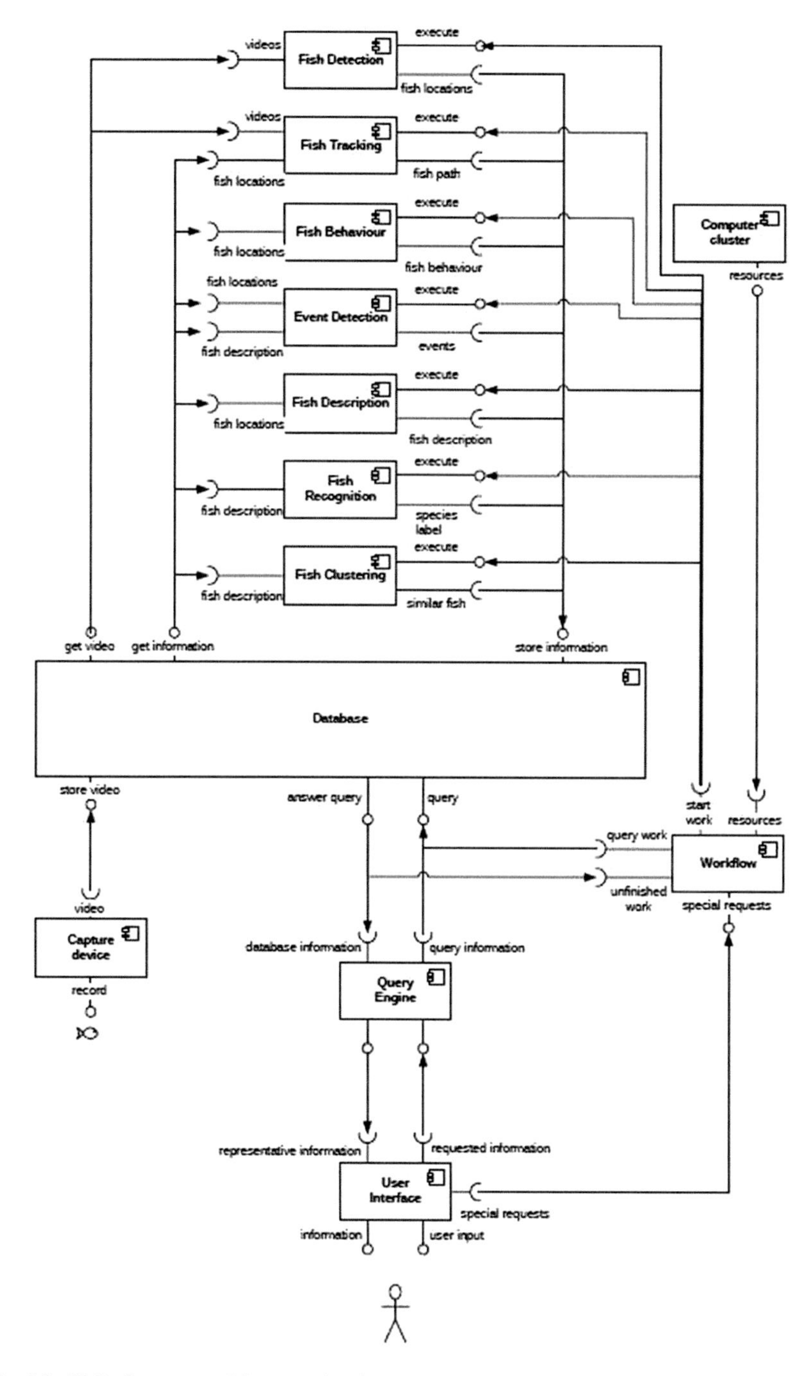

Fig. 6.1 UML Component Diagram, showing the input and output relations of the different components, see Sect. 6.2.4 for a description of the components and interaction

and the fish which have not been processed yet by the Fish Description/Recognition component. The workflow will start these processes if there are resources available (like memory and CPUs). Furthermore, it should be able to handle special requests by the user interface, running different settings of computer vision components (fish detection, tracking, description, recognition and clustering).

In Fig. 6.1, we show a schematic representation (UML Component Diagram) of the entire system. The components basically have interfaces and sockets and the information flow is given by the arrows. This schematic gives a rough overview. In Sect. 6.3, we define the possible inputs and outputs in more details. Notice that most components connect with the database component, because this will both allow to store the output of the components and provide input to the components. The big expectation here is the workflow component, which is able to execute the video and image processing software, but monitors the status of this software using a database table that monitors all the processing.

6.3 Individual Software Components and Their Relations

6.3.1 Fish Detection/Tracking Component

The purpose of this component is to detect the fish in the video stream. The detection will basically locate the fish in each frame. After locating the fish, the contour of the fish will be saved in the database. Another task of this component is to describe the scene. For examples, it will detect if it is dark or light, how much pollution (green, dirt) is in the water and if there are encoding issues. The fish tracking will follow the fish in the video, labeling at which position and in which direction the fish is going. It also provides information on how long the fish was visible in the image. More valuable output can be the interaction of the fish in relation with other fish in the video, like analysis if a fish is pursued by another fish. Other events that can be detected are eating, resting, hiding, fighting, mating, schooling, panic. Notice that we combined the Fish Detection/Tracking components for performance reasons.

6.3.2 Fish Recognition

The Fish Recognition Component will recognize the species to which the fish belongs. Because the fish is visible in multiple frames, these frames might have to be combined to obtain more information about the fish. We also have to select the frame (time and place in video) that contains the best appearance of the fish. This can be done using the contour, but also based on the number of features found by the fish description. If the fish is too far from the camera, it is possible that we can only determine the family to which the fish belongs and not the precise species. In computer vision, recognition of objects is a difficult task with a lot of uncertainties. These uncertainties can be expressed in probabilities or percentage, which allows us to communicate a certainty value (see Chap. 15) that we correctly recognized the fish. For performance reasons, we combined the Fish Description with the Fish Recognition component.

6.3.3 *Fish Clustering*

This software component allows us to make clusters of fish that are very similar to each other. These clusters are used to support the annotation of fish species by marine biologists and non-expert users. Annotation of data is necessary for the Fish Recognition Component to learn the appearance of the different fish species. By clustering groups of fish together that look similar, speed-ups in annotation can be achieved.

6.3.4 *User Interface*

The User Interface Component allows the user to browse the information and to ask specific questions to the system. The User Interface is connected to the query engine, which will retrieve the information for the users. The information provided by the query interface can be linked to other related projects, for instance Taiwan fish database, fishbase.org and Catalogue of Life. The first purpose of the website is to provide an interface to the experts and other visitors to search in a relative easy manner through the enormous amount of data. A special area can be developed for specialists (like marine biologists), so that they can login and that their searches will be remembered. Here, they can also ask for specific features they want to add to the website or special requests to add extra information in the storage facilities. In the project, the query engine is embedded in the user interface, where by improvement of the database design by using summary tables we are able to have fast query performance while still being able to deal with the large amount of data present in the database.

6.3.5 *Work-Flow*

The purpose of the Work-flow component is to organize the work that has to be done. Because the different components have different requirements on CPU, memory and hard-disk consumption, this can be a difficult task. The Work-flow component looks in the database to identify what information can be processed and will execute the appropriate components. The Work-flow component can also get a special request from the User Interface and it will create a processing chain to generate the requested information. The Work-flow component however needs detailed information of the other components, like version, purpose, average memory usage, average CPU demands, average run-time. This information will be contributed by the developer when adding a new software component to the system, so the Work-flow component can handle different kinds of information requests and will schedule the correct components for this. It also needs to detect if a component failed, where it is able to rerun the component on the video or try another component.

6.3.6 Database

The database component allows the different components to store and query information. There will be different kind of information, like videos, images, numerical data and strings. The database definitions were defined at the start of the project (Chap. 7). For both the fish detection and recognition components, there will be a simple interface to query lists of unprocessed videos and fishes. The query interface to these components is relatively simple compared to query interface for the user interface. The database component allows all the other components to share information about which video are processed with each other, but all components must have their unique identifiers when they store information. This allows us to be able to see which component is contributing the information.

6.3.7 Final Overview of the System

An overview of the entire system is given in Fig. 6.2, where the database parts are split into the "Video Storage", "VIP database", "Summary data" and "Processed videos table". The black arrows indicate the information/data flow, where fish are detected/recognized by the Video/Image Processing (VIP) software and afterwards stored in the "VIP database". Because of the large number of fish stored in this database, "Summary data" was created in order to deal with this kind of information, which allows the user to search through this kind of information. The red arrows indicate how the processing of all the data is organized, where a table keeps track of the processing performed on the different videos in the database.

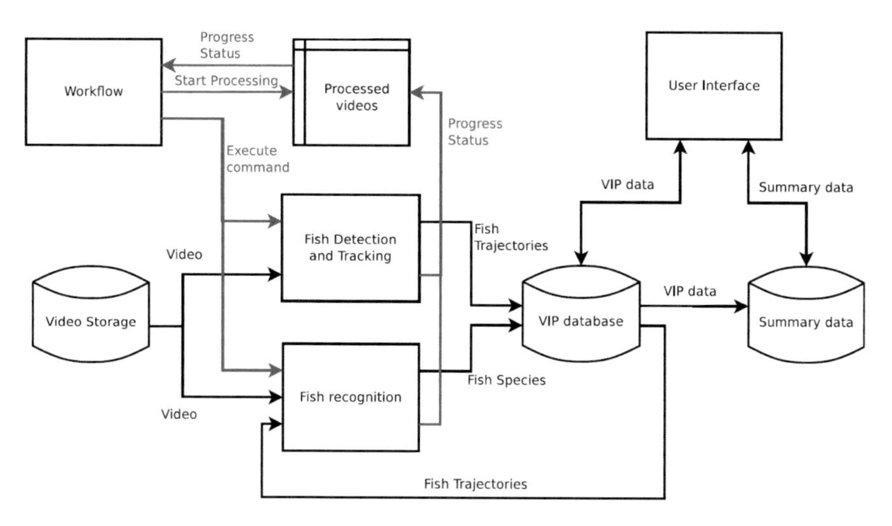

Fig. 6.2 Schematic overview of the Fish4Knowledge system, where the *black arrows* give the information flow in the system, while the *red arrows* show how the software is executed. This system finds the "fish trajectory" (positions of single fish in the multiple frames) in the video and given the fish trajectories determines the most likely fish species. This information is all stored in a Video/Image Processing (VIP) database. The user interface queries the summary data from this VIP database allowing marine ecologists to analyze this information

6.4 Software Development Process Given the Architecture

6.4.1 First Prototype System

The first prototype was working at the end of the second year of the project. It was able to perform video and image processing with multiple software components for fish detection/tracking and fish recognition. For fish detection, we had several different background subtraction methods and different fish tracking methods which could be used for this task. Two versions of the fish recognition software were available where the first version was able to recognize 10 species, and a newer version that recognized 15 species. A simple bulk processing workflow was used to compute the backlog of video data with both the default fish detection and recognition software components. The user interface was able to show statistical information about the processed video data, where we had already a year worth of processed video data stored in this system.

In the first prototype system, the bulk processing workflow needed to be replaced by a workflow that could perform bulk processing as well as respond to user requests allowing users to run different versions of the software to verify, for instance, a hypotheses. Here a connection between the user interface and the workflow was also still necessary.

As of May 27th 2013, the fish detection had processed 70,784 clips of 10 min which was equal to around 983 days of video given the 12 daylight hours we are recording. The fish recognition, which depends on the fish detection component had processed around 67,468 clips of 10 min (937 days of video). In total, we have however 623,472 clips, although there are multiple clips where we have both low resolution and high resolution videos of the same scene (resulting in 528,624 unique clips).

6.4.2 Final System

For the first prototype, we reported that some components were not fully connected with each other (i.e., workflow and interface). This connection has been achieved, allowing marine biologists to run video processing component on the videos in the database. Large improvements in the individual components are made: The fish detection component is able to classify videos into categories like "blurred", "normal", "encoding problem", etc. By looking at the information in the database, we discovered strange results, which by checking the original video were due to for instance "encoding problems" where often more fish are detected then one would normally expect. In the fish recognition components, we are able to recognize more species, going from 15 to 23 species. Also, the recognition component can filter out false positives from the detection stage. The user interface is improved in both the usability and the fact that it can present more views of the data. There is also a connection to the workflow giving marine biologists the ability to process videos with other VIP software.

6.4.3 Data Processing Status

The video and image processing modules analyze the video data detecting and recognizing the fish in the video footage. The video data is saved in 10 min video clips, where in total we have 528,624 distinct video clips. For clarity reasons, we will state the both the number processed by the first prototype (which are measured at May 27th 2013) and the new numbers processed by the second prototype measured at the end of the project (October 2013) and the processing when finishing all valid videos ignoring videos with encoding errors (April 2014) in Table 6.1.

Table 6.1 This table compares the status at May 2013, October 2013 and April 2014, where large difference in the number of processed videos are shown

Measurements	Total	Fish detection			Fish recognition		
		(May 2013)	(Oct 2013)	(Apr 2014)	(May 2013)	(Oct 2013)	(Apr 2014)
Processed videos	528624	70784 (13 %)	530660 (100 %)	530660 (100 %)	67468 (13 %)	243563 (46 %)	456414 (87 %)
Processed videos (class normal)	75806		75806 (100 %)	75806 (100 %)		30807 (93.41 %)	75806 (100 %)
Speed		40 min (std 83 min)	12 min (std 12 min)	12 min (std 12 min)	175 min (std 381 min)	160 min (std 246 min)	160 min (std 246 min)

Lots of videos are blurred or have video encoding effect. The system has processed all videos expect for the videos that contain encoding errors

6.5 Lessons Learned with Current Architecture

6.5.1 Database Definitions

The Database Definitions are very important in this design. In this book, Chap. 7 discusses the exact Database Definitions in more details. In this section, we will focus on the effects the Database Definitions have in the development stage. For this project, the first Database Definitions were created during a meeting, where representatives of the different develop teams came up with an initial design.

During the course of the project, no major changes were necessary however a couple of small changes had to be made (for reasons of storage capacity, added functionality, etc.). It is important that changes can happen, however, it is more important that people are made aware of the changes and that they are communicated before they are made. Another issue is that changes might have an effect on old data which is already stored in a different format. Decisions need to be made if this format and data will still be supported. The further in the developing process, the larger the effect that potential changes can have, it is often good to freeze the Database Definitions until the next development meeting that allows people to propose changes if necessary.

Another potential issue of using Database Definitions to communicate is the growing complexity if more functionality is added to the system. Possible issues are duplicated definitions, unused tables, unknown system capabilities, unstructured design, etc. Software designer has to have a strong background in relational database design to keep the Database Definitions structured. It is important that the developers know what software is responsible for writing to which tables/records because that often defines the functionality of the individual software components.

The last issue related to the Database Definitions is making sure that the exact meaning is understood. An practical example for our database is recording that a fish appearance is in frame 500. Given a C++ implementation, you start counting from zero while matlab starts counting from one. In this case, the developer needs to agree if they use zero or one as starting basis, where often only defining the value as an integer is not enough.

6.5.2 Dependencies

Dealing with dependencies is very important in a research project like this. Although from a software developers point of view, there was no problem with dependencies. In this research project, we discovered that there are some dependencies that had not been solved:
First, in research solutions can be develop that depend on the input data. However, if you have no clue what kind of data you can expect, especially because the partners are also still developing solutions, it will be very difficult to develop data-driven solutions. The only solution here is either to ask partners in the project to develop their software as quickly as possible or ask if they can simulate their expected output. Second, if there is data available the quality of the data becomes important. In this project, we used a lot of automatic analysis software which can make mistakes in analyzing the data. However, multiple decisions from automatic analysis software are chained together and mistakes in the early stages have effects on the later software components. One way of dealing with these problem is having multiple automatic analysis software components that use slightly different methods which allows you to check how consistent the decision are.

6.5.3 Visualization

Visualization is often considered to be important to communicate with the potential users of the system. In the system design the user interface is one of the important parts in the system that should take care of this issue.

In this section, we argue that visualization are not only important for potential users but also for software engineer/scientist to communicate. Developing a system for showing the detected/recognized fish in the video is very difficult due to video encoding, etc. Once we created software that was able to draw contours around the fish in the video and link them to fish names both the computer scientist and the marine biologists had a better feeling for the performance and limitations of the system. Another example is the bulk processing of videos. We created a process that automatically executed the video processing software on the historical videos in the database that had not been processed. The discussions of the stakeholders about which videos to process started when a simple web-interface became available that showed how many videos still needed to be processed in each category.

In this system, we already mentioned that trust is an important issue, which is very difficult to communicate to non-expert users. By creating visualization, users of the system can first of all relate to the different issues in the system and second they know to which level they can to trust the system analysis. The development of good visualization is very difficult especially, because knowledge from two specializations in computer science are often required (human computer interaction and, for instance, computer vision), but new visualizations lead to better understanding of the underlying problems and thus to better solutions for both users and developers.

6.6 Conclusion

The software architecture presented which uses a database for communication and interfacing between software modules was an excellent choice for this kind of data-intensive research project. Part of the success of the project is due to this design decision at the start of the project which allows rapid prototyping. The major challenge of this project: Data-Intensiveness, Flexibility, Dependencies and Trust were solved with this kind of software architecture. Although, this project is successfully finished there are lessons to be learned. First, good maintenance and communication of the database definitions is essential for a project like this. Second, focus not only on database definitions and make sure the quality and content of the data is understood by the different developers as well. Third, start visualizing the stored data as soon as possible, because it helps with communication between developers of different partners and towards the final users especially if trust is a big concern for the end-users.

References

Boom, B.J., J. He, S. Palazzo, P.X. Huang, C. Beyan, H.-M. Chou, F.-P. Lin, C. Spampinato, and R.B. Fisher. 2014. A research tool for long-term and continuous analysis of fish assemblage in coral-reefs using underwater camera footage. *Ecological Informatics* 23(0): 83–97. Special Issue on Multimedia in Ecology and Environment.

Boom, B., and R.B. Fisher. 2011. Fish4knowledge Deliverable D5.1—component interface and integration plan. Technical report, University of Edinburgh.

Chapter 7
Fish4Knowledge Database Structure, Creating and Sharing Scientific Data

Bastiaan J. Boom

Abstract The Fish4Knowledge database is unique in the amount of video analysis information it contains. Because of this, the Fish4Knowledge project decided to make this data as available as possible to the rest of the world. The database is organized as a relation datastore, which can also be accessed through an RDF schema that is linked to the Linked Open Data Cloud. Both schemas are summarized here to explain the data acquired in the project and how this data is organized. Because of the large amount of data, downloading the entire database obtained by this project seems to be impossible. However, the original user interface of the project can display summary information of the database. Information in a more raw format can also be obtain from our website, where we provide both the video and the automatic analysis of fish observed in each video in a readable format (CSV: Comma-Separated Values). This allows future researchers to both use this research and also go beyond the research performed in the Fish4Knowledge project. This data is of particular interest to researchers in Marine Ecology and Image Processing/Computer Vision.

7.1 Introduction

Probably one of the most valuable assets of the Fish4Knowledge project is the final database. This database can be observed as the final output of the project providing marine biologists with a new kind of information which could not be analyzed before the start of the project. In this chapter, we discuss the structure of the database used in the Fish4Knowledge project (Boom et al. 2014), where the database plays a central role in the software design (see also Chap. 6). The database's main function is to share the data between all software components in the system, where software components interface/communicate with each other and share information by means of the database. The database has three main parts, which are connected with each other:

B.J. Boom (✉)
Cyclomedia, Van Voordenpark 1b, 5301 KP, Holland, Zaltbommel, The Netherlands
e-mail: bas.boom12@gmail.com

R.B. Fisher et al. (eds.), *Fish4Knowledge: Collecting and Analyzing Massive Coral Reef Fish Video Data*, Intelligent Systems Reference Library 104,
DOI 10.1007/978-3-319-30208-9_7

(1) The video/image processing part that allows the video/image processing software modules to contribute their analysis of the videos into a database, which other software modules can analyze or which can be represented in a user interface to the end user (marine ecologists).

(2) The workflow part that keeps track of which software modules have processed which videos and also which videos have not yet been processed. It allows us to perform version management, which was needed because during the project newer and better software modules for analysis became available.

(3) The summary part allows us to quickly analyze large parts of the database, where summary tables have been created to speed up the analysis of the large amounts of data stored in the database. For this project, the data of interest to the user is statistical information, which, using summary tables, can be quickly processed.

For the database, two schemas have been defined: a relational one (SQL) and linked-data one using RDF. The data has a clear relational structure, with little or no data integration or data heterogeneity problems, and a relational approach was deemed to be the best solution. In addition, an RDF schema has been defined in order to expose the project data in a Linked Data-compliant solution for Web-scale sharing of resources.

Because of the importance of the database, in addition to the RDF schema, we ensure that there is a partial backup of the database available in a European institute. This data is stored in a format that can be easily used by scientists and contains both the raw data and the analyzed video/image processing. Because of the unique nature of the data, where there are large number of videos that are analyzed by automatic software. This might be currently the biggest database of analyzed video in the world. New projects can both learn from this project or they can take up challenge that are open based on this data. An example is the LifeCLEF contests to which this project contributed a large dataset.

The chapter is organized as follows: the Relational Datastore Schema is given in Sect. 7.2 followed by the RDF schema to provide compliance to Linked Data in Sect. 7.3. The current ways of accessing the data are described in Sect. 7.4, while Sect. 7.5 gives examples of current and future possibilities for this data.

7.2 Relational Datastore Schema

The huge amount of information extracted from the videos is stored in a relational database which is designed to make it easy to retrieve the data typically needed to answer queries by marine biologists (Chap. 2). The datastore schema is described by using the Entity/Relationship model (Chap. 5). The database definitions (SQL) are important in this system, because this is basically the communication mechanism between software modules. More detailed description of the Database Definitions can be found in Chap. 2 and Appendix C, and Deliverable 5.2 (Palazzo et al. 2011) (a design document and changes have been made to the Database definitions/location and interfaces since then).

Table 7.1 This table show the size of the current database hosted by NCHC (Taiwan), where the largest table is the fish_detection table

Table	Number of rows (M)	Raw data size (Gb)	Total data size (Gb)
f4k_db.fish_detection	1445.41	259.37	322.26
f4k_db.detection_video_map	1448.71	101.70	101.70
f4k_db.fish_species	715.33	11.33	26.60
f4k_db.fish	124.28	18.80	21.01
f4k_db.traj_species	101.81	1.61	3.75
f4k_db.frame_class	11.61	1.00	2.65
f4k_db.fish_species_cert	32.55	1.29	1.29
f4k_db.summary_camera_39	7.13	0.56	1.24
f4k_db.summary_camera_46	7.12	0.56	1.24
f4k_db.summary_camera_38	6.31	0.49	1.10
f4k_db.summary_camera_37	4.46	0.35	0.78
f4k_db.summary_camera_42	4.31	0.34	0.75
f4k_db.ui_fish_v2	2.60	0.24	0.69
f4k_db.ui_fish	1.77	0.19	0.64
f4k_db.hmm_fish_detection	1.71	0.34	0.62
f4k_db.summary_camera_44	1.49	0.12	0.26
f4k_db.summary_camera_43	0.83	0.07	0.15
f4k_db.video	0.63	0.09	0.14
f4k_db.processed_videos	0.79	0.07	0.12
f4k_db.video_old	0.52	0.07	0.12
f4k_db.sumary_camera_41	0.63	0.05	0.11
f4k_db.summary_camera_40	0.28	0.02	0.05
f4k_db.video_class	0.53	0.04	0.04

The first column indicates how many records (in millions are presented in the tables, the second column show the amount of raw data in Gb, while the final column give the amount of data that is really necessary for storage in the database because of indexing allowing also quick querying of this information. Notice that is a unique effort in data collection, where not many projects have analyzed such a large amount of video data

Table 7.1 gives an impression of the size of the database (as of October 2013). It shows that the "fish_detection" table is the largest table where the fish detection software had finished processing all the videos in the database. The fish recognition component at that time finished processing half of the video. In Table 7.1, the largest tables in our database are shown, in which we give the number of rows (in millions) and the data size (in Gb) of the table. Beside the "fish_detection" table, other important tables are the "fish_species" table where all the species information is stored and the "fish" table that contains the fish trajectory information. Tables not mentioned in our schema are the "summary_camera_XX" tables that allow fast querying of the database. There are some other experimental tables "detection_video_map" and "traj_species" that are not in the original design (Palazzo et al. 2011). What we observed in this project is that the initial design (Palazzo et al. 2011) still holds but for research and programming purposes other tables have been created to make the life of developers and research easier and the database more scalable.

7.3 Linked Open Data

In addition to the relational scheme, which can be accessed through SQL application programmer interfaces, we also exposed the data as Linked Open Data (LOD) using the RDF data model. There are three categories of RDF data: (1) the relational data describe above is directly mapped to RDF, (2) a taxonomy of Taiwanese coral fish has been published in SKOS (Simple Knowledge Organization System), (3) RDF links between the relational data and the SKOS taxonomy are linked to other relevant datasets.

7.3.1 Direct Mapping to RDF

For the LOD data set, a direct one-to-one mapping between all the relational data and RDF data is used, according to "A Direct Mapping of Relational Data to RDF" Working Draft[1] which is being develop by the W3C RDB2RDF Working Group[2]. A key advantage of this Direct Mapping is that the RDF schema is directly defined from the relational schema. The information is therefore not duplicated. The Fish4Knowledge Linked Open Data derived from the Direct Mapping are (syntax used is Turtle):

Abbreviation	`XXX=http://f4k.project.cwi.nl`
HTML browser entry point	`XXX/lod/`
Namespace URL	`XXX/lod/`
Namespace abbreviation	`f4k:  XXX/lod/`
RDF browser entry point	`f4k:all  XXX/lod/all`
SPARQL end point	`f4k:sparql  XXX/lod/sparql`
SPARQL explorer	`f4k:snorql  XXX/lod/snorql`

The example below shows the RDF description of camera #25. This is the camera 3 in HoBiHu harbor reef given by the Direct Mapping. URLs of the resources which use camera #25 are also returned, where the first of these includes a video fragment #16, that was shot by the camera:

```
@prefix f4k:   <http://f4k.project.cwi.nl/lod/> .
```

[1] http://www.w3.org/TR/2011/WD-rdb-direct-mapping-20110324/.

[2] http://www.w3.org/2001/sw/rdb2rdf/.

```
@prefix vocab: <http://f4k.project.cwi.nl/lod/vocab/resource/> .
@prefix rdf:   <http://www.w3.org/1999/02/22-rdf-syntax-ns#> .
@prefix rdfs:  <http://www.w3.org/2000/01/rdf-schema#> .
@prefix xsd:   <http://www.w3.org/2001/XMLSchema#> .
@prefix d2r:   <http://sites.wiwiss.fu-berlin.de/suhl/bizer/...
                                    .../d2r-server/config.rdf#> .

<http://f4k.project.cwi.nl/lod/resource/cameras/camera_id=25>
   a  <http://f4k.project.cwi.nl/lod/vocab/resource/f4k.cameras>;
      rdfs:label"camera #25 (3@HoBiHu)" ;
      vocab:cameras_camera_id "25"^^xsd:int ;
      vocab:cameras_location "HoBiHu" ;
      vocab:cameras_video_number "3"^^xsd:int .

<http://f4k.project.cwi.nl/lod/resource/videos/video_id=16>
      vocab:videos_camera_id
              <http://f4k.project.cwi.nl/lod/resource/cameras/...
                          .../camera_id=25> .

...
```

The next example shows an entry of the fish_detection table under the Direct Mapping. In this example, we omitted namespace declarations for brevity.

```
f4k:resource/fish_detection/fish_id=261208,video_id=3576, ...
                              ...frame_id=1652
      a vocab:f4k.fish_detection ;
      rdfs:label "fish_detection fish261208/v3576/f1652" ;
      vocab:fish_detection_detection_certainty
              "0.88877"^^xsd:double ;
      vocab:fish_detection_fish_id
              <http://f4k.project.cwi.nl/lod/resource/fish/...
                              .../fish_id=261208> ;
      vocab:fish_detection_frame_id
              "1652"^^xsd:int ;
      vocab:fish_detection_timestamp
              "2010-12-16"^^xsd:date ;
      vocab:fish_detection_tracking_certainty
              "0.79386"^^xsd:double ;
      vocab:fish_detection_video_id
              <http://f4k.project.cwi.nl/lod/resource/videos/...
                              .../video_id=3576> .
```

7.3.2 Taiwanese Coral Reef Fish Taxonomy in SKOS

This RDF data set uses SKOS to provide a detailed taxonomy of all coral reef fish species that live on the Taiwanese reefs. It is based on information from authoritative sources such as the Fish Database of Taiwan by Prof. Dr. K.T. Shao from the Biodiversity Research Center of the Academia Sinica in Taiwan. In September 2014, tentative version was available including 28113 fish images, associated with 2893 species descriptions. These species belong to 1051 genera, 300 families, 61 suborders, 47 orders and 3 (sub)classes. (Note that the taxonomy does not necessarily have a single root as fish are a paraphyletic collection of taxa, of which 3 are potentially relevant to the project).

The Fish4Knowledge SKOS taxonomy plays several roles. First, it provides a species-centric access point to the Fish4Knowledge LOD, in contrast to the relational data described above which primarily provides a detection-centric view. Second, it is a means to publicly share the external resources as LOD, that were used to train the species recognition software modules. Finally, the SKOS vocabulary deals systematically with the variety of names that are associated with fish species, making it understandable for machines.

7.3.3 Interlinking and Alternative Representations of Direct Mapping Data

A vital part of the interlinking is the mapping of the internal database keys of the relational species identifiers to the species that are defined as SKOS concepts in the taxonomy. The species recognition is based on the same authoritative sources that were crawled to create the SKOS taxonomy.

Geographical locations and event type common to LOD datasets such as DBpedia[3] and GeoNames[4] were created, where the number of camera locations is very small allowing us to do this manually.

7.4 Current Accessibility of Data

There are several ways to access the information in our database, the database itself or part of the database in different formats. We give an overview of the possible ways to access the database content. Notice that depending on funding for maintenance not all the data might be available in the future, although we did our best to preserve the data. Currently the database is located at NCHC in Taiwan, but direct access to this server is for security reasons not possible. A web-interface to query this data is provided at the following URL:

http://gleoncentral.nchc.org.tw: The "Private" UI: hosted at NCHC, with access to the full dataset and the workflow, as intended for biologists. Users need an account and a password, which must be created by an administrator. The accounts previously created at this address are still valid.

---

[3] http://ckan.net/package/dbpedia.

[4] http://ckan.net/package/geonames.

http://f4k.project.cwi.nl/demo/ui: The "Public" UI: hosted at CWI, with access to a limited dataset. It is not connected to the workflow. It may be faster when users are connected from Europe (although latency is still high). All users can create an account online.

There is also a website that allows you to query the raw videos which is located at http://gad240.nchc.org.tw/tai/video_query/.

A very large subset of this database has been made available at `groups.inf.ed.ac.uk/f4k/F4KDATASAMPLES/INTERFACE/DATASAMPLES/search.php`, allowing anyone interested in this data to use it. To make this data more accessible for different communities (such as marine biologists, marine ecologists, computer scientists), we store for each video all the processing information in a single table. The tables are stored as CSV files. Each line in the file records the detection of a fish in a given frame in the corresponding video. A typical example of a CSV file is shown in Fig. 7.1.

There are two interesting subsets of the data provided on the UK website:

1. All Years: a 10 min video clip from all working cameras taken at 08:00 every day during Oct 1, 2010–July 10, 2013, giving 5824 video clips (between 7 and 30 Mb). Note, some dates are missing due to broken cameras. This data allows analysis of fish patterns over annual cycles and comparison between sites.
2. Full Day: 690 video clips from the 9 cameras taken from 06:00 to 19:00 on April 22, 2011 (out of 702 possible). This data allows analysis of fish patterns over a full day period and comparison between sites.

The subsets of the provided subset are capture by different cameras, on three different site in Taiwan given below:

- NPP-3: 4 cameras
- HoBiHu harbor: 3 cameras
- Lanyu/Orchid Island—2 cameras

This website also supplies 2 example Matlab script files to demonstrate how to download/use the data files:

- *species_histogram.m* gives a histogram of the number of fish in a given species over time.
- *species_extraction.m* called by species_histogram to download the CSV files from the Fish4Knowledge web site and extract the data. With the interface, you can also download the videos and CSV files directly into your own filespace. You can then adapt the supplied matlab files to analyze the locally stored data.

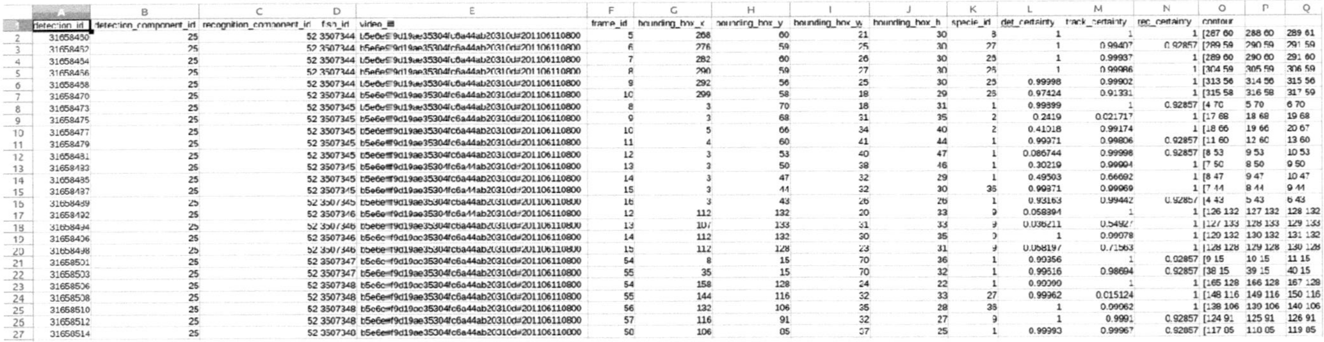

	A	B	C	D	E	F	G	H	I	J	K	L	M	N	O	P	Q
1	detection_id	detection_component_id	recognition_component_id	fish_id	video_id	frame_id	bounding_box_x	bounding_box_y	bounding_box_w	bounding_box_h	specie_id	det_certainty	track_certainty	rec_certainty	contour		
2	31658450	25	52	3507344	b5e6e[illegible]9d19ae35304fc6a44ab20310d#201106110800	5	268	60	21	30	8	1	1	1	[287 60	288 60	289 61
3	31658452	25	52	3507344	b5e6e[illegible]9d19ae35304fc6a44ab20310d#201106110800	6	276	59	25	30	27	1	0.99407	0.92857	[289 59	290 59	291 59
4	31658454	25	52	3507344	b5e6e[illegible]9d19ae35304fc6a44ab20310d#201106110800	7	282	60	26	30	25	1	0.99937	1	[289 60	290 60	291 60
5	31658456	25	52	3507344	b5e6e[illegible]9d19ae35304fc6a44ab20310d#201106110800	8	290	59	27	30	25	1	0.99986	1	[304 59	305 59	306 59
6	31658458	25	52	3507344	b5e6e[illegible]9d19ae35304fc6a44ab20310d#201106110800	9	292	56	25	30	25	0.99998	0.99902	1	[313 56	314 56	315 56
7	31658470	25	52	3507344	b5e6e[illegible]9d19ae35304fc6a44ab20310d#201106110800	10	299	58	18	29	25	0.97424	0.91331	1	[315 58	316 58	317 59
8	31658473	25	52	3507345	b5e6e[illegible]9d19ae35304fc6a44ab20310d#201106110800	8	3	70	18	31	1	0.99899	1	0.92857	[4 70	5 70	6 70
9	31658475	25	52	3507345	b5e6e[illegible]9d19ae35304fc6a44ab20310d#201106110800	9	3	68	31	35	2	0.2419	0.021717	1	[17 68	18 68	19 68
10	31658477	25	52	3507345	b5e6e[illegible]9d19ae35304fc6a44ab20310d#201106110800	10	5	66	34	40	2	0.41018	0.99174	1	[18 66	19 66	20 67
11	31658479	25	52	3507345	b5e6e[illegible]9d19ae35304fc6a44ab20310d#201106110800	11	4	60	41	44	1	0.99971	0.99806	0.92857	[11 60	12 60	13 60
12	31658481	25	52	3507345	b5e6e[illegible]9d19ae35304fc6a44ab20310d#201106110800	12	3	53	40	47	1	0.086744	0.99998	0.92857	[8 53	9 53	10 53
13	31658483	25	52	3507345	b5e6e[illegible]9d19ae35304fc6a44ab20310d#201106110800	13	3	50	38	46	1	0.30219	0.99994	1	[7 50	8 50	9 50
14	31658485	25	52	3507345	b5e6e[illegible]9d19ae35304fc6a44ab20310d#201106110800	14	3	47	32	29	1	0.49503	0.66692	1	[8 47	9 47	10 47
15	31658487	25	52	3507345	b5e6e[illegible]9d19ae35304fc6a44ab20310d#201106110800	15	3	44	32	30	36	0.99371	0.99969	1	[7 44	8 44	9 44
16	31658489	25	52	3507345	b5e6e[illegible]9d19ae35304fc6a44ab20310d#201106110800	16	3	43	26	26	1	0.93163	0.99442	0.92857	[4 43	5 43	6 43
17	31658492	25	52	3507346	b5e6e[illegible]9d19ae35304fc6a44ab20310d#201106110800	12	112	132	20	33	9	0.058394	1	1	[126 132	127 132	128 132
18	31658494	25	52	3507346	b5e6e[illegible]9d19ae35304fc6a44ab20310d#201106110800	13	107	133	31	33	9	0.036211	0.54927	1	[127 133	128 133	129 133
19	31658496	25	52	3507346	b5e6e[illegible]9d19ae35304fc6a44ab20310d#201106110800	14	112	132	30	35	9	1	0.99078	1	[120 132	130 132	131 132
20	31658498	25	52	3507346	b5e6e[illegible]9d19ae35304fc6a44ab20310d#201106110800	15	112	128	23	31	9	0.058197	0.71563	1	[128 128	129 128	130 128
21	31658501	25	52	3507347	b5e6e[illegible]9d19ae35304fc6a44ab20310d#201106110800	54	8	15	70	36	1	0.99356	1	0.92857	[9 15	10 15	11 15
22	31658503	25	52	3507347	b5e6e[illegible]9d19ae35304fc6a44ab20310d#201106110800	55	35	15	70	32	1	0.99516	0.98694	0.92857	[38 15	39 15	40 15
23	31658506	25	52	3507348	b5e6e[illegible]9d19ae35304fc6a44ab20310d#201106110800	54	158	128	24	22	1	0.99990	1	1	[165 128	166 128	167 128
24	31658508	25	52	3507348	b5e6e[illegible]9d19ae35304fc6a44ab20310d#201106110800	55	144	116	32	33	27	0.99962	0.015124	1	[148 116	149 116	150 116
25	31658510	25	52	3507348	b5e6e[illegible]9d19ae35304fc6a44ab20310d#201106110800	56	132	106	35	28	36	1	0.99062	1	[138 106	139 106	140 106
26	31658512	25	52	3507348	b5e6e[illegible]9d19ae35304fc6a44ab20310d#201106110800	57	116	91	32	27	9	1	0.9991	0.92857	[124 91	125 91	126 91
27	31658514	25	52	3507348	b5e6e[illegible]9d19ae35304fc6a44ab20310d#201106110800	58	106	85	37	25	1	0.99993	0.99967	0.92857	[117 85	118 85	119 85

Fig. 7.1 Comma-Separated Values (CSV) file of the information downloadable for marine ecologists of fish behavior over time. This information can be linked to the video, containing both when and where the fish is detected in the video and to which species the fish belongs

7.5 Data Usage and Future Possibilities

We discuss possible future usage of the data provided by the Fish4Knowledge project. Given the current dataset, the two most obvious future possibilities are **usage for marine ecology research** and **creating new challenges in the computer vision domain** which we discuss below. Although, the data might be used for other purposes as well.

- **Usage for marine ecology research**: Currently, we have analyzed a large dataset of videos, which might contain interesting observations/trends that could not previously be studied by marine ecologists. Together with the fish observations, we recorded water conditions such as temperature and water pressure which can be linked/correlated to the fish observations. Other publicly available information can be linked to this information and possible interesting correlations with either behaviors or abundance of species can be discovered. In earlier work, this project for instance looked at the behavior of fish during a typhoon, but another interesting relation can be the behavior of certain species in the presence of another species.
- **Creating new challenges in the computer vision domain**: The data generated by this project is in many respects similar to video surveillance recording of humans. The advantage, however, is that in the case of fish we do not need approval of every recorded subject for research purposes, while in human video surveillance this is necessary. The other advantage is that we have around 3 years of data, where in video surveillance data older than a month often needs to be destroyed. This kind of data gives new challenges in computer vision/image processing, for instance, deblurring of images (Vougioukas et al. 2013) by using previous frames (where the camera is still clean). Another possible research question is how to exploit the fixed geometry of a static background due to a fixed camera, but where the illumination of that background varies widely over short time periods.

The LifeCLEF challenge (Joly et al. 2014) uses a subset of our data for their challenge (Spampinato et al. 2014a), where 4 subtasks were defined based on the Fish4Knowledge project data:

- Subtask 1—Video Based Fish Identification: Four videos fully labeled, 21106 annotations corresponding to 9852 different fish instances.
- Subtask 2—Image Based Fish Identification: 957 videos labeled with 112078 fish annotated
- Subtask 3—Image Based Fish Identification and Species Recognition: 285 videos labeled with 19868 fish (and their species) annotated.
- Subtask 4—Image Based Fish Species Recognition: 19868 fish images annotated.

Although the annotation of this kind of data is still the main challenge, we hope to provide much larger datasets in the future, which can be used for large scale data experiments in, for example, clustering, nearest neighbor search and semi-supervised learning. One of the partly unsolved questions is how to retrieve the rare species in these large databases, where the common fish species are thousands of times more likely to appear than the rare species. This kind of dataset can be a future benchmark for these kinds of problems.

7.6 Summary

This chapter gives an overview of the database structures used in the Fish4Knowledge project. First of all, we showed a solid relational database schema that allows our automatic software to store and analyze the data. This resulted in a very large and unique database of research data to analyze fish behavior over extended periods of time. Second, this data can be made available in the Linked Open Data communities, where we link to already known definitions and resources available. Third, there are multiple ways to access and interact with both parts of the data and the entire database, where this project tried to have multiple backup options available allowing researchers from different fields to access the data even after the project ended. Finally, we indicate possible uses of this data, where both for marine ecology and computer vision/image processing new research areas and challenges can be explored using the data, where one of the important challenges created after the project is the LifeCLEF challenges (Joly et al. 2014; Spampinato et al. 2014a) which uses our data resources.

References

Boom, B.J., J. He, S. Palazzo, P.X. Huang, C. Beyan, H.-M. Chou, F.-P. Lin, C. Spampinato, and R.B. Fisher. 2014. A research tool for long-term and continuous analysis of fish assemblage in coral-reefs using underwater camera footage. *Ecological Informatics* 23(0): 83–97. Special Issue on Multimedia in Ecology and Environment.

Joly, A., H. Müller, H. Goëau, H. Glotin, C. Spampinato, A. Rauber, P. Bonnet, W.-P. Vellinga, and B. Fisher. 2014. Lifeclef 2014: multimedia life species identification challenges. In *Proceedings of CLEF*.

Palazzo, S., C. Spampinato, and J. van Ossenbruggen. 2011. Fish4knowledge Deliverable D5.2 - rdf/rdms datastore definition. Technical Report Del 5.2, Fish4Knowledge Project.

Spampinato, C., R.B. Fisher, and B.J. Boom. 2014a. Lifeclef fish identification task 2014. In *CLEF working notes*.

Vougioukas, K., B.J. Boom, and R.B. Fisher. 2013. Adaptive deblurring of surveillance video sequences that deteriorate over time. *20th IEEE international conference on image processing (ICIP)*, 1085–1089.

Chapter 8
Intelligent Workflow Management for Fish4Knowledge Using the SWELL System

Gayathri Nadarajan, Cheng-Lin Yang and Yun-Heh Chen-Burger

Abstract F4K's workflow component is the chief mediator between the user interface (UI) and the video and image processing (VIP) components that reside in F4K's high performance computing (HPC) platforms. Not only does it decompose high level user queries into lower video-level command-line invocations, it also makes its *own* decisions on which VIP modules and the best parameters to select, which hardware platform to perform the video processing executions on and which fault tolerance strategies to take during the executions so as to optimize the overall system's performance. In this chapter, we describe the workings of F4K's workflow component, SWELL (**S**emantic **W**orkflows with **E**rror Handling for **L**arge Video Ana**l**yses) which comprises a workflow engine and a workflow monitor. We also describe the F4K domain ontologies that have heavily influenced the development of SWELL and have been used for term matching between partner components.

8.1 Introduction

The workflow component of the F4K project is responsible for investigating relevant methodologies to efficiently generate, distribute and monitor processes that constitute the video processing tasks that aggregate to answer higher level (user) queries from

G. Nadarajan (✉)
Biomedical Knowledge Engineering Laboratory (BiKE), #102 Bio
Material Research Building, Seoul National University, 101 Daehak-ro Jongro-gu,
Seoul 110-744, South Korea
e-mail: gaya.nadarajan@gmail.com

C.-L. Yang
School of Informatics, University of Edinburgh, 10 Crichton St,
Edinburgh EH8 9AB, UK
e-mail: s0969605@inf.ed.ac.uk

Y.-H. Chen-Burger
School of Mathematical and Computer Sciences,
Heriot-Watt University, Edinburgh EH14 4AS, UK
e-mail: y.j.chenburger@hw.ac.uk

R.B. Fisher et al. (eds.), *Fish4Knowledge: Collecting and Analyzing Massive Coral Reef Fish Video Data*, Intelligent Systems Reference Library 104,
DOI 10.1007/978-3-319-30208-9_8

marine scientists. More specifically, its task is to take in video data that have been captured by the F4K project partner NARL and analyze and process them in useful ways to answer targeted user queries. The approach that we have chosen for the workflow system is a semantics based one. This approach enables us to separate the problem and domain descriptions, the application computer vision and data analysis software components and the actual workflow enablement component, the workflow engine. These components are understood and connected via a set of knowledge based representations.

This loose-coupling approach facilitates a more flexible mechanism that adapts more easily to a different problem description (or a different problem area when desirable). This is particularly useful, as the problem and domain descriptions evolve over the life time of the F4K project and beyond. It also enables us to make use of new software components that become available over time. Moreover, it allows us to improve the *brain* component, the workflow engine, to become more efficient or richer, as needed, as we make use of high performance computing facilities through NARL. This approach should therefore enable each of these major components of the system to evolve and improve independently, without needing substantial changes to other components.

The problem and domain descriptions are also written using a knowledge-rich language, Prolog, so that logic-based programming techniques can be applied directly. As a result, we are able to make use of planning technology to perform dynamic workflow composition and assist workflow execution. The semantics of the problem and domain descriptions are defined in a set of ontologies that have corresponding knowledge-based representations. This provides semantically more human-understandable and uniform labels to annotate videos and images. Such labels are a translation of the results as produced by the image and video processing software components. This labeling will then be used to assist user-query answering functions that is a part of the workflow system (Nadarajan et al. 2013a, b).

In the sub-sections below, we provide the inner workings of SWELL, namely the workflow engine (Sect. 8.2.1) and the workflow monitor (Sect. 8.2.2). We also report on the collaboration with workflow fault tolerance experts in introducing a measure of "resilience" to evaluate the overall performance of the workflow system.

8.2 SWELL System Design

The F4K workflow management architecture (Fig. 8.1) shows an overview of the components that the SWELL system interacts with, its main functions, and its sub-components. It communicates with the front end and VIP modules using a MySQL database and communicates with NARL's HPC platforms using command line interfaces with resource schedulers. As can be seen there are two workflow management sub components: (1) Workflow Engine and; (2) Workflow Monitor. The user (on-demand) queries can be one of the following:

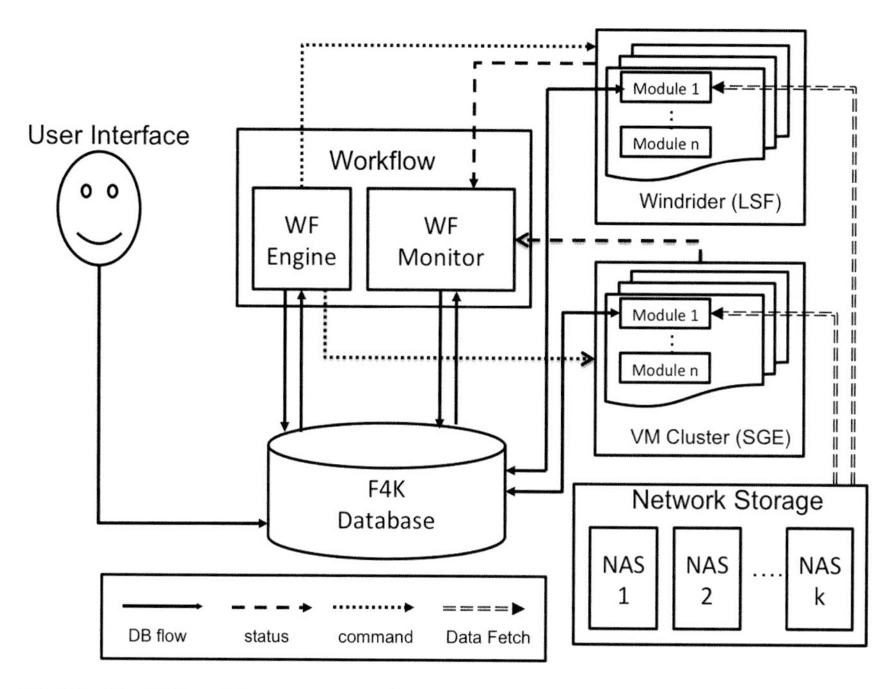

Fig. 8.1 The F4K workflow component, SWELL, communicates with partner components using a central database and command line invocations. It consists of a *workflow engine* which deals with the generation of abstract and concrete workflow instances from user requests, and a *workflow monitor* which deals with error handling and user query statistics

Q1 Detect and track all fish in a given date range and set of camera locations.
Q2 Identify all fish species in a given date range and set of camera locations.
Q3 Estimate how long a fish detection (Q1) or species recognition (Q2) query will take to produce results, without sending the query for execution.
Q4 Abort a query (Q1–Q3) that is currently being processed.

Internal batch queries are those that are invoked by the workflow system itself. These are predominantly batch tasks on newly captured video clips that involve fish detection and tracking (Q1 above) and fish species recognition (Q2 above). These batch queries are given low priority and are scheduled at "quiet" times, i.e., when on-demand queries are least likely to be processed.

The computing environment consists of nine nodes on a cluster of virtual machines with a total of 66 CPUs (called VM cluster) and two nodes on a supercomputer with a total of 96 CPUs (called Windrider). The workflow system resides in the master node of the VM cluster and makes use of both platforms to process queries. It deals with two different resource schedulers: Grid Engine (SGE) (Oracle 2014b) on the VM cluster and Load Sharing Facility (LSF) (IBM 2014) on Windrider (Fig. 8.2). The queues that are used by F4K workflow are *monos01*, *serial*, *short*, *medium* and *long*. Each queue has its own capacity and restrictions, as governed by its administrator.

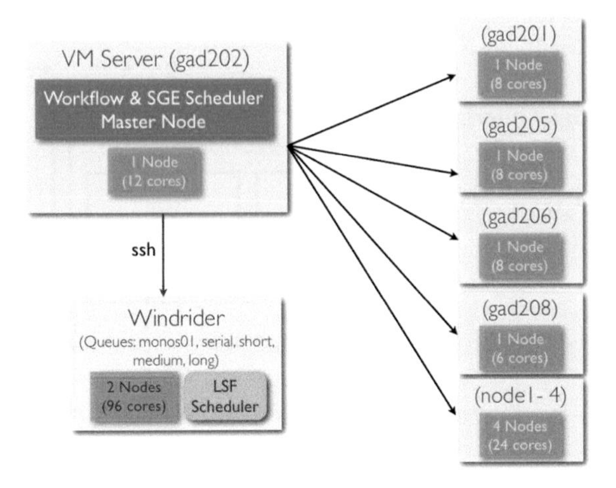

Fig. 8.2 The F4K computing environment is a heterogeneous platform made up of a group of virtual machines (VM cluster) and a supercomputer (Windrider)

For example, *monos01* is dedicated to the F4K user on Windrider. However, all other queues are shared with other Windrider users, therefore optimal queue usage is not within our control. *Serial* only takes single-core jobs on a first-come, first-served basis which resembles a non-parallel queuing system. For *short*, *medium* and *long* only two jobs can execute at a time and there is a minimum interval between the completion of a job and the execution of the next job. These three queues can take multiple-core jobs, i.e., parallel jobs.

The database tables that are used for communication between the SWELL system and other partner components are *query_management*, *job_monitoring* and *processed_videos*. Their definitions are provided in Appendix B. The workflow system and the UI communicate using the *query_management* table. Queries are inserted by the UI and processed by the workflow. The workflow engine and workflow monitor communicate using the *job_monitoring* table. Finally when the VIP modules are executed, they update the shared table *processed_videos*. We now look at the workflow engine in more detail.

8.2.1 Workflow Engine

The workflow engine is a standalone component which constantly listens for new queries and processes them. It integrates with other components in the F4K system via database tables; it communicates with the user interface via the *query_management* table, video processing modules via the *processed_videos* table and with the workflow monitor via the *job_monitoring* table.

When the workflow detects a query, it first recognises the type of query: whether it is compute-intensive (Q1 and Q2) or not (Q3 and Q4). Example queries are given in Sect. 8.2. For compute-intensive queries (Q1 and Q2), the workflow will need to

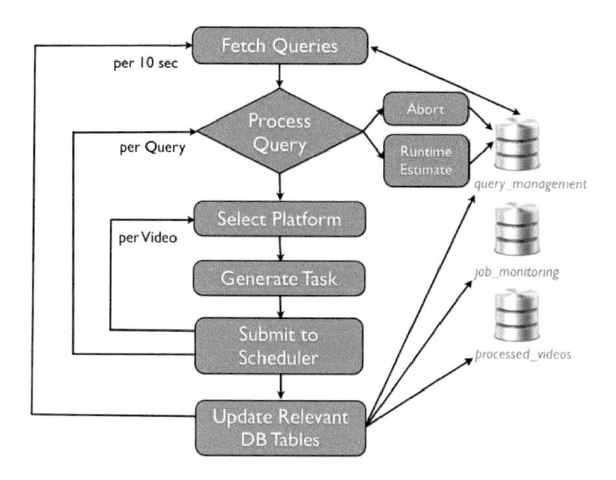

Fig. 8.3 The workflow engine is responsible for processing on-demand user queries and batch queries. It composes and schedules jobs for the queries and updates relevant database tables to pass control to the workflow monitor

select a platform for execution. Two types of such queries can exist: high priority (user) queries and low priority (batch) queries. At present all high priority jobs are scheduled on the Windrider platform, while the low priority jobs are split equally between the VM cluster and Windrider (Fig. 8.3).

Batch queries are initiated internally by the workflow as follows. On a daily interval, the workflow looks for new *unprocessed* videos over the last 24 hours from the database. If no new unprocessed videos are present, it looks for hourly unprocessed *historical* videos and creates fish detection and tracking queries on them, which will be caught by the workflow engine. Failing that, it looks for half hourly unprocessed videos and creates fish detection and tracking queries. Missing videos are skipped and a record of their timestamps and location are kept in the database.

After selecting the appropriate platform for a query, the workflow engine retrieves each video associated with that query. An example query is "What is the overall fish population in the Nuclear Power Plant (NPP-3) station between January and March 2012?". The user interface component deals with presenting the query and communicating the results to the user. For each video, the command line call including the invocation of the resource scheduler and the selection of appropriate algorithms for that query type will be generated by the workflow engine. This is done via a planning-based workflow composition mechanism (Nadarajan et al. 2013a). Using this mechanism, new software modules with enhanced algorithms can be easily added, detected and selected by the workflow engine. The planner plays an important part in performance-based selection of video processing software modules. The command line call that is sent to the scheduler to process each video is known as a **job**.

Another important feature of the workflow engine is in dealing with job dependencies. This scenario applies to fish species recognition jobs (Q2). A fish recognition module can only be applied to a video when a fish detection module has *already* processed it. The workflow engine deals with a fish species recognition job on a video using the following control logic:

- If fish detection has been completed, then run fish recognition only.
- If fish detection has not been started, run fish detection and fish recognition, specifying a dependency flag between them.
- If fish detection has been started but not completed yet, run fish recognition with a dependency flag on the running fish detection job.

Currently, the workflow schedules a maximum of 300 batch jobs (i.e., processes 300 videos) every 24 hours and listens for new on-demand queries every 10 seconds. Implementation-wise, there are two daemon processes for (i) managing the queries (every 10 seconds); and (ii) creating batch jobs (every 24 hours). When the jobs are queueing and executing, the workflow monitor oversees their executions for successful completion, or deals with errors that occur. The workflow monitor can handle various scenarios on Windrider and the VM cluster. The workings of the workflow monitor are presented next.

8.2.2 *Workflow Monitor*

Although LSF and SGE resource schedulers provide basic job status monitoring functions, they are not intelligent enough to handle different types of job executing scenarios that need more sophisticated methods to be dealt with. For example, a low priority job could end up starving if it has been waiting in the queue which is constantly packed with high priority ones. Therefore, it is essential to build a smart job status monitoring and error handling system to tackle different job executing scenarios.

Another benefit of having a job status monitoring and error handling system is to provide real-time and detailed statistics of the progress of each query. It allows the user to be able to track up-to-date status of her/his request. Currently, the system is able to compute the following statistics:

1. The percentage of completed jobs of each user query.
2. The percentage of successful job executions of each user query.
3. The percentage of abandoned job executions (by user/system) of each user query.
4. The percentage of pending jobs of each user query.

The job status monitoring and handling system is designed to run in parallel with the workflow engine to support the workflow system, by regularly checking status of scheduled jobs in the queue. Figure 8.4 shows the high-level design architecture of the system. It communicates with the workflow engine via the *job_monitoring* table; it updates job statistics via the *query_management* and *job_monitoring* table.

There are two major components in workflow monitoring system: (1) Job status classifier and (2) Event handler. Job status classifier constantly inspects a job's status and classifies it into one of five categories: Pending, Running, Suspending, Failed (abnormal) and Done (success). Each job category has its own event handler that was carefully designed to handle the possible execution scenarios described below.

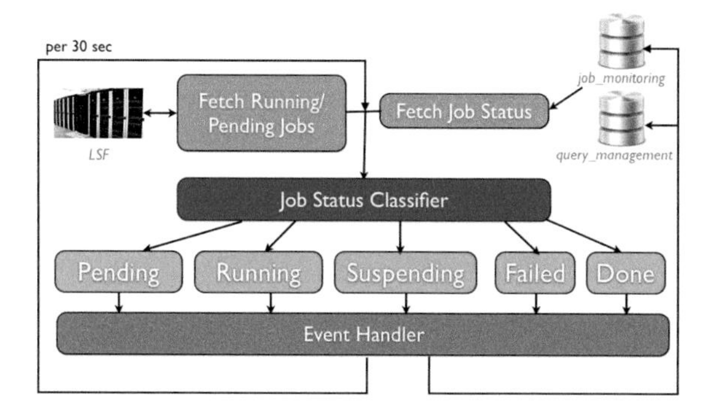

Fig. 8.4 The workflow monitoring system is responsible for handling different execution scenarios. It updates each job status and statistics to relevant database tables

8.2.2.1 Pending Jobs

The workflow engine may schedule hundreds or thousands of high priority jobs per on-demand query. It also processes up to 300 low priority background jobs on a daily basis. Therefore, it is possible that an individual job will be waiting in the queue for a long time. The event handling procedure has different pending time thresholds based on the job priority. High priority jobs have much shorter thresholds than low priority jobs. If a high priority job, which needs to be processed immediately, has been pending longer than its waiting threshold, the event handler will check if there is any low priority job in the queue and suspends it to make sure that the high priority job can be executed shortly.

If the event handler detects a low priority job that has been pending longer than its waiting threshold, it will kill the job and resubmit it with the highest priority to make sure it will be executed and not starved in the queue. The main reason that the event handler uses the *kill-and-resubmit* action is that the supercomputer platform disallows the user to alter the job priority of submitted jobs. The only solution to increase the priority is to kill the pending job and resubmit it with a higher priority.

8.2.2.2 Prolonged Jobs

If a job is running longer than expected, it might have encountered unexpected problems and valuable computing resource could be wasted. The average execution time for a job varies with the software module (which has a unique component identifier) that is selected for execution (see Table 8.4). The maximum running time limit has been set as 3 times the average execution time of the assigned component identifier of the job. If the job running time is longer than its threshold, it will be considered as a failed job (Sect. 8.2.2.4). The event handler will be triggered to kill and resubmit the job to the scheduler. When the event handler resubmits the job, it also keeps track of the number of times the job has been submitted to the scheduler. If a job has been submitted to the scheduler twice, it will be killed and marked as a failed job. This prevents further wasting of computing resource.

8.2.2.3 Suspended Jobs

A job may be suspended by the workflow engine or the scheduler itself. A mechanism exists to ensure that a job is not suspended for longer than a predefined threshold. The event handler will be triggered when this threshold is exceeded, where it will resume the suspended job.

8.2.2.4 Failed Jobs

An executing job might fail due to various reasons such as computing node failure, missing data or internal interconnection issue. This will be relayed as an exit status by the scheduler. To increase the successful completion chances of such failed jobs, the event handler adopts *resubmit policy*. If a running job exits unexpectedly or finishes with an error code, the event handler will be triggered, where it resubmits the job back to the queue. In order to have a balance between job resubmission frequency and effective computing resource usage, the failed job is resubmitted at most twice.

8.2.2.5 Completed Jobs

If a job is completed successfully, the event handling will update the final job execution statistics to *job_monitoring* and *query_management* table.

8.2.2.6 Dependent Jobs

Although job dependencies are dealt by the workflow engine as mentioned in Sect. 8.2.1, the workflow monitoring system still applies the same policies (Sects. 8.2.2.1–8.2.2.5) above to each dependent job. When the jobs that have other jobs depending on them fail or face unforeseen circumstances, the workflow monitor will handle them as any other job, while any effect that will have on the dependent jobs will be handled internally by the scheduler.

8.2.3 Workflow Evaluation

As the workflow system is complex, it is not trivial to perform unit testing on each component. Furthermore the performance rate of the workflow is not easily known due to other factors, such as damaged data, failed infrastructure and other system components which could cause jobs to fail. Hence we have taken two criteria into consideration when evaluating the workflow's performance: the impact on the overall system when the workflow is not present and its quality of resilience. These criteria will be explained and described next.

8.2.3.1 Impact on Overall System Without Workflow

In order to understand the performance of the workflow system with respect to the handling of different execution scenarios outlined in Sect. 8.2.2, we decided to study its impact on the overall system when the workflow system is not used. We outline the list of errors that can occur within the F4K workflow execution environment and observe the impact on the system when the workflow is used and not used. Table 8.1 summarizes how each scenario is handled in the presence and absence of the workflow. Additionally, it states the possible effects on the system when the job monitoring and error handling system is not used.

In summary, we can conclude that when the system does not make use of the workflow, optimal resources and queues are not being selected as no fault tolerance mechanism is in place. Jobs that fail are not rerun and in extreme cases jobs can starve. All these factors can affect the overall system performance drastically, justifying the role and contributions of the workflow.

8.2.3.2 Quality of Resilience

In the workflow community, "Quality of Resilience" (QoR) Tolosana-Calasanz et al. (2011) is a metric that identifies *how* robust a given workflow *is likely to be* prior to its enactment. While workflow Quality of Service (QoS) tries to characterize service

Table 8.1 Summary of possible impacts on overall system with and without workflow error handling

Scenario	System handling using workflow	System handling without workflow	Possible effect(s) without using workflow
Successful Job	Finished	Finished	All jobs are waiting in the same queue without utilising full system capability
Failed Job	Re-run once	Exit directly	The failed job will not be detected until a manual check is performed
Job dependency	With dependency handling	Without error handling	The dependent job could be queueing forever
High priority job waiting too long	Suspend low priority job to release resources	Job waits in the queue	High priority jobs can be held for a long time
Low priority job waiting too long	Resubmit with higher priority	Job waits in the queue	Low priority jobs can be starving in the queue

levels such as performance or cost from a users' perspective, and to monitor and to maintain the agreed service levels, QoR aims at specifying the *likelihood of failure* of a workflow instance. Whereas significant work in workflow enactment has focused on performance (often measured as the workflow makespan) and the associated QoS metrics, limited attention has been given to resilience. In F4K, the workflow is generated through a planner, which is able to take a user query and combine services that can carry this out. The concept of resilience also includes availability, reliability, and fault tolerance as outlined in Tolosana-Calasanz et al. (2011).

Fault tolerance in F4K consists of three main phases: fault detection, identification, and correction—orchestrated by the workflow monitor. Consider a workflow in F4K to detect, track (Q1, Sect. 8.2) and identify (Q2) all fish species over a given date range and set of camera locations. The planner will generate a workflow template consisting of two data dependent steps, a first step ($t1$) for detecting & tracking and a second ($t2$) to identify fish species. 4 candidate tools (or **abstract tasks** in QoR terms) for $t1$ and 2 for $t2$. The planner uses detection accuracy and performance (measured in terms of the execution time) as a criteria for selection between them. F4K has registered over 6,00,000 executions of this query. This enables us to use this data to understand the QoR associated with this workflow (Rana et al. 2013).

Table 8.2 summarizes QoR values for all possible instances of $t1$ and $t2$. Based on this table, metric $m1$ gets the number of alternatives for a given abstract task—3 for task $t1$ and 1 for $t2$. Metric $m2$ counts the number of inputs for each task—$t1$ has no previous input tasks, $t2$ has $t1$ as input. Metric $m3$ measures the number of resources involved in the execution—all tasks in F4K require a single computational resource. Metric $m4$ considers the number of failed executions out of the overall number of executions. Metric $m5$ gets the execution time per task. With 4 different variants for $t1$ and 2 for $t2$, up to 8 different workflow configurations can be generated ($t1_1 - t2_1$: $wf_1, t1_1 - t2_2$: $wf_2, \ldots, t1_4 - t2_2$: wf_8). For each of the 8 workflow configurations, the following QoR metrics at workflow level can be computed (see Table 8.3): metric $m6$ accounts for the average number of alternative tasks per task. Metric $m8$ measures the number of tasks of each workflow that have input dependencies. Metric $m9$ considers the number of failed executions compared to total executions of a workflow. Metric $m10$ reflects the workflow execution time (makespan). Finally, metric $m11$ extracts the number of computational resources required by a workflow.

Table 8.2 F4K quality of resilience metrics at task-level

Quantitative QoR metrics classification: task level							
Metric	Description	$t1_1$	$t1_2$	$t1_3$	$t1_4$	$t2_1$	$t2_2$
m1	Number of alternatives tasks	3	3	3	3	1	1
m2	Number of input tasks	0	0	0	0	1	1
m3	Number of resources	1	1	1	1	1	1
m4	Task failure rate	3.02	4.12	6.0	2.1	21.7	12.3
m5	Task execution time (s)	397	411	1596	1342	4984	13134

Table 8.3 F4K Compilation of quality of resilience metrics at workflow level

Quantitative QoR metrics classification: workflow-level									
Metric	Description	wf_1	wf_2	wf_3	wf_4	wf_5	wf_6	wf_7	wf_8
m6	Average number of alternatives tasks	2	2	2	2	2	2	2	2
m8	Number of task joins	1	1	1	1	1	1	1	1
m9	Wf failure rate	12.36 %	7.66 %	12.9 %	8.21 %	13.85 %	9.15 %	11.9 %	7.2 %
m10	Wf execution time (s)	2550	5244	5371	4747	7857	9726	5531	14643
m11	Total numbers of resources	2	2	2	2	2	2	2	2

In Table 8.2, it can be seen that abstract task $t1_1$ has the lowest execution time among the four available tasks, hence the best candidate for selection. For $t2$, workflow planning will select $t2_1$ for the same reason. Nevertheless, this solution does not take into account the likelihood of failure using the combination of $t1_1 - t2_1$ (denoted as $wf1$ in Table 8.3). Besides, it is important to note that a failure in a workflow configuration leads to recovery strategies that, in the best case, also affect performance—due to their inherent overhead. In the worst case, however, these failures require human intervention. Therefore, a resilience aware planner would take into account the QoR information gathered, so that the choice for the workflow configuration would consider performance, accuracy and also QoR. For instance, in Table 8.3, a planner enriched with such QoR information would be aware of the fact that combining $t1_1 - t2_1$ (with the lowest execution time) has a higher failure rate (12.36 %) than using $t1_1 - t2_2$ (7.66 %).

We have shown how QoR metrics can be associated with multiple workflow alternatives, that enable users to make more informed decisions about which alternative to choose. A decision made exclusively in terms of performance and accuracy may lead to configurations with poor QoR, which may lead to worse performance due to the overhead in the recovery of the fault tolerant techniques. The QoR analysis of F4K workflows has paved an avenue for the development of a data mining model that would allow QoR computation on the fly.

8.3 F4K Domain Ontologies

We created a set of domain ontologies that are based on user requirements for the Fish4Knowledge (F4K) project—goal, video description and capability. The roles of the ontologies are to (1) support the development of appropriate functions of the project's workflow system, and (2) serve as a communication media to interface with other F4K components. The ontologies were designed in collaboration with image processing experts, marine biologists and user interface experts to capture the domain knowledge succinctly.

Based on a mapping between the user requirements and a high level abstraction of the capabilities of the VIP modules, we have constructed the Goal Ontology (Sect. 8.3.1). The Video Description Ontology (Sect. 8.3.2) contains the environmental factors related to the videos, among others. The Capability Ontology (Sect. 8.3.3) describes details of the VIP tools and techniques. For ontology development and visualization purposes, OWL 1.0 (McGuinness and van Harmelen 2004) was generated using Protégé version 4.0. Where applicable, ontological diagrams were derived using the OntoViz plugin. To date, the Goal Ontology contains 52 classes, 85 instances and 1 property, the Video Description Ontology has 24 classes, 30 instances and 4 properties and the Capability Ontology has been populated with 42 classes, 71 instances and 2 properties.

8.3.1 Goal Ontology

The Goal Ontology contains the high level questions posed by the user, interpreted by the system as VIP tasks (termed as **goals**) and the constraints to the goals. The high level queries were derived from mapping the 20 questions posed by F4K's marine biologists to the goals that they fall into based on the types of tasks that are required to be carried out to address these questions (Nadarajan and Chen-Burger 2012). A VIP task as indicated here is a high level goal in our Goal Ontology that can be understood by an image processing expert, who writes low level programs or software to solve it.

The main relationship between classes shown here is the sub-class relationship, *is-a*. Instances and properties are not shown due to the limitation of the visualization tool. The 'Goal' class (Fig. 8.5) is the umbrella concept that includes the eight main

Fig. 8.5 Goal ontology denoting the main classes of goals that were applicable to the 20 questions posed by F4K's marine biologists

VIP tasks that can be found in Fig. 8.5. We also created a level of intermediate classes between the goal class and the lower level VIP goals to increase the flexibility and readability of the ontology. Under these general concepts, more specific goals may be defined, for example 'Fish Detection', 'Fish Tracking', 'Fish Clustering', 'Fish Species Classification' and 'Fish Size Analysis'. The principle behind keeping the top level concepts more general is also to allow the ontology to be easily extended to include (new) tasks as appropriate.

'Constraint on Goal' (Fig. 8.6) refers to the conditions that restrict the video and image processing tasks or goals further. In F4K's context, the main constraint for a VIP goal is the 'Duration', a subclass of 'Temporal Constraint'. Each task may be performed on all the historical videos, or a portion specified by the user—within a day, night, week, month, year, season, sunrise or sunset (specified as instances of 'Duration'). Other constraint types include 'Control Constraint', 'Acceptable Error' and 'Detail Level'. The control constraints are those related to the speed of VIP processing

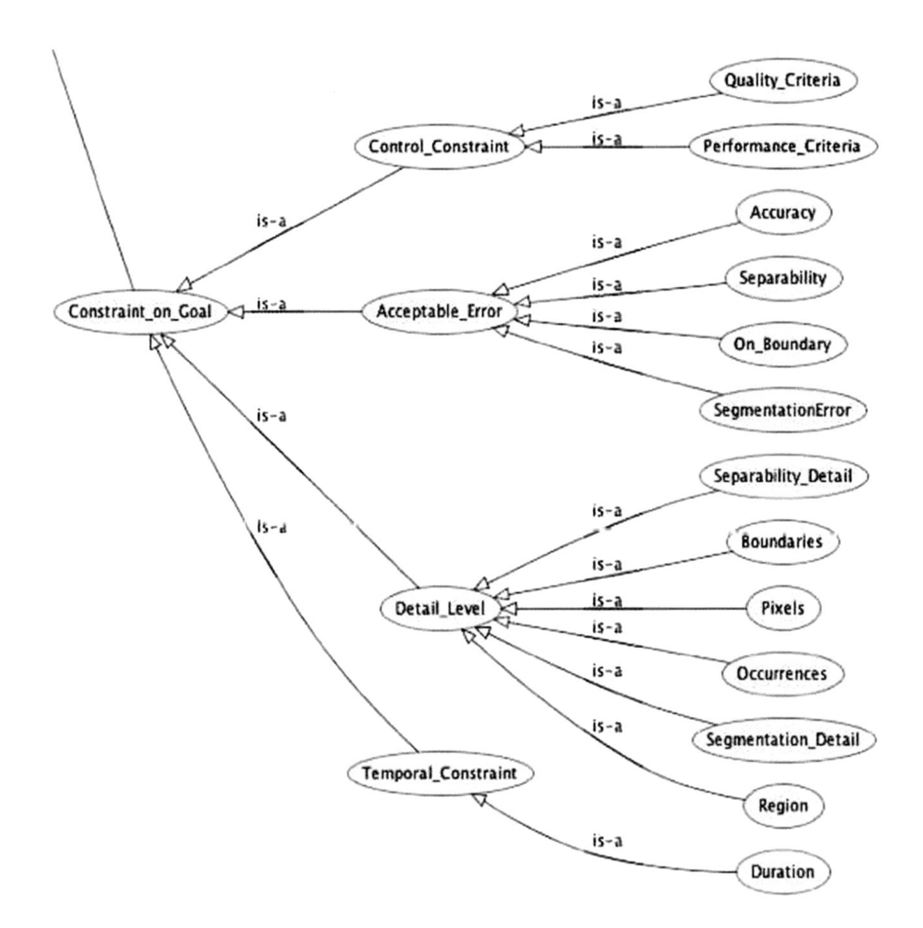

Fig. 8.6 Snapshot of the constraints of the goal ontology that consist of four types of constraints

and the quality of the results expected by the user. Acceptable errors are the threshold for errors that the user may want. For example, 'Accuracy' level may include false positives (or not) when detecting objects. The class 'Detail Level' contains constraints that are specific to particular details, for example detail of 'Occurrence' is used for detection tasks to restrict the number of objects to be detected.

8.3.2 Video Description Ontology

The Video Description Ontology describes the concepts and relationships of the video and image data, such as what constitutes video/image data, the acquisition conditions such as lighting conditions, color information, texture, **environmental conditions** as well as spatial relations and the range and type of their values. Figure 8.7 gives a pictorial overview of the main components of the Video Description Ontology. The upper level classes include 'Video Description', 'Descriptor Value', 'Relation', and 'Measurement Unit'.

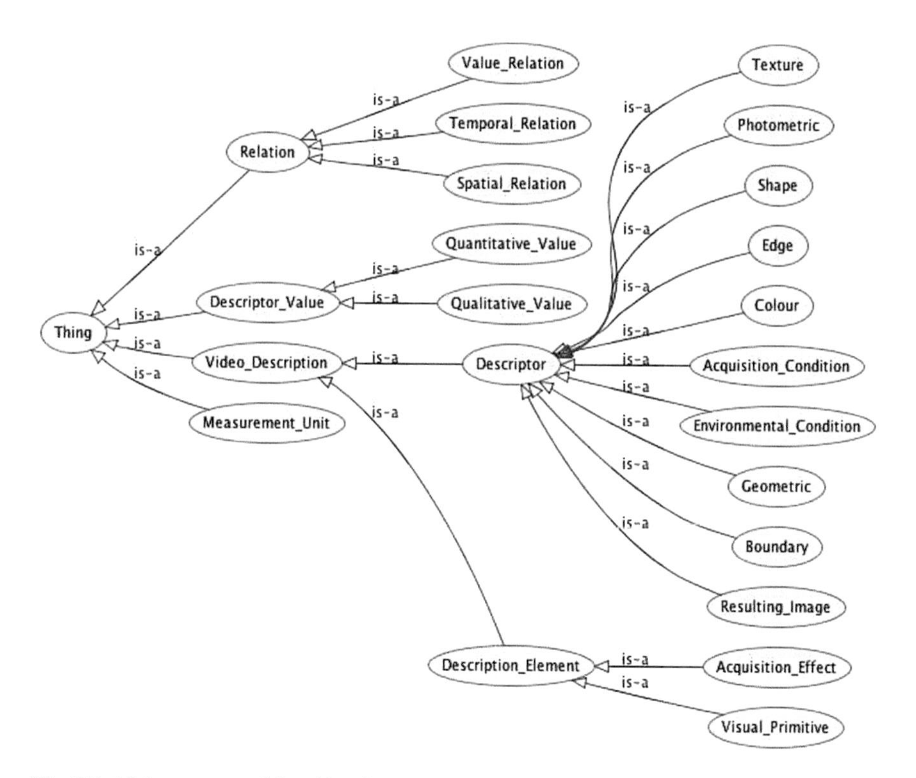

Fig. 8.7 Main concepts of the video description ontology

The main class of this ontology is the 'Video Description' class, which has two subclasses—'Description Element' and 'Descriptor'. A description element can be either a 'Visual Primitive' or an 'Acquisition Effect'. A visual primitive describes visual effects of a video/image such as observed object's geometric and shape features, e.g., size, position and orientation while acquisition effect descriptor contains the non-visual effects of the whole video/image that contains the video/image class such as the brightness (luminosity), hue and noise.

The descriptors for the description elements are contained under the 'Descriptor' class and are connected to the 'Description Element' class via the object property 'hasDescriptionElement'. Typical descriptors include shape, edge, color, texture and environmental conditions. Environmental conditions, which are acquisitional effects, include factors such as current velocity, pollution level, water salinity, surge or wave, water turbidity, water temperature and typhoon. These values that the descriptors can hold are specified in the 'Descriptor Value' class and connected by the object property 'hasValue'. For the most part, qualitative values such as 'low', 'medium' and 'high' are preferred to quantitative ones (e.g., numerical values). 'Qualitative' values could be transformed to quantitative values using the 'convertTo' relation. This would require the specific measurement unit derived from one of the classes under 'Measurement Unit' and conversion function for the respective descriptor. This ontology is used to describe the videos and external effects on it such as environmental conditions.

8.3.3 Capability Ontology

The Capability Ontology (Fig. 8.8) contains the classes of video and image processing (VIP) tools, techniques and performance measures of the tools with known domain heuristics. The main concepts intended for this ontology have been identified as 'VIP Tool', 'VIP Technique' and 'Domain Descriptions for VIP Tools'. Each VIP technique can be used in association with one or more VIP tools. A VIP tool is a software component that can perform a VIP task independently, or a function within an integrated computer vision library that may be invoked with given parameters. 'Domain Description for VIP Tool' represent a combination of known domain descriptions (video descriptions and/or constraints to goals) that are recommended for a subset of the tools. This will be used to indicate the suitability of a VIP tool when a given set of domain conditions hold at a certain point of execution. At present these domain descriptions are represented as strings and tied to VIP tools.

The main types of VIP tools are video analysis tools, image enhancement tools, clustering tools, image transform tools, basic structures and operations tools, object description tools, structural analysis tools and object recognition and classification tools. The class 'Object Description Tool' specifies tools that extract features such as color, texture, size and contour, while image transform tools are those concerned with operations such as point, geometric and domain transformations. 'VIP Technique' is a class that contains technologies that can perform VIP operations. For now, two

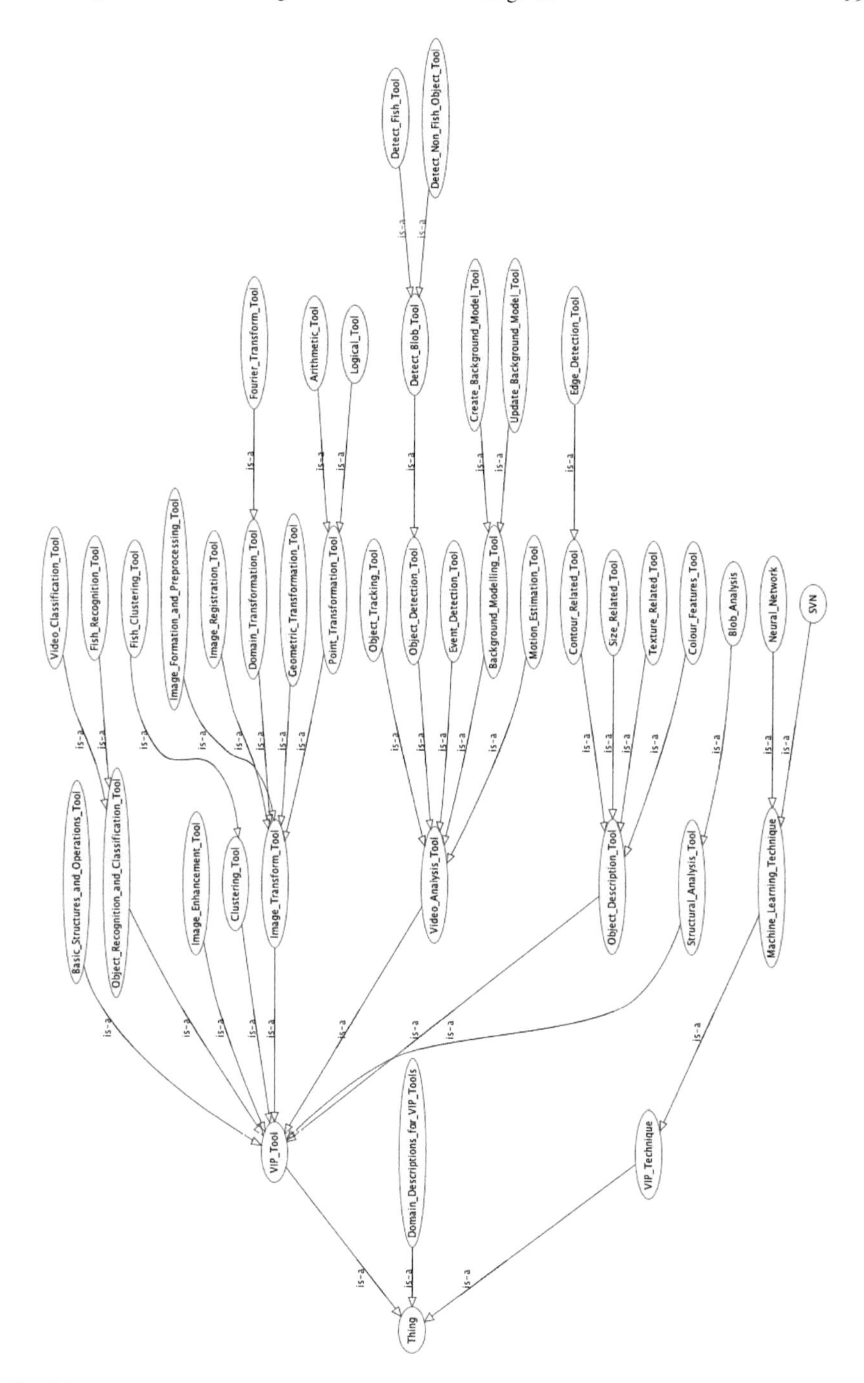

Fig. 8.8 Main concepts of the capability ontology

Table 8.4 The performance metrics of the stable software components for fish detection and tracking (IDs 135, 141 and 142) and fish species recognition (IDs 128 and 135) in F4K

Component ID & Type	Average execution time (s)	Average queue time (s)	Maximum execution time (s)	Minimum execution time (s)	Average wait time (s)
128 recognition	8796 (~2.4 h)	6164 (~1.7 h)	355381 (~4 days)	15	68
135* detection	734 (~12 min)	90 (~1.5 min)	19604 (~5.4 h)	0	93
136* recognition	9902 (~2.75 h)	42655 (~11.5 h)	344113 (~4 h)	16	32
141 detection	892 (~14.7 min)	31460 (~8.7 h)	2845 (~47 min)	10	4
142 detection	11336 (~3.15 h)	53205 (~4.8 h)	28107 ~7.8 h)	180	11

Note * denotes default component

types of machine learning techniques have been identified. These techniques could be used to accomplish the task of one or more VIP tools. For example, neural networks can be used as classifiers as well as detectors.

The F4K domain ontologies were enhanced with performance metrics capabilities in the final year of the project (Nadarajan et al. 2013b). In particular, we included hardware- and software-related measures that would inherently improve the overall performance of the workflow system when considered appropriately. This enhancement was implemented in the form of extensions to the Capability Ontology. The terminologies in the ontologies were used by the workflow and partner components for consensual knowledge and term sharing. Most of these terms are communicated and used through the database.

At F4K's production run, we have gathered statistics for the performance metrics related to specific software components. Table 8.4 shows the aggregation for all the stable software components that have been used to run fish detection and tracking and fish species recognition tasks. Note that the default components were selected based on a combination of their performance and accuracy levels. Component 135 was the most optimal choice for the workflow, given that it has the fastest average execution time and shortest average queueing time among the three available detection components. For species recognition, however, component 136 was preferred over component 128 as it was the most accurate of the two, despite having higher average execution and queueing times.

8.4 Concluding Remarks

The F4K's workflow management system combines automatic composition, scheduling and monitoring within a complex problem environment. We have described how F4K's SWELL system has played an integral part in connecting partner components and efficiently making use of HPC resources for solving data- and compute-intensive video analyses. We have also presented F4K's domain ontologies which are instrumental in providing shared terminologies for marine biologists, image processing experts, user interface designers and workflow experts. We are continuing our efforts in improving overall system performance by collaborating with workflow fault tolerance experts in designing a data mining workflow evaluation framework.

References

IBM. 2014. Load sharing facility (lsf). http://www-03.ibm.com/systems/platformcomputing/products/lsf/. Accessed 11 October 2014.

McGuinness, D. and van Harmelen, F. 2004. OWL web ontology language. World wide web consortium (W3C).

Nadarajan, G., and Y.-H. Chen-Burger. 2012. Goal, video description and capability ontologies for fish4knowledge domain. In *Proceedings of the 12 Special Session on Intelligent Workflow, Cloud Computing and Systems, KES-AMSTA*.

Nadarajan, G., Y.H. Chen-Burger, and R.B. Fisher. 2013a. Semantics and planning based workflow composition for video processing. *Journal of Grid Computing* 11(3): 523–551.

Nadarajan, G., Yang, C.-L., and Y.-H. Chen-Burger. 2013b. Multiple ontologies enhanced with performance capabilities to define interacting domains within a workflow framework for analysing large undersea videos. In *Proceedings of the 5th international conference on knowledge engineering and ontology development (KEOD 2013)*.

Oracle. 2014b. Open grid scheduler (sge). http://gridscheduler.sourceforge.net/. Accessed 11-October 2014.

Rana, O.F., R. Tolosana-Calasanz, G. Nadarajan, C.L. Yang, and Y.H. Chen-Burger. 2013. Analysing quality of resilience in fish4knowledge video analysis workflows. In *Proceedings of the International Workshop on Clouds and (eScience) Applications Mandagement—CloudAM*

Tolosana-Calasanz, R., Lackovic, M., Rana, O., Banares, J., and D. Talia. 2011. Characterizing quality of resilience in scientific workflows. In Proceedings of the 6th workshop on workflows in support of large-scale science, 117–126.

Chapter 9
Fish Detection

Daniela Giordano, Simone Palazzo and Concetto Spampinato

Abstract Fish detection is the upstream module of the whole Fish4Knowledge system and, as such, it needs to be as accurate and fast as possible. Driven by these needs, several state of the art and new approaches for object segmentation in videos have been developed and tested. We opted for background modeling—based approaches as they fit better with the underwater domain peculiarities. In particular, kernel density estimation methods, modeling colors, texture and spatial information of both the background and the foreground, proved to be the best performing ones not only in underwater video sequences but also in other complex scenarios. To provide more robustness to fish detection, we also developed a post-processing layer (added on top of the background modeling one) able to filter out effectively false detections by using "real-world" object properties. Despite the low-quality (low frame rate and spatial resolution) of the processed underwater videos, the achieved results can be considered satisfactory especially considering that most of the state of the art approaches failed. This chapter provides, therefore, an overview on the development and deployment of fish detection module for the Fish4Knowledge system. It includes a detailed analysis of the challenges of underwater video analysis, the limitations of the existing approaches, the devised solutions and the experimental results.

9.1 Introduction

In the Fish4Knowledge project, fish detection in videos has been carried out through background modeling approaches as opposed to template matching (not applicable because of the large variability of fish appearance) and motion analysis

D. Giordano · S. Palazzo · C. Spampinato (✉)
Dipartimento di Ingegneria Elettrica, Elettronica E Informatica,
Universita' di Catania, Viale A. Doria 6, 95125 Catania, Italy
e-mail: cspampin@gmail.com

D. Giordano
e-mail: dgiordan@diit.unict.it

S. Palazzo
e-mail: simopal6@gmail.com

R.B. Fisher et al. (eds.), *Fish4Knowledge: Collecting and Analyzing Massive Coral Reef Fish Video Data*, Intelligent Systems Reference Library 104,
DOI 10.1007/978-3-319-30208-9_9

(low-resolution videos do not allow an effective estimation of a fish motion model) methods. Background modeling aims at building an estimated image of the scene without objects of interest. This model is then compared to each new video frame for identifying foreground objects. The most popular background modeling approaches are the density-based ones, where the temporal distribution of each background pixel is modeled by either a probability density function (e.g., Gaussian) (Stauffer and Grimson 1999) or non-parametric kernel density estimation (Sheikh and Shah 2005). In early years, these approaches used mainly historical variations of pixel colors while, recently, there is a trend towards background modeling methods that employ spatial and texture information, i.e., methods that build the background model by taking into account textures computed on neighboring pixels. In addition, methods that estimate the background and the foreground separately have been favored to the ones relying only on the background model, since the latter do not account for the spatial and temporal changes that may happen in the foreground. While excluding foreground modeling and spatial/texture information might work and make the algorithms run effectively in simple scenarios, there are evident limitations when more complex scenes need to be modeled, as in the case of the underwater environment. Indeed, the underwater domain shows several difficulties that make the fish detection task challenging and all the strategies adopted within the F4K project have been influenced by the following factors (affecting the performance of most of the existing background modeling approaches):

- **Sudden light changes**: mainly due to the light propagation in water as affected by the water surface shape;
- **Multimodal backgrounds and periodic movements**: (e.g., plants affected by flood-tide and drift) which may lead to misclassification of background areas as target objects and vice versa;
- **Low-quality videos**: in terms of image resolution and video frame rate, due to bandwidth limitations between the cameras and the storage servers;
- **Image quality**: atmospheric phenomena (e.g., typhoons, storms), murky water and bio-fouling generally affect the quality of video frames, thus making the video analysis components more prone to errors. Image compression errors also affected many videos;
- **Appearance model**: as fish have three degrees of freedom and undergo erratic movements, their shape is subject to sudden changes (further amplified by the low video frame rate);
- **Motion model**: besides the difficulty introduced by the low video frame rate (which caused fish to move by a significant amount of pixels between two consecutive pixels), fish motion patterns are typically hard to understand and predict.

The underwater case is, therefore, an extreme case for background modeling since it may show a combination of above factors. For instance, dynamic or multimodal backgrounds, abrupt lighting changes, and radical and instant water turbidity changes can all be found in the same scene. To complicate even more the situation, the underwater environment shows two almost exclusive characteristics with respect to other domains: three degrees of freedom and erratic movements of objects (i.e., fish).

Moreover, beside implementing accurate fish detection approaches able to deal with all the aforementioned difficulties, in the Fish4Knowledge project, algorithms' efficiency and processing times were of primarily importance. Indeed fish detection had to deal with continuously-recording videos and with a huge amount of previously-recorded videos (dating back to 2009). In addition fish detection was the upstream module of the entire system, so it needed to be as fast as possible. This aspect greatly influenced the choice of the methods as to balance the trade-off between accuracy and efficiency.

In this chapter we will describe the methods employed to deal with all the above aspects. We, first, describe the most recent work on background modeling, highlighting strengths and weaknesses. Then, a description of the approaches devised for fish detection, together with their performance, is given. Since background modeling is not able to solve per se the fish detection problem, a post-processing approach, exploiting real-world object properties, has been developed with the goal to correct background modeling failures. Exhaustive performance analysis on several manually annotated underwater videos is finally presented. The last section draws the conclusion and provides indication for future developments.

9.2 Related Work

The objective of background modeling and subtraction algorithms is to create a model of the scene without objects and then to detect the foreground by comparing the current frame with the estimated background. This model should be able to cope with noise and illumination variations, and at the same time, it should also identify and remove object shadows, which technically are part of the foreground but can affect the performance of higher level applications (from object tracking to behavior understanding). Over the years several background modeling approaches of increasing complexity have been proposed. Among these, so far, the most popular are the *density-based* ones, which model each background pixel through a probability density function (pdf) based on learned visual cues. Such *pdf* is then updated adaptively according to the input frames. The most common used *pdf* is the Gaussian one and it is considered as the "de facto" standard baseline method for background modeling (Zhang et al. 2011). For example, Wren et al. (1996) modeled each pixel using a single Gaussian distribution in the YUV color space. The main downside of these approaches is that they do not perform well with dynamic natural environments which, instead, involve the use of multimodal density functions. Gaussian mixture models demonstrated to work reliably in such cases (Stauffer and Grimson 1999), although they lack of effective strategies to update adaptively the components in mixtures. To avoid the difficulty in identifying the appropriate shape of the *pdf*, non-parametric methods, e.g., kernel density estimation approaches, have been adopted for background modeling (Mittal and Paragios 2004) with a fair success. They build the background model by accumulating the values from the pixel's history, thus requiring many samples for accurate model estimation. This makes non-parametric methods computationally expensive, thus inappropriate for real-time applications.

Recently, in (Barnich and Van Droogenbroeck 2011) the authors classified background and foreground pixels by comparing, for each, its current value to a history of recent values. They state that the main differences/advantages of their method with respect to the state-of-the-art approaches lie in the randomness introduced to the background update policy: at each frame, a pixel's model is updated only with probability p (e.g., 1/16), and the choice of which history value to remove is random rather than according to a *FIFO* policy. However, due to the exploitation of pixel colors only, this method shows several limitations, namely, (1) the impossibility to handle luminosity changes and shadows, (2) correlation between features of a multidimensional feature space and (3) the lack of effective mechanisms for including structural variations of pixels' neighbors.

Recent research has, instead, proved that using pixel spatial information to build the background model, by joint domain-range density estimation (Sheikh and Shah 2005), achieves higher accuracy. Also, combining properly visual features (color, texture and/or motion) when modeling temporal and spatial pixel variations improves performance further (Han and Davis 2012).

Most of these methods employ complex texture features (more robust to luminosity changes than color features) to model the background; for example, (Heikkila and Pietikainen 2006) have successfully applied the local binary pattern (LBP) texture operator, because of its tolerance to illumination variations. Similarly, scale-invariant local ternary patterns (SILTP) (Liao et al. 2010) have been also used with success thanks to their insensitivity to light changes and shadows in the scene. (Zhang et al. 2011) argued that both LBP- and SILTP-based approaches are not able to model temporal pattern variations. Therefore, they maintain a background model using a spatial-temporal feature, the texture pattern flow, to compute inherent motion information. This feature is able to reflect the fact that the foreground moves as a block in a certain direction. While this assertion is generally true for humans, it does not hold for fish because they tend to move erratically with frequent direction changes.

Motion cues have been largely used (Mittal and Paragios 2004; Monnet et al. 2003) though being rather computationally expensive and failing with low resolution images because of the difficulty to estimate reliably motion. Other approaches adopt multidimensional kernel density estimators but do not perform effectively if there is dependency between features. To address this issue, (Han and Davis 2012) do not use a multi-feature background modeling, but they adopt multiple single dimensional models, which are then combined through Support Vector Machines. However, this approach in practice is not viable because it requires a training phase for each different scenario it has to deal with.

Furthermore, although in many cases modeling only the background has proved successful, it shows a major shortcoming: if object features are similar to the background, the object will unavoidably be detected as part of it. This consideration leads us to model explicitly the foreground and methods which keep both models (i.e., one for the background and one for the foreground) improved the performance considerably (Gallego et al. 2009).

All the discussed approaches are pixel-based and often result in misclassification errors because of noise, which often affects isolated pixels (Barnich and Van Droogenbroeck 2011). For this reason, block-based methods (Seki et al. 2003; Tsai and Lai 2009), which assume an inter-dependency between background models of neighboring pixels, have been also proposed. For example, (Seki et al. 2003) model the background by principal component analysis (PCA) for each image block: the foreground image is created by comparing the current image and the back-projection of its PCA components into the image space. The main shortcoming of block-based approaches is the inaccurate (in terms of object shapes and boundaries) object segmentation, requiring a consistent post-processing phase to improve the algorithm's output.

Summarizing, the key efforts of the recent literature on background modeling are:

- pixel-wise background model with increasingly sophisticated probabilistic models;
- inclusion of complex texture features, e.g., (Liao et al. 2010; Heikkila and Pietikainen 2006), as they are more robust against illumination changes and shadow;
- explicit foreground model (Sheikh and Shah 2005).

All of the above solutions have been tested and devised for human-centered contexts, but they are likely to fail in complex scenarios, such as the underwater one, where the images suffer from low contrast mainly because of light attenuation (caused by absorption and scattering), which strongly limits the object visibility (Raimondo and Silvia 2010). Also the targets to be detected, i.e., fish, move erratically and fast, complicating greatly the background and/or foreground modeling process. (Porikli 2006a) compared different algorithms (both for detection and tracking) under extreme conditions (that resemble somewhat the ones present in underwater scenes) such as erratic motion, sudden and global light changes, presence of periodic and multimodal backgrounds, arbitrary changes in the observed scene, low contrast and noise. However, the best performing methods on the scenarios described in Porikli (2006a) failed on the underwater context (Kavasidis and Palazzo 2012).

9.3 The Fish Detection Approaches

9.3.1 Background

Driven by the computation time constraint and by the features of the underwater domain, we first implemented and tested (see experimental result section) several pixelwise state of the art approaches, which were previously tested under conditions recalling the ones present in the Fish4Knowledge underwater videos

(Porikli 2006b). In particular, two algorithms based on mixtures of probability density function, e.g., the well-known "Adaptive Gaussian Mixture Model" (AGMM) (Stauffer and Grimson 1999) and the "Adaptive Poisson Mixture Model" (APMM) (Faro et al. 2011), were first used. Both methods showed good performance with multi-modal backgrounds but they failed in cases of frequent or abrupt lighting changes.

Since real-world physics often induces near-periodic phenomena in the environment, frequency decomposition-based representations of the background have been proposed (Porikli and Wren 2005). These algorithms detect objects by comparing the frequency transform representations of the background and the current scene. This requires to keep a fairly large number of past frames in order to accurately model the current frequency map against which the background model is compared in the F4K context, this kind of approaches performed well with repetitive scenes and low-contrast colors but was not able to deal adequately with erratic/fast fish movement and sudden lighting transitions.

A technique devised to handle issues related to noise and illumination changes consists of representing the scene using intrinsic images, where each image is obtained as a multiplication between its static and dynamic components (Porikli 2005). This approach has been used as a basis for the "Intrinsic Model" approach, which performs the multiplicative background-foreground estimation under uncontrolled illumination using intrinsic images. Every image is split into two components: the reflectance image (static component) and the illumination image (dynamic component). The former is invariant to lighting changes and will be almost identical under any light conditions. The background is modeled by calculating the temporal median of these components. Although it allowed for improved performance in cases of light changes, fast and erratic fish movements, it suffered when multi-modal backgrounds (e.g., algae) were present.

All the above approaches initially showed encouraging performance, but when more complex scenes were taken into account, their performance dropped dramatically leading us to investigate other solutions. In detail, their main downsides were in the adopted background model and background update mechanism which were not suitable to deal with the peculiarities of the underwater domain. As a consequence, other solutions were investigated; in particular, the VIBE approach (Barnich and Van Droogenbroeck 2011) that models background pixels with a set of neighborhood samples instead of with an explicit pixel model. One of the main peculiarities of this method is that spatial influence of neighboring pixels is integrated into the background model and foreground is also modeled (though with a very simple method), thus allowing for more robust detection. Since this approach showed a good compromise between efficiency and accuracy (see Performance Analysis section), it was employed for processing the whole set of historical videos.

9.3.2 A Texton-Based Kernel Density Estimation for Video Object Segmentation

Although the VIBE approach was adopted for the production run, more accurate approaches for enhancing detection performance were sought. In particular, following the current research trends in background modeling, we have developed a domain-range kernel density estimation approach modeling not only the background pixels but also the foreground ones. The method also exploits, for the modeling process, information on neighboring pixels and textures robust to illumination changes. Performance evaluation showed a significant improvement in accuracy not only in the underwater domain but also in other scenarios, outperforming the most recent approaches. Of course, the increase in accuracy was achieved at the expenses of efficiency as the new method was about one hundred time slower than VIBE. In the following the new approach is explained more thoroughly.

9.3.2.1 A Joint Domain-Range Model for Fish Detection

The joint domain-range model proposed (Spampinato et al. 2014c) extends the work in Sheikh and Shah (2005) by integrating texture features (computed using textons) into the model. In detail, for each pixel p at location (x, y) we build a feature vector v_p consisting of the (x, y) coordinates, the (s_1, s_2, s_3) color channels in a given color space (we tested RGB, HSV and Lab) and the energy of the texton image in a region $R = w \times w$ centered on p.

Our approach starts with assessing the magnitude of the maximum gradient image G within the region $R = w \times w$ by computing the eigenvector corresponding to the greatest eigenvalue of the matrix $J \times J^T$ with J being the Jacobian of the color space vector (Di Zenzo 1986). Afterwards the texton image is obtained from image G. Textons have been extensively adopted in texture analysis, especially for image retrieval (Julesz 1981) and can be defined as sets of patterns shared over the image (Julesz 1981). We adopted the texton images proposed in (Liu and Yang 2008). Generally, a smaller number of textons is used (usually 4); we, instead, make use of 7 textons because of the need to capture even the slightest texture variations within the considered region R and a smaller number of textons would prevent us from achieving this goal. The texton image is obtained by detecting textons on the image G and then setting to zeros all the pixels that do not belong to the detected textons. Figure 9.1 shows an example of how a texton image is derived using only four textons (for simplicity) on a 7×7 image.

As a global texture feature describing a pixel (x, y) and its neighbors in the region $R = w \times w$, we compute the energy e_T of the texton image T of the region R defined as:

$$e_T = \sqrt{\sum_{(i,j) \in R} T(i,j)^2} \tag{9.1}$$

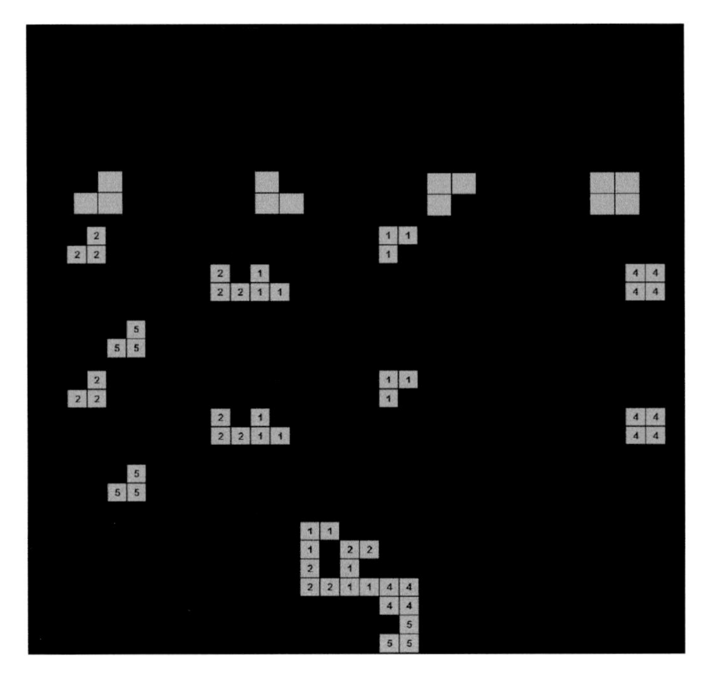

Fig. 9.1 **Texton Image Extraction**. *First row*: an example 7 × 7 image; *second row*: used textons; *third row*: texton detection in the previous image; *fourth row*: components of the texton image; *fifth row*: texton image

Therefore, each pixel $\boldsymbol{p}$ is then represented through the feature vector $v_p = (x, y, s_1, s_2, s_3, e_T)$ (with s_1, s_2, s_3 being the color values), and the joint domain-range model consists of the corresponding 6-dimensional space, on which the *pdf*s of the background and foreground models are built.

This is performed by means of Kernel Density Estimation (*KDE*) (Rosenblatt 1956): given the sets $\psi_b = \{\boldsymbol{b}_1, \boldsymbol{b}_2, \ldots, \boldsymbol{b}_n\}$ and $\psi_f = \{\boldsymbol{f}_1, \boldsymbol{f}_2, \ldots, \boldsymbol{f}_m\}$, containing all background and foreground samples, respectively. The corresponding *pdf*s can be approximated as:

$$P(\boldsymbol{p}|\psi_b) = \frac{1}{n} \sum_{i=1}^{n} \varphi\left(\frac{\boldsymbol{p} - \boldsymbol{b}_i}{H}\right) \tag{9.2}$$

$$P(\boldsymbol{p}|\psi_f) = \frac{1}{m} \sum_{i=1}^{m} \varphi\left(\frac{\boldsymbol{p} - \boldsymbol{f}_i}{H}\right) \tag{9.3}$$

where $\varphi(\boldsymbol{x})$ is a KDE kernel function with the usual properties of unitary integral, symmetry, zero-mean and with identity covariance and H is the kernel bandwidth. In our case we used the common multivariate Gaussian (as the novelty of the approach relies on integrating textons in the joint-domain range model not in the adopted kernel function), defined as:

$$\varphi(\mathbf{x}) = \frac{1}{\sqrt{2\pi^N |H|}} e^{\left(-\frac{1}{2}(\mathbf{x}-{}^{-})^T H^{-1}(\mathbf{x}-{}^{-})\right)} \tag{9.4}$$

with $\mathbf{x}$ being the variables, H the bandwidth matrix ($6\times$ 6 diagonal matrix because our model is six-dimensional) and $\boldsymbol{\mu}$ the mean vector. This representation allows to achieve two objectives: first of all, a spatial dependency relationship between pixels is introduced due to the joint domain-range model; secondly, the foreground model is managed separately from the background one.

In order to reduce the dimensionality of the model matrices and the frame processing time, the Binned KDE (Hall and Wand 1996) is used, i.e., the models are quantized into a more compact $X \times Y \times S_1 \times S_2 \times S_3 \times E$ space. Practically, the models are two matrices: P_b and P_f, representing at all times the current values of $P(\boldsymbol{p}|\psi_b) = P_b(v_p)$ and $P\left(\boldsymbol{p}|\psi_f\right) = P_f(v_p)$, respectively. Initially, the first N video frames are used to build the background model. For each pixel p in each frame, the $v_p = (x, y, s_1, s_2, s_3, e_T)$ feature vector is computed—appropriately quantized for the Binned KDE—and the discrete KDE kernel is applied at its location, thus increasing the $P_b(v_p)$ model cells by 1 and the neighboring cells also increased by a 0.5. The P_b matrix is then normalized by the total number of pixels used for the initialization. The foreground model P_f is initialized (although no foreground pixels have been detected yet) as follows: it is set to $P_f(v) = \gamma$, for each v cell in the model, where γ is a low value (0.1 in this work), accounting for the possibility of observing any uniformly distributed pixel value at any locations. The background update procedure, which will take into account the properties of the objects appearing in the following frames, is later described.

9.3.2.2 Foreground Detection

As new video frames are available, the current appearance of the observed scene is analyzed to identify areas which present (non-background) motion. In particular, the probabilities that each pixel belongs to either the background or the foreground are computed. Thanks to the discrete KDE representation of the models, such computation is straightforward, since the probability that pixel $\boldsymbol{p}$ with $v_p = \{x, y, s_1, s_2, s_3, e_T\}$ belongs to the background or the foreground models are simply $P_b(v_p)$ and $P_f(v_p)$, respectively. A motion binary map $M(x, y)$ is then built where each pixel is classified according to the log-likelihood ratio:

$$M(x, y) = \begin{cases} 0 \text{ if } -\ln \frac{P_b(v_p)}{P_f(v_p)} > T \\ 1 \text{ otherwise} \end{cases} \tag{9.5}$$

where $\boldsymbol{v_p}$ is the pixel feature vector at location (x, y); 0's and 1's in the output motion map represent background and foreground pixels, respectively, in the current video frame. The threshold T is set to balance sensitivity to foreground and background changes and robustness to noise.

9.3.2.3 Background and Foreground Model Update

After pixel classification has been completed, background and foreground models are updated with values from the current video frame.

The background update procedure consists of integrating the current frame's classification results into the KDE estimation, namely the $P(\boldsymbol{p}|\psi_b) = P_b(v_p)$ (given the binned pdf representation as described above) function with $v_p = (x, y, s_1, s_2, s_3, e_T)$ and ψ_b representing the current background KDE support points. In order to take into account the possibility that new objects appear in the scene (or that background pixels are misclassified), we update the background model with all the background pixels (referred as $\psi_{b,\mathrm{curr}}$) in the current image. The $P(\boldsymbol{p}|\psi_{b,\mathrm{curr}}) = P_{b,\mathrm{curr}}(v_p)$ function is computed from the $\psi_{b,\mathrm{curr}}$. The new background model $P_{b,\mathrm{new}}(v_p)$ is computed as a weighted mean between the current background model $P_b(v_p)$ and $P_{b,\mathrm{curr}}(v_p)$:

$$P_{b,\mathrm{new}}(v_p) = \alpha P_{b,\mathrm{curr}}(v_p) + (1 - \alpha) P_b(v_p) \tag{9.6}$$

The foreground model is recomputed every time from the ψ_f set of pixels detected as foreground in the current frame. As for the background, KDE is applied to estimate the $P_f(\boldsymbol{p}|\psi_f) = P_f(v_p)$ *pdf*. Similarly to what was done for the model initialization, a γ value is added to $P_f(v_p)$ to account for the appearance of new objects in the frame (Sheikh and Shah 2005).

9.4 Improving Detection Performance

As described in the previous section, several video object segmentation approaches have been developed and tested on the F4K underwater videos. Despite some approaches perform better than others, none of them is able to deal effectively with false positives that can be due to: (1) scene changes (e.g., light changes, camera movements, etc.) not absorbed by the background models, and (2) correct detection of non-interest objects (e.g., plant and algae movements). While in the former case, background modeling approaches should be able to cope with scene changes, in the latter case higher-level modules are needed to select only the objects of interest (fish in our case).

This led us to add a post-processing level (Spampinato and Palazzo 2012) to the fish detector in order to recover from potential failures. In detail, to discriminate fish from other background objects we adopted a set of specific features of real-world objects. The set of considered features exploits two main concepts: (1) the "human perceptual organization model" to discriminate blobs that are most likely produced by the motion of a biological object from blobs that may arise due to changes in the background (i.e., luminosity), (2) the "motion objectness" to compute the probability that a change detected by the above algorithms is due to fish movement instead of background movement (e.g., corals or algae).

9.4.1 Perceptual Organization Model Features

The ability of humans to identify objects, and more in general structures, without a priori knowledge of their contents is known as perceptual organization, which is governed by the four Gestalt laws that identify some basic principles of whole objects such as proximity, similarity, continuity, symmetry and convexity (Liu et al. 1999), i.e., real-world objects tend to have convex shape, tend to be symmetric with respect to a reference axis, etc. To measure quantitatively the Gestalt laws in real-word applications, we used the boundary energy function proposed by (Cheng et al. 2012):

$$E\left[\partial R\right] = \frac{-\int\int_R f\left(x,y\right)dxdy}{L\left(\partial R\right)} \tag{9.7}$$

where ∂R is the object's contour, $L\left(\partial R\right)$ the contour's length, and $f\left(x,y\right)$ is a weight function for each point belonging to the object. The first step for the evaluation of $f\left(x,y\right)$ is a superpixel segmentation (Felzenszwalb and Huttenlocher 2004) of the object's region into homogeneous patches. For each pixel (x,y), belonging to patch i, the corresponding weight is computed as:

$$f\left(x,y\right) = e^{-\theta \bullet \eta (S_i - S_a)} \tag{9.8}$$

In this formula, S_i is a two-component vector $[B_i \ \ C_i]$, where B_i and C_i represent, respectively, the boundary complexity of patch i and its cohesiveness with the other patches which make up the object. S_a is a reference vector computed at the largest patch of the object (Cheng et al. 2012). θ is a weight vector and η is vector element-by-element absolute value.

9.4.2 Motion Objectness

To distinguish between moving blobs produced by fish movement and blobs due to background object movements, we extended the "Objectness" concept (Alexe et al. 2012), used to identify generic class objects on still images, to the "Motion Objectness" one. Motion Objectness, particularly, takes into account the fact that fish is characterized by specific intraframe and interframe properties which make its movement different from background objects.

We use the term "blob" to indicate a particular area closed by a contour as provided by the fish detection processes.

- **Intraframe Properties**. To describe the peculiarities of a fish within an image, we took under consideration the following characteristics:
  - *Closed boundary in space*. This property aims at evaluating how the blob's contour matches the object's boundary. To measure it, we assess the density of edges (ED) included in a blob. Let δ_b and b^{θ}_{in} be, respectively, the contour of the considered blob b and the inner blob obtained by shrinking b by a factor θ. The edge density is given by:

$$ED = \frac{\sum_{p \in b^{\theta}_{in}} M_{ED}(p)}{A_{b \setminus \delta_b}} \tag{9.9}$$

    where $M_{ED}(p)$ is a binary edgemap and indicates if the pixel p is classified as edge by a Canny edge detector. $A_{b\setminus\delta_b}$ is the area of the blob b minus its contour δ_b ($\setminus$ is the set difference symbol).
    Moreover, we also used the percentage of superpixels intersecting (SI) the blob's contour, computed as follows:

$$SI = 1 - \frac{\sum_{s \in \mathscr{S}} \min(|s \setminus b|, |s \bigcap b|)}{|b|} \tag{9.10}$$

    where $\mathscr{S}$ is the set of superpixels computed as in (Felzenszwalb and Huttenlocher 2004) and $|b|$ the blob's area.
  - *Appearance difference from surrounding areas*. The dissimilarity of an object to its surrounding area is estimated by analyzing the color contrast (CC) along the object's boundary. Let b^{θ}_{out} and b^{θ}_{in} be the outer and the inner blob obtained, respectively, by dilating and shrinking the original blob b by a factor θ (empirically set to 2 in our implementation), the color contrast along the boundary of a blob b is computed as the Chi-square distance ($\chi^2(;)$) between the LAB histograms of the two rings (outer and inner) surrounding the object's boundary:

$$CC = \chi^2(h(b \setminus b^{\theta}_{in}), h(b^{\theta}_{out} \setminus b)) \tag{9.11}$$

  - *Internal homogeneity of color and texture*. Most fish appear to have a limited number of colors and a uniform texture (due to the low resolution of the video), especially when compared with complex background objects (e.g., algae, corals, rocks, etc.). The internal homogeneity of a blob has been assessed by computing the average color value and the average texture of all the superpixels in a blob. The more similar these average results are, the more likely the detected blob is actually a fish. The average color homogeneity (HC) is given by:

$$HC = 1 - \frac{\sum_{s \in \mathscr{S}} ||C_s - \overline{C}||}{Dim(\mathscr{S})} \quad (9.12)$$

where $\mathscr{S}$ is the same set of superpixels described above, C_s is the average color within each superpixel s, $\overline{C}$ is the average color within the whole blob and $Dim(\mathscr{S})$ is the number of superpixels. Analogously, the measurement of the internal texture homogeneity is performed by averaging the outputs of a bank of Gabor filters applied to each superpixel in the detected blob.

- *Preferred positions*. The spatial coordinates of the blob's centroid are also used to measure "objectness", since we assume that some positions are more likely than others to contain objects.

- **Interframe Properties**. Any fish holds the *motion coherence* property that allows us to distinguish it from the rest of the scene. To measure this property, we propose two cues based on the object's motion vector (calculated according to Bouguet 2000):

  - *Difference of motion vectors at object boundary*: Let $M_{b,in}$ and $M_{b,out}$ be the average motion vectors computed, respectively, in the ring just inside and the ring just outside the blob's boundary, the motion difference at the boundary Δ_{MV} is assessed as the Chi-square distance between the motion histograms assessed in the two rings (whose size was set empirically set to 3 in our implementation):

$$\Delta_{MV} = \chi^2(h(M_{b,out}), h(M_{b,in})) \quad (9.13)$$

  - *Internal motion homogeneity*. This cue is based on the assumption that the internal motion vectors of a correctly-detected fish are more uniform than the ones of a false positive. The detected blob is split into a set of superpixels (as above) and the average of motion vector's magnitude in each superpixel is compared with the global one. The internal motion homogeneity MH is computed as follows:

$$MH = 1 - \frac{1}{Dim(\mathscr{S})} \sum_{s \in \mathscr{S}} \frac{1}{Dim(\mathscr{V}_s)} \left| \sum_{p \in \mathscr{R}_s \cap \mathscr{V}_s} |M_p| - \overline{M} \right|^2 \quad (9.14)$$

where $\mathscr{S}$ is the set of superpixels in the analyzed blob and $Dim(\mathscr{S})$ its dimension, $\mathscr{R}_s$ is the union between the superpixel's current bounding box $R_s(t)$ and the bounding box of its last appearance $R_s(t-1)$, $\mathscr{V}_s$ is the set of valid points in $\mathscr{R}_s$, i.e., the points whose displacement project them inside $\mathscr{R}_s$ and $Dim(\mathscr{V}_s)$ is the number of valid points in $\mathscr{V}_s$. Finally, M_p is the motion vector (two components: x and y) describing the displacement of pixel p in two consecutive appearances and $\overline{M}$ is the average motion vector between all superpixels.

Fig. 9.2 Example of estimated probability of the blobs (for simplicity only the bounding *boxes* are drawn) to be a fish

The feature vector, containing the above objectness measures and the perceptual organization energy value of each detected blob, is then given as input to a naive Bayes classifier with two classes: *"object of interest" (OI)* and *"false positive" (FP)*, which computes the probability that the considered blob is a fish or not. Figure 9.2 shows an example of estimated probabilities of some detected blobs to be fish.

9.5 Performance Analysis

The evaluation of the fish detection performance was carried out by testing the performance, first, of the background modeling approaches which identify the foreground pixels that are then grouped together to form objects (fish), and, then, of the post-processing module which, instead, aims at filtering out the false positives due to errors in the previous step.

9.5.1 *Fish Segmentation in Underwater Videos*

To test our background modeling approaches, we used a dataset consisting of 17 "real-life"underwater videos. This dataset[1], in detail, contains about 3500 fish masks, manually labeled using the tool in (Kavasidis et al. 2013a), equally distributed in seven video classes (classification performed according to the typical features of underwater videos): "Blurred" (smoothed and low contrasted images), "Complex Background Texture" (background featuring complex textures), "Crowded" (lots of fish), "Dynamic Background" (background movements, e.g., plant movements, etc.),

[1] http://f4k.dieei.unict.it/datasets/bkg_modeling/.

Fig. 9.3 Underwater video dataset. From *top-left* to *bottom-right*: (1) Blurred, (2) Complex background texture, (3) Crowded, (4) Dynamic background, (5) Luminosity change, (6) Camouflage foreground object

"Hybrid" (more than one features: e.g., plant movements together with luminosity changes), "Luminosity Change" (videos affected by transient and abrupt luminosity changes), "Camouflage Foreground Object" (e.g., fish with colors similar to the background) and examples are shown in Fig. 9.3. We did not use any post-processing to improve detection results but removed the connected components having area less than 15 pixels. The accuracy was measured in terms of precision/recall curves and F-measure F_1, defined as:

$$Precision = \frac{TP}{TP + FP}, \; Recall = \frac{TP}{TP + FN}, \; F_1 = \frac{2 \times Precision \times Recall}{Precision + Recall} \tag{9.15}$$

where TP, FP and FN are, respectively, the true positives, the false positives and the false negatives measured when comparing, on a pixel basis, the ground truth binary masks and the output masks of the background modeling approach.

We tested the performance of several state-of-the-art approaches, compared to our kernel-density estimation approach described in Sect. 9.3.2, namely:

- P-Finder (Wren et al. 1996) which models the background with only one single Gaussian *pdf* (Gaussian):
- Two methods that exploit a mixture of Gaussians, namely, the original Gaussian Mixture Model by (Stauffer and Grimson 1999) (*GMM*) and its improvement by Zivkovic (*ZGMM*) (Zivkovic and van der Heijden 2006);
- The Eigenbackground (*EIGEN*) Subtraction method (Oliver et al. 2000),
- Two non-parametric kernel density estimation approaches: Sheikh's method (Sheikh and Shah 2005) (*KDE-RGB*), which uses only color features for modeling the background, and the MultiLayer background model (*ML-BKG*) by (Yao and

Odobez 2007), which, instead, employs also texture features computed via Local Binary Patterns.

- *VIBE* (Barnich and Van Droogenbroeck 2011) (background modeling through actual pixel values instead of using a predefined *pdf* shape) which has been previously applied to the underwater domain (Kavasidis and Palazzo 2012).

To avoid any implementation bias in the performance evaluation, we used only methods for which the original code was available.[2]

The performance of the different methods are reported as F-Measure values and illustrated in Table 9.1, which shows that combining color and texture features (as in our method and in *ML-BKG*) enhanced the background modeling's performance also in complex scenarios where targets and background had similar texture features. *ML-BKG* performed well mainly with complex backgrounds, while our approach shows good accuracy also in the cases of smooth regions (e.g., Blurred class) because of the ability of textons to describe also tiny texture variations.

The other information that can be derived from these results is that methods relying on a *pdf* with a predefined form (e.g., Gaussian) are not suitable to deal with complex scenes, where, instead, non-parametric methods perform much better. A qualitative comparison of our approach, *VIBE* and *ML-BKG* is presented in Fig. 9.4, which shows that our approach had good qualitative performance when compared to the other two.

Table 9.1 F-Measure scores (in percentage) for different methods on our underwater dataset

Video Class	P-finder	*GMM*	*ZGMM*	*EIGEN*	*ML – BKG*	*KDE – RGB*	*VIBE*	*Our method*
Blurred	75.3	83.3	77.8	81.7	70.3	92.6	85.1	**93.3**
Complex background	75.6	67.0	75.9	74.8	83.7	**87.5**	74.2	81.8
Texture crowded	71.2	**85.2**	74.4	73.8	79.8	82.5	84.6	84.2
Dynamic background	51.0	62.0	64.3	71.5	**77.5**	59.1	67.0	75.6
Hybrid	74.6	62.7	75.5	80.7	72.2	**85.7**	79.8	82.6
Luminosity change	48.1	63.1	59.1	70.4	**82.7**	72.1	70.4	73.0
Camouflage foreground object	72.4	66.3	70.0	70.2	73.5	54.1	76.3	**82.2**
Average	66.9	69.9	71.0	74.7	77.1	76.2	76.8	81.8
Standard deviation	11.1	9.2	6.5	4.4	4.9	13.7	6.4	6.0

[2] Most of the methods are available at https://code.google.com/p/bgslibrary/. The code of the remaining methods were made available by the authors and reference to the code can be found in the corresponding papers.

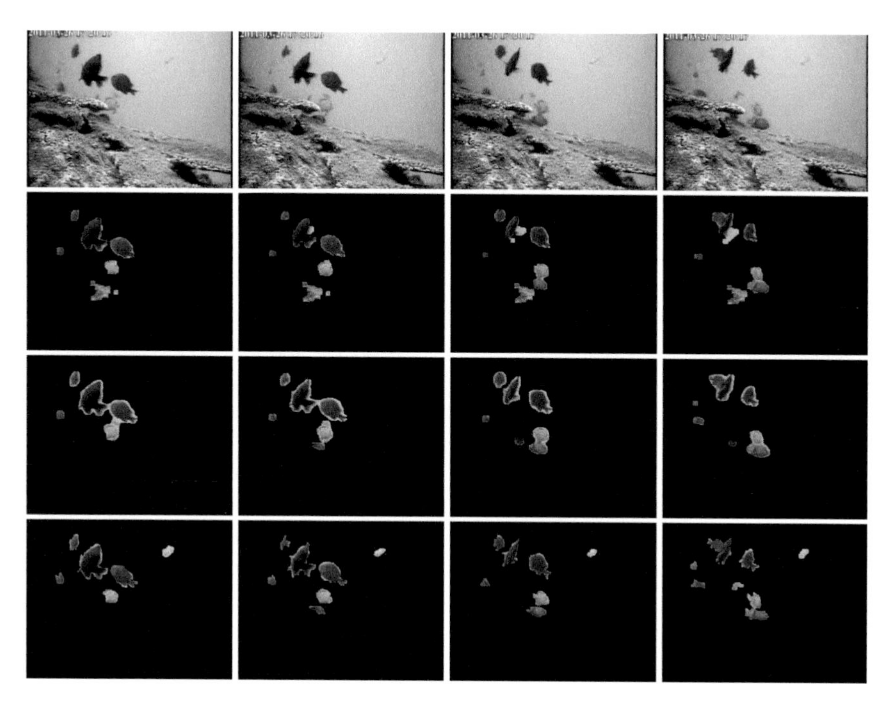

Fig. 9.4 Qualitative Comparison: background subtraction results with (from *top* to *bottom*) (1) $VIBE$ (second row) which detects parts of rocks as fish because of *light* changes, (2) $ML-BKG$ (third row) which outperforms $VIBE$ in term of false positive reduction as it did not detect the rocks; and (3) our approach (last row) that reduces false positives (rocks not detected) but is also able to detect tiny fish (*bright* spot on the *right* hand side) which looks like the background

Table 9.2 Processing Times (frames/sec) of the employed background modeling algorithms

Algorithm	320 × 240	640 × 480
P-Finder	250	60
GMM	200	50
VIBE	100	25
ZGMM	100	25
EIGEN	30	10
ML-BKG	20	3
Our approach	1.5	0.3

Table 9.2 reports the processing times (frames/sec) of the background modeling algorithms (C++ implementation) on a PC powered by an Intel i7 3.4 Ghz CPU and 16 GB RAM.

9.5.2 *Post-Processing*

Fish detections are extracted from the binary motion masks (i.e., the output of the background modeling process) by searching for connected regions of foreground pixels. However, due to the complexity of the monitored environment (lighting changes, sunlight gleaming, plant movements), many false alarms, i.e., image regions mistakenly identified as fish, may be detected. In order to reduce these errors, a detection post-processing (see Sect. 9.4) was employed to filter out objects which do not show motion and appearance traits typical to fish. A Naive Bayes classifier has been used to discriminate between good and bad detections; the training and test sets were built by manually labeling 852 samples, equally divided between images representing single fish (as positive samples) and background portions (as negative samples). Each sample consisted of the frame at time t, in which the object appears, the corresponding binary mask and the frame at time t + 1 (used for the computation of optical flow descriptors). The precision and recall scores achieved by the module were respectively 89.4 and 97.3 % (overall misclassification rate of 6.8 %).

By filtering out detections with low scores (lower than 0.5) we were able to increase fish detection performance of about 3–4%, of course, at the expense of the processing time, which increased by, at least, one order to magnitude.

9.6 Conclusions

In the Fish4Knowledge project, several state-of-the-art as well as new devised background modeling approaches have been developed and tested both in the underwater domain and in other complex environments. All these methods were greatly influenced by the low underwater video quality: in fact, the low frame rate impeded to estimate a reliable fish motion model, while the low spatial resolution had an impact on the fish appearance computation. Despite all these difficulties, the achieved results can be considered satisfactory.

In conclusion, statistical methods modeling texture, motion and appearance of both the background and the foreground outperformed other approaches in extremely complex environments. However, identifying objects of interest by optimizing a classification objective built upon low level visual features will never solve the problem, but, instead, a knowledge-based representation of the visual and semantic information will be needed, thus also allowing for actual visual reasoning as humans do.

References

Alexe, B., Deselaers, T., and V. Ferrari. 2012. Measuring the objectness of image windows. *IEEE Transactions on PAMI*, 99(PrePrints).

Barnich, O., and M. Van Droogenbroeck. 2011. ViBe: A universal background subtraction algorithm for video sequences. *IEEE Transactions on Image Processing* 20(6): 1709–1724.

Bouguet, J.-Y. 2000. Pyramidal Implementation of the Lucas-Kanade Feature Tracker Description of the Algorithm.

Cheng, C., A. Koschan, C.-H. Chen, D.L. Page, and M.A. Abidi. 2012. Outdoor scene image segmentation based on background recognition and perceptual organization. *IEEE Transactions on Image Processing* 21(3): 1007–1019.

Di Zenzo, S. 1986. A note on the gradient of a multi-image. *Computer Vision, Graphics, and Image Processing* 33(1): 116–125.

Faro, A., D. Giordano, and C. Spampinato. 2011. Integrating location tracking, traffic monitoring and semantics in a layered its architecture. *IET Intelligent Transport Systems* 5(3): 197–206.

Felzenszwalb, P., and D. Huttenlocher. 2004. Efficient graph-based image segmentation. *International Journal of Computer Vision* 59(2): 167–181.

Gallego, J., Pardas, M., and G. Haro. 2009. Bayesian foreground segmentation and tracking using pixel-wise background model and region based foreground model. In *2009 16th IEEE international conference on image processing (ICIP)*, 3205–3208.

Hall, P., and M.P. Wand. 1996. On the accuracy of binned kernel density estimators. *Journal of Multivariate Analysis* 56(2): 165–184.

Han, B., and L. Davis. 2012. Density-based multifeature background subtraction with support vector machine. *IEEE Transactions on Pattern Analysis and Machine Intelligence* 34(5): 1305–1312.

Heikkila, M., and M. Pietikainen. 2006. A texture-based method for modeling the background and detecting moving objects. *IEEE Transactions on Pattern Analysis and Machine Intelligence* 28(4): 657–662.

Julesz. B. 1981. Textons, the elements of texture perception, and their interactions. *Nature*.

Kavasidis, I. and S. Palazzo. 2012. Quantitative performance analysis of object detection algorithms on underwater video footage. In *Proceedings of the 1st ACM international workshop on multimedia analysis for ecological data*, 57–60. New York: ACM.

Kavasidis, I., Palazzo, S., Di Salvo, R., Giordano, D., and C. Spampinato. 2013a. An innovative web-based collaborative platform for video annotation. *Multimedia Tools and Applications*, 1–20.

Liao, S., Zhao, G., Kellokumpu, V., Pietikainen, M., and S. Li. 2010. Modeling pixel process with scale invariant local patterns for background subtraction in complex scenes. In *2010 IEEE conference on computer vision and pattern recognition (CVPR)*, 1301–1306.

Liu, G.-H., and J.-Y. Yang. 2008. Image retrieval based on the texton co-occurrence matrix. *Pattern Recognition* 41(12): 3521–3527.

Liu, Z., D.W. Jacobs, and R. Basri. 1999. The role of convexity in perceptual completion: Beyond good continuation. *Vision Research* 39(25): 4244–4257.

Mittal, A. and N. Paragios. 2004. Motion-based background subtraction using adaptive kernel density estimation. In *Proceedings of the 2004 IEEE computer society conference on computer vision and pattern recognition, 2004. CVPR 2004*, 2: II-302–II-309.

Monnet, A., Mittal, A., Paragios, N., and V. Ramesh. 2003. Background modeling and subtraction of dynamic scenes. In *Proceedings of the ninth IEEE international conference on computer vision, ICCV '03*, - vol. 2:1305–1313. Washington: IEEE Computer Society.

Oliver, N., B. Rosario, and A. Pentland. 2000. A bayesian computer vision system for modeling human interactions. *IEEE Transactions on Pattern Analysis and Machine Intelligence* 22(8): 831–843.

Porikli, F. 2005. Multiplicative background-foreground estimation under uncontrolled illumination using intrinsic images. In *Seventh IEEE workshops on application of computer vision, 2005. WACV/MOTIONS '05 Volume 1*. 2: 20–27.

Porikli, F. 2006a. Achieving real-time object detection and tracking under extreme conditions. *J. Real-Time Image Processing* 1(1): 33–40.

Porikli, F. 2006b. Achieving real-time object detection and tracking under extreme conditions. *Journal Real-Time Image Processing* 1(1): 33–40.

Porikli, F. and C. Wren. 2005. Change detection by frequency decomposition: Wave-back. In *Proceedings of the workshop on image analysis for multimedia interactive*.

Raimondo, S. and C. Silvia. 2010. Underwater image processing: state of the art of restoration and image enhancement methods. *EURASIP Journal on Advances in Signal Processing*.

Rosenblatt, M. 1956. Remarks on some nonparametric estimates of a density function. *The Annals of Mathematical Statistics* 27(3): 832–837.

Seki, M., Wada, T., Fujiwara, H., and K. Sumi. 2003. Background subtraction based on cooccurrence of image variations. In *Proceedings of the 2003 IEEE computer society conference on computer vision and pattern recognition 2003* 2: II-65–II-72 .

Sheikh, Y., and M. Shah. 2005. Bayesian modeling of dynamic scenes for object detection. *IEEE Transactions on Pattern Analysis and Machine Intelligence* 27(11): 1778–1792.

Spampinato, C. and S. Palazzo. 2012. Enhancing object detection performance by integrating motion objectness and perceptual organization. In *2012 21st international conference on pattern recognition (ICPR)*, 3640–3643

Spampinato, C., S. Palazzo, and I. Kavasidis. 2014c. A texton-based kernel density estimation approach for background modeling under extreme conditions. *Computer Vision and Image Understanding* 122: 74–83.

Stauffer, C., and W.E.L. Grimson. 1999. Adaptive background mixture models for real-time tracking. *IEEE Computer Society Conference on Computer Vision and Pattern Recognition* 2: 246–252.

Tsai, D.M., and S.C. Lai. 2009. Independent component analysis-based background subtraction for indoor surveillance. *IEEE Trans Image Process* 18(1): 158–167.

Wren, C., Azarbayejani, A., Darrell, T., and A. Pentland. 1996. Pfinder: Real-time tracking of the human body. In *Proceedings of the second international conference on automatic face and gesture recognition 1996*, 51–56.

Yao, J. and J.-M. Odobez. 2007. Multi-layer background subtraction based on color and texture. In *IEEE Conference on Computer Vision and Pattern Recognition 2007. CVPR'07*, 1–8. New York: IEEE.

Zhang, B., Y. Gao, S. Zhao, and B. Zhong. 2011. Kernel similarity modeling of texture pattern flow for motion detection in complex background. *IEEE Transactions on Circuits and Systems for Video Technology* 21(1): 29–38.

Zivkovic, Z., and F. van der Heijden. 2006. Efficient adaptive density estimation per image pixel for the task of background subtraction. *Pattern Recognition Letters* 27(7): 773–780.

Chapter 10
Fish Tracking

Daniela Giordano, Simone Palazzo and Concetto Spampinato

Abstract Object tracking is an essential step of a video processing pipeline. In the Fish4Knowledge project, recognizing fish trajectories allows to provide information to higher-level modules, such as behavior understanding and population size estimation. However, video quality limitations and appearance/motion characteristics of fish make the task much more challenging than in typical human-based applications in urban contexts. To solve this problem, robust appearance and motion models must be employed: this chapter describes an approach devised to tackle the fish tracking problem in this project, and presents and evaluation of the tracking algorithm in comparison with state-of-the-art techniques.

10.1 Introduction

In computer vision and video processing, object tracking is the problem of following an object across frames; in other words, a tracking algorithm should be able to recognize that two image regions in two different video frames represent the same object. The key to solve this problem is to adopt a descriptive model of the object which is consistent in time and robust to pose changes, illumination changes, temporary occlusions between objects or between an object and scene elements, and so on. Unfortunately, no such globally-working model has been developed, since domain-specific characteristics and requirements make the definition of a general model unfeasible; this is particularly true for underwater videos, due to the nature

D. Giordano · S. Palazzo (✉) · C. Spampinato
Dipartimento di Ingegneria Elettrica, Elettronica e Informatica,
Università degli Studi di Catania, Viale A. Doria 6, 95125 Catania, Italy
e-mail: dgiordan@diit.unict.it

S. Palazzo
e-mail: simopal6@gmail.com

C. Spampinato
e-mail: cspampin@gmail.com

R.B. Fisher et al. (eds.), *Fish4Knowledge: Collecting and Analyzing Massive Coral Reef Fish Video Data*, Intelligent Systems Reference Library 104,
DOI 10.1007/978-3-319-30208-9_10

of the targets (fish) and the surrounding environment, if compared to other typical tracking applications where targets are people or vehicles.

In the scope of the Fish4Knowledge project, fish tracking is a fundamental task for providing information on trajectory-based behavior understanding and population statistics generation. However, the specific application context makes the task very challenging: in addition to the difficulties mentioned above, the technical limitations related to link the underwater cameras to the mainland storage and processing servers affected both image quality, which makes appearance model more difficult as details are lost and noise is added, and video frame rate, which makes a fish move a larger distance in consecutive frames, making tracking harder in cluttered environments.

In the following, we will describe the way we tackled the fish tracking task: after presenting a brief review of state-of-the-art tracking algorithms, we will show in detail the solution we devised, how it exploits information coming from the object detection module, and how tracking quality is evaluated, both offline and at runtime.

10.2 Literature on Fish Tracking

Many different approaches have been studied on how to solve object tracking (Yilmaz et al. 2006), from the widely-used algorithms based on Kalman filters or particle filters (Doucet et al. 2001) to recent ones such as Babenko et al. (2011), Yilmaz et al. (2004), Porikli et al. (2005).

One of the simplest ways to see the tracking problem is as an estimation of the probability density function of a state representing an object's position and appearance, given the set of all measurements up to that moment. When the measurement noise is assumed to be Gaussian, Kalman filters (Doucet et al. 2001) provide an optimal solution, in a least-square sense, to the general problem of estimating the state of a linear discrete-time dynamical system. Each cycle of the algorithm is divided into two steps: a prediction step, in which the current state and the error covariance are projected forward to estimate the prior for the following step; and a correction stage, in which new measurements (i.e., information extracted from new frames) is incorporated to obtain a better estimate.

A more general class of state-space-based filters is represented by particle filters, where the current state distribution is modeled as a set of weighted randomly-distributed samples which are updated as new measurements become available. Particle filter tracking is used in several applications (Bouaynaya et al. 2005; Zhou et al. 2003) and provide robust performance, however it may become impractical because of the size of the state vector and of the large number of particles.

A computationally efficient and very popular approach is mean-shift tracking (Comaniciu and Meer 2002), which models the object's probability density in terms of a color histogram, and moves the object region in the largest gradient direction, in order to maximize the similarity between the reference and candidate object regions, measured with the Bhattacharyya coefficient or the Kullback–Leibler divergence criteria. This technique is more efficient than particle filters, however it has worse performance and fails in the case of occlusions and quick appearance changes, when the color distribution of the background is too similar to that of the target object or when the object moves outside of the kernel search area.

The CAMSHIFT (Bradski 1998) algorithm is basically a variant of the mean-shift. The main limitations of the mean-shift algorithm concern the absence of a model-update mechanism, in terms of both the histogram values and the size of the searched object. CAMSHIFT applies mean-shift to find the best-matching region for a target, then updates the size of the object according to the zeroth moment, i.e., the sum of probability contributions of the current window. To the best of our knowledge, before the F4K project, CAMSHIFT had been the only algorithm tested for tracking and counting fish in real-life unconstrained scenarios, as described in Spampinato et al. (2008).

Other approaches, rather than finding a global representation of an object and looking for regions which resemble them, extract local point features from an object (chosen in such a way as to make the description invariant to affine or projective transformations) and try to find correspondences in the following frames. The algorithm described in Shi and Tomasi (2008) follows this strategy. However, this technique presents a few disadvantages, especially with smooth object surfaces (for which it is difficult to extract distinguishing feature points) or when objects undergo pose changes, intersections and severe deformations—which is common for fish.

In recent years new tracking algorithms have been proposed by the scientific community: some of them are inspired by or are variants of classic algorithms, while others represent completely new approaches in the field. One which caught our attention for its suitability to the task at hand is the one proposed in Porikli et al. (2005), which employs covariance models to describe fish appearance. Covariance models (which have recently gained attention also in the object recognition field, e.g., San Biagio et al. 2013) represent objects as the covariance matrix of pixel-based features, such as location, color, gradient intensity; this representation allows to encode both spatial and statistical information in a concise and elegant way, and proved to be particularly useful for deformable objects.

10.3 Underwater Object Tracking

As mentioned in Sect. 10.1, fish tracking suffers several complications with respect to most human-centered tracking approaches, which mainly focus on people and vehicles.

First of all, being able to state that "object in frame X is the same object as in frame Y" implies that the two objects share a very similar appearance: while recognizing this is obvious and immediate for humans, it may be quite a challenge for a computer, depending on the characteristics of the target object. For example, although the anatomy of people is of course the same for everyone, wearing clothes greatly simplifies tracking, since they provide useful visual clues on a person's appearance model.

In the case of fish, an additional complication is introduced by the shape and deformability of their bodies: following the human analogy, whereas we can exploit the fact that human silhouette does not change while moving (we could even say it is almost isotropic with respect to the vertical rotation axis), the same does not hold for a fish, since its two-dimensional shape (i.e., as seen from a camera) can change greatly depending on the direction it is swimming. Fish speed also accentuates the difficulty of appearance-based matching in different frames, especially in a low-frame rate (5–8) scenario as the F4K one, since the time interval between two consecutive frames becomes large enough that fish shape does not change gradually, but may be very different from one position to the next. Finally, the absence of distinctive visual clues for fish belonging to the same species makes it impossible to identify them individually or to understand if it is the same fish or another one which went out of the camera view and back in.

Fish tracking must also take into account the way fish move. A few seconds of a video may be enough to estimate a target's motion model when we are dealing with people or cars, since both are characterized by a typical linear motion pattern without sudden accelerations; the same does not apply for fish, since they move in an erratic pattern which may be difficult and computationally expensive to model mathematically. To further complicate matters, humans mostly move in two dimensions, whereas fish have three degrees of freedom and interact in a full 3D space (which also has an effect on appearance, which can change depending on the distance from the camera).

These problems are reflected in the design of the tracking algorithms in two ways:

- appearance model: how can we describe how a fish looks, in such a way that the description does not change even when fish pose changes?
- motion model: given the locations of a fish in the previous video frames, how can we predict where it will be in the next frame?

Both questions are basically the core of any algorithm for object tracking; unfortunately, no globally optimal solution is available, as the choice of both models is closely related to the specific problem at hand. In our case, it was important that the chosen model encode appearance, pose and spatial information: the approach which seemed to better suit these requirements was the covariance-based modeling approach in Porikli et al. (2005). However, the authors of that paper employed no motion model, relying on a window search around an object's previous location to find its new position, which was too inefficient for our purposes and performed almost randomly in case of occlusions between fish of the same species. Therefore, after employing in the first stage of the project a simple motion model based on linear motion, we finally resolved to adopt a modified particle filter framework, which allowed to both handle erratic movements and to integrate information coming from the object detection module in a natural way.

10.4 Tracking with Covariance Modeling

In this section, we will describe two versions of the tracker used in the F4K project: the "covariance-based tracker" (also shortened as *COV*) and the "covariance-based particle filter" (*COVPF*). Although the particle filter version was the final product of our research for a suitable algorithm, it is strongly based on the more "basic" covariance-based tracker; for this reason, it is interesting to see how the first version of the tracker was designed, how it performed, and what its limitations were, before going on to explain the modifications introduced by adding a particle filter framework, and how it solved the problems left open by the initial *COV* tracker.

10.4.1 Covariance-Based Tracker (COV)

In the first stage of the work, the approach adopted to perform fish tracking consisted of the following steps:

- Identification of objects of interest, performed by the fish detection algorithms in the previous processing stage (see Chap. 9).
- Computation of a mathematical model describing the appearance of each detected object in the current frame.
- Association between previously-tracked objects and current detections, according to the best matches between appearance models.

As we mentioned earlier, we adopted a covariance-based appearance model, which represents objects (i.e., a certain area in a frame) as the covariance matrix of pixel-based feature vectors, extracted from all pixels which make up the object. In order to represent different aspects of a region and to capture the correlation between different kinds of features, each pixel descriptor is a vector consisting of the following components: x and y coordinates, RGB color values, hue value (from the HSV color space) and gray-scale intensity gradient. Given the set of all such 8-dimensional vectors, we compute the corresponding covariance matrix, which has size 8×8 and will be used as the appearance descriptor for the image. In our tests, we experimented using other kinds of features (such as local histogram, higher-order derivatives and different color spaces), however the chosen configuration allowed to better recognize object similarity under different pose and lighting conditions.

Speaking of similarity, the immediate step after defining an appearance model is choosing a way to compare the models of two regions. Unfortunately, covariance matrices do not lie on a Euclidean space (for example, the space of covariance matrices is not closed under multiplication by a negative factor), which prevents from using the Euclidean distance as a similarity measure. To solve the problem, we employed the distance measure proposed by Förstner and Moonen (1999):

$$d(C_1, C_2) = \sqrt{\sum_{k=1}^{d} \ln^2 \lambda_k(C_1, C_2)} \tag{10.1}$$

where C_1 and C_2 are two covariance matrices (i.e., two region descriptors), d is the order of the matrices (i.e., the number of pixel features), and $\{\lambda_k(C_1, C_2)\}$ is the set of generalized eigenvalues of C_1 and C_2, computed as:

$$\lambda_k C_1 x_k - C_2 x_k = 0 \tag{10.2}$$

where x_k are the generalized eigenvectors, and $k = 1 \dots d$.

Another important concept to explain is that of *search area*. Given a tracked object, our aim is to find its new location in the current frame. Typically, searching for all possible locations is too time-expensive, and approaches based on the estimation of an object's future motion are employed to identify a *search area*, where we assume the object will be found in the next frame. The size of a search area represents a trade-off between the risk of missing an object and that of making a wrong association: if the search area is too small, the object might move outside of it, and the tracker would mistakenly mark it as "missing"; on the other hand, if the search area is too large, the risk of assigning a wrong detection increases. In our case, the low video frame rate makes the decision more complex, because fish move by a fairly large number of pixels at each frame, which would require a large search area, thus increasing the risk of ambiguity.

Due to the complexity and relative unpredictability of fish motion patterns, we decided not to use an explicit mathematical model to estimate a fish's location given its past displacements, but to adopt a heuristic based on an object's size (in terms of pixels; this is useful because it is loosely related to the distance between the object and the camera and to its apparent speed) and on the information gathered from all other tracked objects. In practice, the following is the procedure for the estimation of the search area for object O:

- If O was already present in previous frames, let as compute its average displacement a_O, that is the average amount of pixels it has moved in each previous frame.
- If we have previous information on other objects, compute the average displacement of all objects with similar size as O, where an object's size is the number of pixels it is made up of. We define "size similarity" by quantizing objects' size in steps of 400 pixels: two objects will have "similar" sizes if they belong to the same quantization group. Let us call the average displacement a_s (for "size").
- Set O's estimated motion d_O as follows:
  - If both a_O and a_s are available, set d_O to the average between the two.
  - If only one between a_O and a_s is available, set d_O to that value.
  - If neither is available, set d_O to 20.

- If O has been missing for t frames, increase d_O by $5 \times t$ pixels.
- At this point, define our current search area as a square centered at O's previous location, with side $2 \times d_O$.
- If we have a previous location for O, compute its latest velocity vector v_O, then move the search area along v_O's direction by $d_O/2$ pixels. For objects moving at relatively high speed, this allows to give more importance in the search to the direction of motion, rather than considering all directions equiprobable.
- Given the final search area S_O and a detection D, we will have a match if D's region intersects S_O.

All numeric values have been optimized to suit video and fish motion characteristics.

Now that all basic ideas have been explained, we can describe the tracking algorithm in detail:

- Let $\{T_1, T_2, \ldots, T_n\}$ be the set of tracked objects up to video frame f. Each T_i consists of a $\{L_i, v_i, t_i, C_i, \{c_1, c_2, \ldots, c_B\}\}$ structure, where L_i is the most recent location of the object, v_i is its current velocity vector, t_i is the number of consecutive frames on which the object could not be found (so it is 0 if the object was found at frame $f - 1$), C_i is the covariance matrix representing the object's appearance, and c_1 to c_B are the covariance matrices of the last B regions where the object was found (in our implementation, $B = 10$). Also, let $\{D_1, D_2, \ldots, D_m\}$ be the set of objects returned by the object detection module at frame f.
- For each tracked object T_i, use L_i, v_i and t_i to compute its search area S_i.
- Process the set of moving objects in frame f (provided by the motion detection module), and select all objects which intersect S_i; let the resulting set be $\{D_1, \ldots, D_m\}$, with corresponding covariance matrices $\left\{C_{D_1}, \ldots, C_{D_m}\right\}$.
- Assign to object T_i the detection D_j which minimizes the covariance distance $d_c(C_{D_j}, C_i)$. Update T_i's structure fields according to D_j.
  - In particular, update the set of $c_1, \ldots, c_B$ matrices by removing the oldest one and inserting C_{D_j}. Compute the new appearance model of T_i as an intrinsic mean of the covariance matrices in the set, by exploiting the property that symmetric matrices have Lie group structure (the mathematical details are out of the scope, and can be found in Porikli et al. 2005).
- Remove objects which have been missing from the scene for more than 6 frames.
- Create new tracked-object structures for unassigned detections.

The results obtained with this algorithm (see Sect. 10.6) show that it can achieve good tracking results, under the hard conditions imposed by the videos. However, the accuracy of the algorithm is strongly linked to that of the detection algorithm, since it assumes that all and only moving objects will be provided by the underlying motion algorithm; for this reason, tracking may fail because of detection inaccuracy.

10.4.2 Covariance-Based Particle Filter (COVPF)

The second phase in the design of the tracking algorithm aimed at dealing with the main limitations shown by the original *COV* tracker:

- As we said earlier, since the algorithm only associated objects identified by the fish detection module, its accuracy was necessarily dependent on the latter's. A failure in the fish detection algorithm necessarily propagates to the fish tracking algorithm.
- The computation of the search area for a fish in a new frame was performed through a heuristic method, based on the history of objects in the video, rather than on a more generic and mathematically rigorous model.
- The algorithm could not handle occlusions, since two fish "touching" each other (in the two-dimensional image projected on the camera) would be identified as a single blob by the fish detection algorithm. This caused the tracking algorithm to associate the aggregated blob to one of the two fish (thus missing the other one), and also caused the model of the associated fish to be disrupted by the inclusion of pixel information belonging to the other fish.

In order to tackle these issues, we decided to modify the tracker towards a *particle filter* framework, while keeping the covariance modeling approach. Particle filters (Doucet et al. 2001) model location and motion information of an object through a set of state vectors (called *particles*) which typically represent a hypothesis about the location, shape, velocity, etc. of an object. Each particle is assigned a certain weight, based on the similarity between the object's current appearance model and the visual features of the image region defined by the particle's state; the state vector obtained through the weighted average of all particles is used to estimate the object's location and motion properties. During the execution of the algorithm, at every video frame each particle's location is modified according to its previous motion information (and to the motion model used), plus some random noise, which allows to avoid deterministic decisions and to "explore" the search space.

The main difference with respect to the previous algorithm is that it uses information from the upstream detection module only as a hint for where fish might have moved, rather than as the only source of possible locations. The advantages of this modification can be found in the way the tracker behaves when the fish detection algorithm fails:

- If a camouflage effect hides a fish due to its similarity to the background, the tracker still tries to locate it, in spite of the absence of detections.
- If two fish occlude, the particle filter can tear them apart, whereas the previous tracker only would only see them as a single blob.

Nevertheless, completely ignoring the information provided by the motion detection module would be as bad as relying exclusively on them. In fact, the *COV* tracker does work well enough, which means that the objects identified as moving blobs are very likely to be correct detections. In order to incorporate this kind of information into the particle filter framework, we introduced two mechanisms: (1) at each frame, each moving blob is added to the set of particles for each object; (2) the weight of each particle depends on the percentage of pixels where motion was detected.

In practice, the covariance-based particle filter algorithm works as follows:

- Let $\{T_1, T_2, \ldots, T_n\}$ be the set of tracked objects up to video frame f. Each T_i consists of a $\left\{\left\{p_{i1}, p_{i2}, \ldots, p_{iq}\right\}, t_i, C_i, \{c_1, c_2, \ldots, c_B\}\right\}$ structure, which differs from the definition in the previous paragraph by the presence of the particles p_{ij}, each of which defines its own weight w_{ij}, velocity v_{ij}, bounding box b_{ij}, covariance matrix C_{ij} and a binary rectangular mask m_{ij}, having the same size as the bounding box, which specifies a more precise segmentation of the hypothetical fish shape within b_{ij}. In our implementation, the number of particles q is set to 20. As in the previous algorithm, let $\{D_1, D_2, \ldots, D_m\}$ be the set of objects returned by the object detection module at frame f.
- At each frame, the algorithm manages a simple motion history model used to limit the random exploration of particles. Given the initial motion history mask $M(0) = 255_{h \times w}$ (i.e., a matrix of size $h \times w$, being the image dimensions, whose elements are all set to 255), the motion history mask $M(f)$ is computed according to the following:

  - Apply a subtracting decay constant (set to 85) to motion mask $M(f-1)$ in order to "forget" areas where motion appeared in past frames but not in recent ones. Obtain mask $M_d(f-1) = M(f-1) - 85_{h \times w}$ (negative values are set to zero).
  - Compute the grayscale frame difference $D(f) = I(f) - I(f-1)$, where $I(x)$ is the grayscale image at video frame x.
  - Threshold $D(f)$ by setting to 0 and 255 all pixels respectively smaller and larger than 5, thus obtaining D_{thr}.
  - Compute the new motion mask as $M(f) = M_d(f-1) + D_{\text{thr}}$ (saturate mask values to 255).
  - Smooth the resulting mask by applying the morphological closure operator (i.e., a dilation followed by an erosion).

- For each tracked object T_i, initialize a new particle set $\left\{p'_{i1}, \ldots, p'_{iq}\right\}$ by randomly selecting particles from the original set. The probability the particle p_{ij} is selected is proportional to w_{ij}, and each particle can be selected multiple times.
- Move particles according to their current velocity, and modify their bounding boxes by adding Gaussian noise ($\sigma = 5$) to each dimension. Compute the average velocity v_{avg}.
- For each particle p'_{ij}, update its binary mask m'_{ij} by setting it to the subregion of $M(f)$ (the motion history mask) defined by b'_{ij} (the particle's bounding box).
- Add particles initialized from the blobs returned by the motion detection algorithm.
- Compute the particles' covariance C'_{ij} using only "active" pixels from the binary mask m'_{ij}. If the covariance matrix cannot be computed (for example because the number of active pixels is too small, which is the case when the intersection between the particle's bounding box and the motion history contains too few pixels,

i.e., the particle has moved to a static region), remove the corresponding particle from the set.

- Remove particles whose covariance distance from C_i (T_i's model) is greater than 3 (chosen empirically).
  - If no valid particle is found, mark T_i as missing for the current frame.
- Update each particle's weight as $w'_{ij} = w'_{\mathrm{cov},ij} - w'_{\mathrm{bkg},ij} - \frac{1}{2}w'_{\mathrm{mot},ij}$, where
  - $w'_{\mathrm{cov},ij}$ is the covariance distance between the new particle's covariance matrix C'_{ij} and C_i (the object's current model).
  - $w'_{\mathrm{bkg},ij}$ is the fraction of pixels inside b'_{ij} which belong to moving blobs returned by the fish detection module (bkg refers to "background modeling").
  - $w'_{\mathrm{mot},ij}$, similarly, is the fraction of pixels inside b'_{ij} which are "active" in the motion history mask $M(f)$.
- Update the estimated object location and corresponding covariance model.
- Update particle velocities as the mean between its previous velocity and the object's global velocity (computed from its most recent estimated locations), plus some Gaussian noise with standard deviation $\sigma = 5$ pixels.
- Remove objects which have been missing from the scene for more than 6 frames.
- Create new tracked-object structures for unassigned detections.
  - In particular, for each new tracked object create a random particle set. Each particle is initialized from the blob's bounding box (plus Gaussian noise) and a global velocity initialization constant set to 20 pixels (plus Gaussian noise).

The particle filter extension allows to solve problems which could not be handled through a detection-based approach, while at the same time giving enough importance to locations where motion happened.

In order to provide a quantitative assessment of the differences between the two tracking approaches, in the following we will present the evaluation methodology and results.

10.5 Assessing Tracking Quality Online

Evaluation is a fundamental requirement for all computer visions algorithms, as it provides evidence on the quality of a technique in practice. However, the generation of evaluation data or *ground truth*, that is the set of annotations which allow to compare an algorithm's performance to the expected results, is time-consuming, error-prone and tedious to users, who—in the case of tracking algorithms—have to analyze manually each frame of a video and label each association between objects. For this reason, research communities have been developing self-evaluation approaches, which typically evaluate tracking decisions by analyzing short-term (i.e., in a small number of frames) regularity of object motion (for example, a sudden and disproportionate change of direction of an object is considered an indication of bad tracking) and appearance changes (e.g., big variations in the shape ratio or in the histogram may indicate that the algorithm lost the object and is following a wrong one). Existing approaches can be classified into three main categories: (1) *feature-based* (Chau et al. 2009) that analyze the internal state or output (shape ratio, area, speed, color and direction variations) of tracking algorithms, (2) *hybrid* (Erdem et al. 2001; Wu and Zheng 2004) that combine several temporal and non-temporal features to get an assessment of each tracking decision and (3) *trajectory-based* (Wu et al. 2010) that exploit intrinsic information of the generated trajectories to measure the quality of a track. Of course, while such approaches cannot replace regular ground-truth-based evaluation, they still provide useful information which can be used and computed at runtime, for example to perform online filtering based on tracking quality.

As part of our evaluation procedure, we devised an on-line method to test tracking algorithms without ground-truth data by analyzing the regularity of motion, shape and appearance of each tracking decision and combining this information into a probability-like score representing the overall evaluation of that tracking decision. Results show how this approach is able to reflect the performance of tracking algorithms on different target motion patterns, besides fish.

In detail, for each tracking decision (i.e., an association between an object in frame t and one in frame $t+1$) the following features are computed and combined into a single score as an average.

- *Difference of shape ratio between frames*: this score detects rapid changes in the object's shape, which might indicate tracking failure. This value is high if the shape ratio ($R = \frac{W}{H}$, with W and H, respectively, the width and the height of the bounding box containing the object) between consecutive frames $t-1$ and t keeps as constant as possible.

$$R_{max} = max\left\{R_t, R_{t-1}\right\}$$
$$R_{min} = min\left\{R_t, R_{t-1}\right\}$$
$$shape\_ratio\_score_t = \alpha_{SR}\frac{R_{min}}{R_{max}}$$

- *Histogram difference*: this feature evaluates the difference between two consecutive appearances of the same object by comparing the respective histograms (analyzing independently the three RGB channels and the grayscale versions of the two objects). Given histograms H_t and H_{t-1}, the corresponding score is proportional to the ratio between the intersection and the union of the two histograms:

$$histogram\_difference_t = \alpha_{HD}\sum_{i=0}^{255}\frac{min\left\{H_t\left(i\right), H_{t-1}\left(i\right)\right\}}{max\left\{H_t\left(i\right), H_{t-1}\left(i\right)\right\}}$$

- *Direction smoothness*: although fish movements are erratic, we can safely assume that a fish trajectory is as good as it is regular and without sudden direction changes in the short term (i.e., in few consecutive frames). This value keeps track of the direction of the object computed from the last three frames and checks for unlikely changes in the trajectory. It is computed as:

$$direction\_smoothness_t = 1 - \frac{|\theta_1 - \theta_2|}{180}$$

where θ_1 and θ_2 are the angles (with respect to the x axis) of the last two displacements of the object: θ_1 is computed from the object's locations in frames $t-2$ and $t-1$, and θ_2 from its locations in frames $t-1$ and t. For simplicity, we use $\theta_1 - \theta_2$ in the formula, although the actual implementation handles the case of angles around the $0°/360°$ boundary.

- *Speed smoothness*: similarly to the previous feature, this value checks whether the current speed of the object (i.e., the displacement between the previous position and current one) is similar to the average speed in the object's history. Let P_t and P_{t-1} be positions of the object at frames t and $t-1$, we compute $s_t = ||P_t - P_{t-1}||$, so that s_t represents the last displacement (speed) of the object, and compare it with the average speed $\bar{s}$ in order to compute $speed\_smoothness$ as:

$$\begin{aligned} s_{max,t} &= max\,\{s_t, \bar{s}\} \\ s_{min,t} &= min\,\{s_t, \bar{s}\} \\ speed\_smoothness_t &= \alpha_{SS}\frac{s_{min}}{s_{max}} \end{aligned}$$

- *Texture difference*: texture features (mean and variance of several Gabor filters) are computed from two consecutive appearances and compared. Given two feature vectors v_{t-1} and v_t, this value is computed as:

$$texture\_difference_t = 1 - \sqrt{\sum_{i=1}^{n}(v_{t-1}(i) - v_t(i))^2/\alpha_{TD}} \quad (10.3)$$

- *Temporal persistence*: this value is the number *TP* of frames in which a given object appears.

All constants α included in the above formulas were experimentally identified to approximately normalize the resulting values between 0 and 1. The overall quality score q_S is computed as follows:

- Compute the average μ of the above-described values over all frames, except the temporal persistence *TP*.
- If $TP > N$, return μ, where N (set to 10) is the number of frames in which an object has to appear in order to be considered a reliable track;
- Else, return $\mu \cdot \left(0.9 + \frac{TP}{10 \times N}\right)$ (i.e., if an object appears in less than N frames, limit the maximum score it can get).

Figure 10.1 shows a few sample trajectories with their average q_S scores; notice that the trajectory of the top-left image is unrealistic (and is caused by a temporary

Fig. 10.1 Examples of fish trajectories: **a** an erroneous path (average q_S score 0.63) due to a failure of the tracking algorithm; **b** a correct path (average q_S score is of 0.91); **c** a complex but correct fish path with an average q_S score of 0.81; **d** trajectory with an average q_S score of 0.71 due to a fish-plant occlusion

mis-association between objects) and its average score, computed as average of the scores at all tracking decisions, was 0.63, whereas the image at the top-right side shows a correct trajectory whose average score is 0.91. The two bottom images show, from left to right, a complex but correct trajectory (with a 0.81 score) and a trajectory which is correct up to a certain point, before an occlusion happened, so its total tracking score is 0.71.

10.6 Results

The comparison between the original covariance-based tracker and the covariance-based particle filter was performed on a set of 10 ground truth videos. Each video is 10 min long, sampled at 8 frames per second; the resolution is 320×240 (for 7 videos) and 640×480 (for 3 videos). The ground truth annotation process consisted in manually drawing the contours of all fish in each frame and associating detections in different frames as instances of the same fish. In total, the annotated dataset contains 35,391 single detections for 2218 trajectories. Figure 10.2 shows the kind of scenes included in the ground truth.

Ground-truth tracking accuracy is shown in terms of the following indicators:

- Correct Counting Ratio (*CCR*): percentage of correctly identified ground-truth fish. This ratio provides information not only on the tracking algorithms but also on the overall system performance from background modeling to fish detection to tracking.
- Average Trajectory Matching (*ATM*): average percentage of common points between each ground-truth trajectory and its best matching tracker trajectory;
- Correct Decision Rate (*CDR*): let a "tracking decision" be an association between a fish at frame t_1 and a fish at frame t_2, where $t_1 < t_2$; this tracking decision is correct if it corresponds to the actual association, as provided by the ground truth. The correct decision rate is the percentage of correct tracking decisions, and gives an indication of how well the algorithm performs in following an object, which is not necessarily implied by the average trajectory matching (see Fig. 10.3).

Fig. 10.2 The scenes included in the tracking ground truth video set

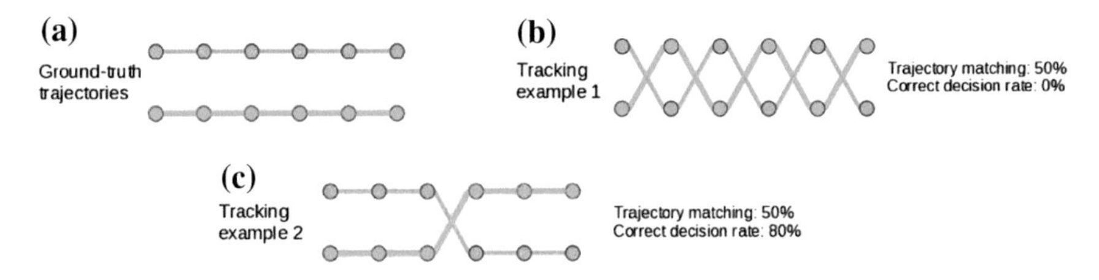

Fig. 10.3 Difference between the trajectory matching score and the correct decision rate. **a** shows two ground truth trajectories of two fish, whereas the other two images represent two examples of tracking output. In **b**, although the tracker fails at each tracking decision the trajectory matching score is 50 %, whereas the correct decision rate is 0. Differently, in **c** the tracker fails only at one step and the trajectory matching score is 50 % (as the previous case) whereas the correct decision rate is 80 % (4 correct associations out of 5, for each object)

Table 10.1 Evaluation of the original covariance-based tracker and the new one using the particle filter approach

Video	Objects	*COV*				*COVPF*			
		ATM	CCR	CDR	OE	ATM	CCR	CDR	OE
1	1058	0.75	0.70	0.74	0.83	0.50	0.68	0.93	0.85
2	3072	0.92	0.51	0.81	0.84	0.84	0.53	0.93	0.89
3	16,321	0.66	0.67	0.77	0.82	0.56	0.65	0.65	0.72
4	1927	0.73	0.56	0.80	0.87	0.69	0.55	0.89	0.78
5	1284	0.64	0.59	0.67	0.74	0.48	0.59	0.78	0.77
6	1656	0.70	0.55	0.66	0.75	0.56	0.52	0.87	0.80
7	5477	0.66	0.72	0.75	0.79	0.71	0.74	0.77	0.79
8	820	0.95	0.90	0.73	0.71	0.80	0.80	0.75	0.71
9	1447	0.88	0.66	0.73	0.78	0.84	0.63	0.83	0.73
10	1903	0.84	0.57	0.70	0.71	0.80	0.53	0.75	0.73
Avg		*0.77*	*0.64*	*0.74*	*0.78*	*0.68*	*0.62*	*0.82*	*0.78*

For each video we show the number of objects included in the corresponding ground truth and the three tracking evaluation indicators for the two algorithms: the first three scores (ATM, CCR and CDR) are ground-truth based, and the fourth score (OE) is computed using the online evaluation method described in Sect. 10.5. The last row shows the average values of all scores

Table 10.1 shows the results of the original covariance-based algorithm (labeled *COV*) and the new covariance-based particle filter (labeled *COVPF*) on each of the 10 videos in the dataset, and on average for all videos.

It is possible to notice a decrease in the ATM score, which can be explained by the presence of more false positives in the trajectory (due to the particle filter's attempts at finding detections for a fish in the absence of motion data), a general constancy in the CCR score, and an increase in the CDR score. This last result reflects the improvements introduced by the particle filter: when a fish swims "alone", i.e., without fish nearby (which may disturb the tracking association process), the only source of tracking decision errors is missing fish detections; instead, in the case of occlusions, the probabilities of misassociations increase, as the tracker has multiple candidates to associate to each tracked fish. In order to evaluate the effect of the videos' low frame rate on the quality of the tracker, we also tested the algorithms on a set of preliminary videos from the AQUACAM project[1] (in collaboration with the F4K Research Consortium). Such videos have a much higher resolution (1920 × 1080 pixels) and frame rate (29 frames per second) than the videos available for this project; an example frame is shown in Fig. 10.4. Our results, run on 3 such videos containing 461 fish trajectories, are shown in Table 10.2, and show the benefits that both trackers obtain from the higher frame rate and video quality. In particular, the particle filter tracker showed the largest improvements: this is due to the fact that smaller motion between frames yields a more stable motion model, which allows to

[1]http://c-fish.org/what-we-do/aquacam-research-programme/.

Fig. 10.4 Example of scene from the AQUACAM video dataset.

Table 10.2 Evaluation of the algorithms on the HD-quality AQUACAM videos

Video	Objects	*COV*				*COVPF*			
		ATM	CCR	CDR	OE	ATM	CCR	CDR	OE
1	344	0.86	0.85	0.84	0.76	0.86	0.85	0.88	0.81
2	260	0.84	0.85	0.83	0.73	0.85	0.85	0.88	0.79
3	121	0.75	0.71	0.81	0.69	0.80	0.76	0.83	0.74
Avg		*0.81*	*0.80*	*0.83*	*0.73*	*0.83*	*0.82*	*0.86*	*0.78*

distribute the particles more effectively in the search region; this results in higher quality tracking, as is shown by the corresponding values for trajectory matching and correct decisions.

10.7 Conclusions

Fish tracking is a fundamental part for the Fish4Knowledge system, as it provides the basis for behavior analysis and population counting. In this chapter we presented the two approaches we employed for the task: a covariance-based tracker and a covariance-based particle filter, designed to solve several of the issues which could not be handled by the former due to its strong dependence on the motion detection module. Our evaluation showed that the approaches were able to satisfactorily track fish in a context as hard as unconstrained underwater videos. Nevertheless, there is still much room for improvement: for example, fish trajectories, though intrinsically three-dimensional (much more so than humans' or cars') are still captured from a 2D plane, therefore losing important information which could help not only on the tracking itself, but also on the behavior understanding part based on trajectory analysis. In this sense, an extension to this work is being carried out in the context of the AQUACAM project, with which the Fish4Knowledge Consortium has an ongoing collaboration.

References

Babenko, B., M.-H. Yang, and S. Belongie. 2011. Robust object tracking with online multiple instance learning. *IEEE Transactions on Pattern Analysis and Machine Intelligence* 33(8): 1619–1632.

Bouaynaya, N., W. Qu, and D. Schonfeld. 2005. An online motion-based particle filter for head tracking applications. In *Proceedings of the IEEE International Conference on Acoustics, Speech and Signal Processing*.

Bradski, G.R. (1998). Computer vision face tracking for use in a perceptual user interface.

Chau, D.P., F. Bremond, and M. Thonnat. 2009. Online evaluation of tracking algorithm performance. In *The 3rd International Conference on Imaging for Crime Detection and Prevention*.

Comaniciu, D., and P. Meer. 2002. Mean shift: A robust approach toward feature space analysis. *IEEE Transactions on Pattern Analysis and Machine Intelligence* 24(5): 603–619.

Doucet, A., N. De Freitas, and N. Gordon. 2001. *Sequential Monte Carlo methods in practice*. Statistics for engineering and information science New York: Springer.

Erdem, C.E., A.M. Tekalp, and B. Sankur. 2001. Metrics for performance evaluation of video object segmentation and tracking without ground truth. *Proceedings of International Conference on Image Processing* 2: 69–72.

Förstner, W., and B. Moonen. 1999. A metric for covariance matrices. Technical report, Department of Geodesy and Geoinformatics, Stuttgart University.

Porikli, F., O. Tuzel, and P. Meer. 2005. Covariance tracking using model update based on Lie algebra. In *Proceedings of the IEEE Conference on Computer Vision and Pattern Recognition*.

San Biagio, M., M. Crocco, M. Cristani, S. Martelli, and V. Murino. 2013. Heterogeneous autosimilarities of characteristics (HASC): Exploiting relational information for classification. In *International Conference on Computer Vision*, 809–816.

Shi, J., and C. Tomasi. 2008. Good features to track. In *Proceedings of the IEEE International Conference on Computer Vision and Pattern Recognition*, 593–600.

Spampinato, C., Y.-H. Chen-Burger, G. Nadarajan, and R. Fisher. 2008. Detecting, tracking and counting fish in low quality unconstrained underwater videos. In *Proceedings of the 3rd International Conference on Computer Vision Theory and Applications (VISAPP)*, vol. 2, 514–519.

Wu, H., and Q. Zheng. 2004. Self-evaluation for video tracking systems. In *Proceedings of the 24th Army Science Conference, USA*.

Wu, H., A.C. Sankaranarayanan, and R. Chellappa. 2010. Online empirical evaluation of tracking algorithms. *IEEE Transactions on Pattern Analysis and Machine Intelligence* 32(8): 1443–1458.

Yilmaz, A., X. Li, and M. Shah. 2004. Object contour tracking using level sets. In *Proceedings of the Asian Conference on Computer Vision*.

Yilmaz, A., O. Javed, and M. Shah. 2006. Object tracking: a survey. *ACM Computing Surveys* 38(4): 13.1–13.45.

Zhou, S., R. Chellappa, and B. Moghaddam. 2003. Visual tracking and recognition using appearance-based modeling in particle filters. In *Proceedings of the International Conference on Multimedia and Expo*.

Chapter 11
Hierarchical Classification System with Reject Option for Live Fish Recognition

Phoenix X. Huang

Abstract This chapter presents a Balance-Guaranteed Optimized Tree with Reject option (BGOTR) for live fish recognition in a non-constrained environment. It recognizes the top 15 common species of fish and detects new species in an unrestricted natural environment recorded by underwater cameras. This system can assist ecological surveillance research, e.g., obtaining fish population statistics from the open sea. BGOTR is automatically constructed based on inter-class similarities. We apply a Gaussian Mixture Model (GMM) and Bayes rule as a reject option after hierarchical classification—we estimate the posterior probability of being a certain species and then filter out less confident decisions. The proposed BGOTR-based hierarchical classification method achieves significant improvements compared to state-of-the-art techniques on a live fish image dataset of 24,150 manually labeled images from the south Taiwan sea.

11.1 Introduction

Live fish recognition in the open sea has been investigated to help understand the marine ecosystem, which is vital for studying the marine environments and promoting commercial applications. This recognition task is fundamentally challenging because of its complex situation where the illumination changes frequently. Prior research is mainly restricted to constrained environments (fish in the tank or on a conveyor system) or dead fish, and these machine vision systems have only explored applications for a limited number of fish species. These methods perform worse when they deal with unconstrained fish in a real-world underwater environment, especially when the dataset is greatly imbalanced.

P.X. Huang (✉)
Google, 1600 Amphitheatre Parkway, Mountain View, CA 94043, USA
e-mail: forestrocket@gmail.com

R.B. Fisher et al. (eds.), *Fish4Knowledge: Collecting and Analyzing Massive Coral Reef Fish Video Data*, Intelligent Systems Reference Library 104,
DOI 10.1007/978-3-319-30208 9_11

In contrast, our work investigates novel techniques to perform effective live fish recognition in an unrestricted natural environment and presents an application of machine vision and learning for free swimming fish. This so-called Balance-Guaranteed Optimized Tree with Reject option (BGOTR) system adopts a hierarchical classification that is based on inter-class similarities to improve the normal hierarchical method and to integrate computer vision techniques and marine biological knowledge. Multiclass classifier and feature selection are built together into a hierarchical tree and optimized to maximize the classification accuracy of grouped classes. BGOTR exploits a novel rejection mechanism to re-classify samples that tend to be confusable with other classes. Meanwhile, trajectory voting combines temporal information with the classification results so that majority results of the same species are preserved while potential outliers produced by occasional illumination changes or fish postures are eliminated. Conflicting decisions resulting from several confusable species are effectively dealt with by voting using each fish detection that appears in multiple frames of a video shot. The reject option after hierarchical classification is conducted by applying the Gaussian Mixture Model (GMM) method to model the feature distribution of the training images. Low confidence decisions of test samples are rejected so that a substantial proportion of classification errors and new species are thrown out although a small number of correctly recognized fish are also removed due to incorrect rejection. After forward sequential feature selection and training each Support vector machine (SVM), Individual feature selection based SVM (IFS-SVM) classifies each test sample by counting votes that are optimized for every pair of specific classes. Tested on a manually labeled fish dataset of 24,150 images, which is the largest and most varied dataset used for fish species recognition, BGOTR demonstrates better accuracy averaged both by all images and by all classes, compared with other previous research. This is the first time that the hierarchical classification method with reject option has been implemented in a live fish recognition system. A figure of the whole recognition system is shown in Fig. 11.1.

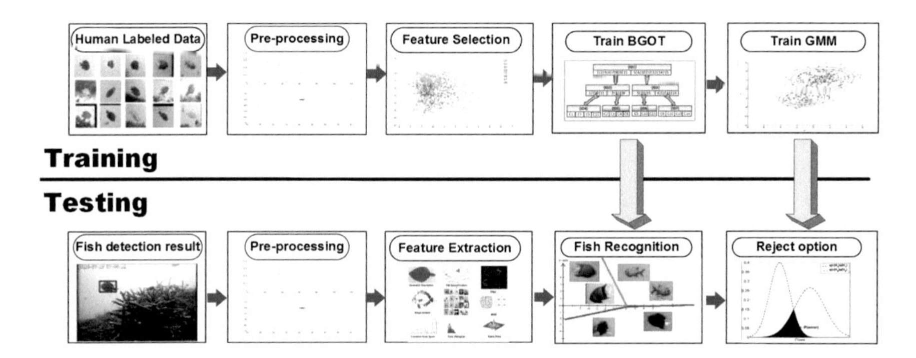

Fig. 11.1 The fish recognition system, an overview framework

11.2 Related Work

Traditionally, marine biologists have employed many tools to examine the appearance and quantities of fish. For example, they cast nets to catch and recognize fish in the ocean. They also dive to observe underwater environment, using photography (Caley et al. 1996). Moreover, they combine net casting with acoustic (sonar) (Brehmer et al. 2006). Nowadays, much more convenient tools are employed, such as hand-held video filming devices. Embedded video cameras are also used to record underwater animals (including insects, fish, etc.), and observe fish presence and habits at different times (Nadarajan et al. 2011). This equipment has produced large amounts of data, and it requires informatics technology like computer vision and pattern recognition to analyze and query the videos. Statistics about specific oceanic fish species distribution, besides an aggregate count of aquatic animals, can assist biologists resolving issues ranging from food availability to predator-prey relationships (Rova et al. 2007). Unlike the simple and constrained environments found in the majority of previous work (e.g., fish tanks (Lee et al. 2004; Ruff et al. 1995), conveyor belts (Strachan 1993), dead fish (Larsen et al. 2009)), we investigate the recognition task of more fish species in a more complex and fundamentally challenging natural environment. We use underwater camera to record and recognize fish, where the fish can move freely and the illumination levels change frequently both locally from caustics arisen from the ocean surface waves and globally due to the sun and cloud positions (Toh et al. 2009). Recently, Duan et al. (2012) used fine-grained method to closely related categories like classify animal species by choosing relevant local attributes. However, the fine-grained method requires high standard about the quality of input images, which is not always met in our dataset. Instead, we designed some species-specific features for fish recognition (e.g., white tail for *Chromis margaritifer*, color stripe for *Amphiprion clarkii*).

In general, fish recognition is an application of multi-class classification. A common multi-class classifier could be considered as a flat classifier because it classifies all classes at the same time (Carlos and Alex 2010). A critical drawback is that it does not consider certain similarities among classes; these classes could be better separated by specifically selected features. One solution is to integrate domain knowledge and construct a tree to organize the classes hierarchically (Deng et al. 2010). This method, called hierarchical classification, has significant advantages by grouping similar classes into certain subsets and selecting specific subsets of features to distinguish them at a later stage (Gordon 1987). However, one problem of the hierarchical classification method is error accumulation. Each level of the hierarchical tree has some classification errors and these compounds as one goes deeper down the tree. As a result, realistic applications usually require rejection to eliminate the accumulated errors from hierarchical classification (Wang and Casasent 2009). In fish recognition, especially when our database is extremely imbalanced, misclassified samples are passed into deeper layers and reduce the average accuracy of the final recognition performance. Furthermore, false detections (e.g., non-fish objects, blurred images) and fish from an unknown species are also input to the recognition process. We introduce rejection into hierarchical classification by calculating the Bayesian posterior density. A GMM model is applied at the leaves of the hierarchical tree as the reject option. It evaluates the posterior probability of the test samples and rejects low probability samples. Using a reject option produces a lower false positive rate, but at the price of a slightly lower true positive rate due to incorrect rejections.

11.3 Feature Extraction

We observe fish images from underwater telerecording streams. These fish images record the illumination values (RGB) of pixels over the observing range. Unfortunately, the appearance of the fish are not constant due to the various conditions of, e.g., illuminations, reflections, shadows, etc. However, computers can only distinguish the fish from digital numeral data of extracted features. For example, in fish recognition, some species of fish have specific colors, fin shapes, stripes or texture. Computer vision techniques exploit these similarities, and present them by similar feature density distributions.

This section describes the feature extraction methods that are implemented for fish recognition in unconstrained circumstances since the quality of underwater video streams affect the recognition accuracy by adding distortions and noise to the original image. The pre-processing procedures are undertaken to improve the quality of features, including a Grabcut method for better segmentation of the fish inside the bounding box, a novel fish rotation algorithm to align the fish into the same direction. Afterwards, we give the technical details about our feature extraction algorithms and idiosyncratic fish descriptors. A combination of color, shape and texture properties in different parts of the fish such as tail, head, top and bottom are extracted.

11.3.1 Image Pre-processing

The pre-processing is undertaken to improve the quality of features. Firstly, the detection and tracking software described in Spampinato et al. (2014b) is used to obtain the fish and mask images. Then the Grabcut algorithm (Rother et al. 2004) is employed to segment fish from the background, similar to Edgington et al. (2006), Cline and Edgington (2010)). Given prior information such as reference frame or pre-label foreground area, the graph cut solution gives each pixel a weight between foreground (source) and background (sink), and solves the segmentation problem with a minimum cost cut method to divide the source from the sink. The solution finds the global energy optimum. This approach converts an image processing problem into a graph energy minimization problem, and there is a universal algorithm to tackle the graph cut question. The optimization procedure is based on the similarity between a pixel and its local neighbors. This method can overcome normal image distortion, such as additional noise and water reflection, which triggers segmentation errors in other algorithms. We then add padding around the detected fish to ensure that the whole fish is included. The padding may extend outside the input frame if the fish is close to the edge of the frame. An example of a detected fish is provided in Fig. 11.2, where most parts of the key feature (white tail) are preserved by the segmentation algorithm.

After acquiring the fish bounding boxes, we align the fish images in the same direction before further processing. We rotate their bodies by an estimated angle so that fish from the same species are facing the same directions. Thereafter, we can divide the fish into several parts and extracts specific features (e.g., focus on the white tail part for *Chromis margaritifer*). The rotation angle is estimated by using a heuristic method inspired by the streamline hypothesis that a fish's head part is smoother than its tail part because it needs a more frictional tail (caudal fin) to swim and keep its body balanced. As a result, the centroid of the curvature value on the fish contour is located on the tail part.

More specifically, the curvature value of each boundary pixel is defined as follows (Mokhtarian and Suomela 1998; He and Yung 2004):

$$\kappa(u,\sigma) = \frac{X_u(u,\sigma)Y_{uu}(u,\sigma) - X_{uu}(u,\sigma)Y_u(u,\sigma)}{(X_u(u,\sigma)^2 + Y_u(u,\sigma)^2)^{\frac{3}{2}}} \tag{11.1}$$

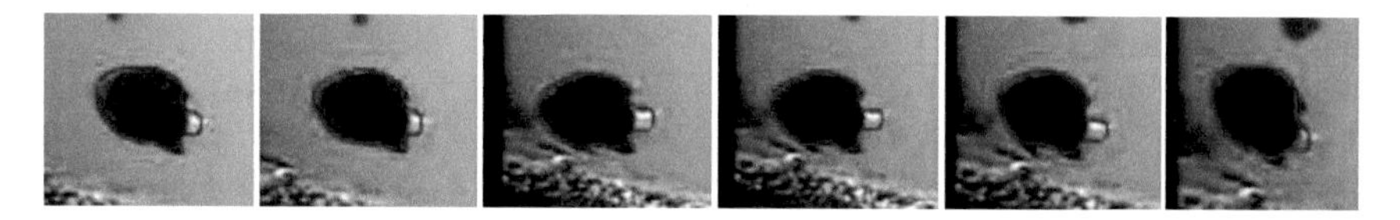

Fig. 11.2 An example of fish detection from a whole trajectory of *Chromis margaritifer*. This species of fish has a noteworthy white tail. This feature is essential for discriminating it from other species of fish, especially *Dascyllus reticulatus*. These images have successfully maintained most parts of the white tails

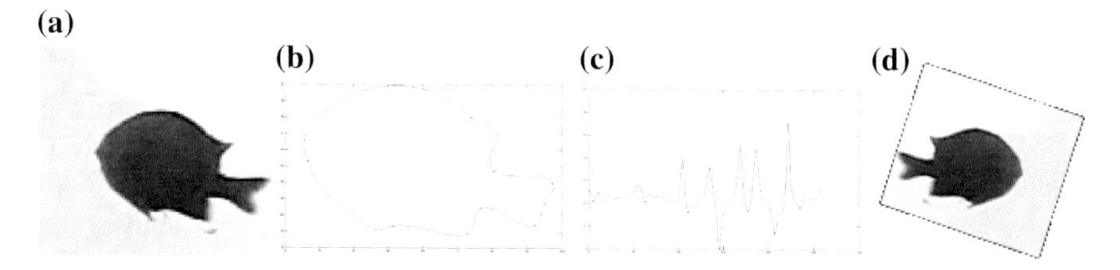

Fig. 11.3 Fish orientation demonstration: **a** input image of *Dascyllus reticulatus* fish; **b** fish boundary after Gaussian smoothing, with small spines eliminated since we are only interested in substantial fluctuations; **c** curvature levels along fish boundary, where the x-axis is the index of pixels of the contour starting from the top part of the fish and counting anti-clockwise, and the y-axis shows the degree of curvature; **d** oriented fish image for further processing. This method helps to divide fish in a constant way and extracts specific features (e.g., the white tail of *Chromis margaritifer*)

where $X_u(u,\sigma)$, $X_{uu}(u,\sigma)$ and $Y_u(u,\sigma)$, $Y_{uu}(u,\sigma)$ are the first and the second derivative of $X(u,\sigma)$ and $Y(u,\sigma)$, respectively; $X(u,\sigma)$ and $Y(u,\sigma)$ are the convolution result of 1-D Gaussian kernel function $g(u,\sigma)$ with fish boundary coordinates $x(u)$ and $y(u)$. We fix σ so that κ depends only on u. A typical fish orientation procedure is illustrated in Fig. 11.3. Considering the first image (Fig. 11.3a) as input, we first smooth the contour image with a Gaussian filter to eliminate spines, which generate pulses in curvature and should be excluded since we only care about substantial components (Fig. 11.3b). The degrees of curvature of fish contour are illustrated in Fig. 11.3c, where the x-axis is the index of pixels of contour starting from the top part of the fish and passing anti-clockwise and the y-axis stands for the curvature degree. The curvature degree fluctuates more severely on the right side than on the left since the curvature is concentrated at the rear half of the fish. In order to refine the estimation of tail direction, we fit the fish boundary into an ellipse shape, and then use the deflective angle for minor trimming. Figure 11.3d shows the final result, where the *Dascyllus reticulatus* is rotated horizontally and faces right. The fish orientation method achieves 95 % correct fish orientation $\pm 15^\circ$ using 1000 manually labeled fish images.

11.3.2 Feature Extraction

The procedure of feature extraction is often considered as a black box in object recognition applications. However, the quality of features is critical in the following classification step. Feature engineering work aims at obtaining discriminative characteristics of input data. In this section, we propose a set of effective low level visual descriptors for fish images. We treat this as an incremental process, where new features are designed to improve on the accuracy achieved by appropriate combinations of existing features. More specifically, we put all existing features into a pool for selection, and the algorithm chooses the candidate features which maximize the averaged classification accuracy over all species. Sixty nine types of feature are extracted. These features are a combination of color, shape and texture properties in different parts of the fish such as tail/head/top/bottom, as well as the whole fish. We use normalized color histogram in the Red&Green channel and the Hue component in HSV color space. These color features are normalized to minimize the effect of illumination changes. In order to equalize the color histogram and create a more uniform distribution for the whole dataset to maximize contrast, we calculate the average distribution of the whole dataset and use it as the global probability function for histogram equalization. We also introduce a set of new features which help distinguish fish species that tend to be misclassified, including projected color density, tail/head and tail/body area ratios. These features are designed to integrate computer vision techniques with marine knowledge. Those fish that have the same ancestors share similar synapomorphic characteristics. They indicate the distinction between species, for example, the presence or absence of components, specific number, and so on. Some of these synapomorphic characteristics can be obtained from the video frame, mostly from the shape of the fish contour. Firstly, we exploit the projected color density, which describes the color variations of fish body changes in both horizontal and vertical directions and generates a density histogram by calculating the mean value of color along the axis. This feature is useful for describing the significant surface marks such as the colorful tail, stripes, and spots of fish. The mean and standard deviation of the projected density are stored as idiosyncratic fish features.

In order to describe the fish texture, we calculate the Gray-Level Co-occurrence Matrix (GLCM), Fourier descriptor and Gabor filter. The GLCM describes the co-occurrence frequency of two gray scale pixels at a given distance d. The frequency is calculated for four angles ϕ: $0°$, $45°$, $90°$, and $135°$. The offset distance ranges from 1 to 10. We computed the GLCM for the multi-spectral image and produced inter-plane combinations of the co-occurrence matrix where six combinations (RR, RG, RB, GG, GB, and BB) are concatenated. We compute 12 features of each normalized GLCM introduced by Soh and Tsatsoulis (1999), Haralick et al. (1973): contrast, correlation, energy, entropy, homogeneity, variance, inverse difference moment, cluster shade, cluster prominence, maximum probability, auto-correlation, and dissimilarity.

Histogram of oriented gradients and Moment Invariants, as well as Affine Moment Invariants, are employed as the shape features. Furthermore, some specific features like tail/head area ratio, tail/body area ratio, etc. are also included.

These descriptors are found to be effective. They are designed to integrate domain knowledge with machine vision methods and considered together as a pool for feature selection in the classification step. This pool is incrementally constructed so that additional features are designed and introduced after analyzing the experimental results. As discussed before, we propose 69 groups of features (2626 dimensions) to recognize fish. Example and more details are included in Huang (2014). These features are a combination of the color, shape, and texture properties of different parts of the fish such as the tail/head/top/bottom as well as the whole fish. All features are normalized by subtracting the mean and dividing by the standard deviation (z-score normalized after 5 % outlier removal).

11.4 Fish Recognition

The Balance Guaranteed Optimized Tree with reject option (BGOTR) is based on the inter-class similarity among fish species, and it groups similar classes at the upper levels of the tree to distinguish them at a later stage. BGOTR is a recursive hierarchical structure using a multiclass decision (here using SVM) at each tree node. The feature selection method chooses particular subsets of features to maximize the accuracy over all subsets at each node. Discussion of multiclass classifiers is presented in this section, which compares the normal flat classifier approach to the hierarchical classification method. The latter method uses a divide and conquer strategy, and organizes candidate classes into multiple levels. In a greatly imbalanced dataset, the less common classes are grouped with other classes and this strategy helps ease the imbalance of data. The hierarchical classification method also exploits the correlations between classes and finds similar groupings. Unlike biological hierarchical classification methods like the taxonomy tree, which aims to systematize animals into their pre-defined hierarchical categories, the BGOTR method chooses an optimal binary split of the given classes at every node. It improves the normal hierarchical method by arranging more accurate classifications at a higher level and keeping the hierarchical tree balanced. The reject function evaluates the posterior probability of the tested samples given the recognition result. This is a post-recognition step and the rejection is independent of the recognition since it is applied only to the recognition results. The "rejection" term targets the specific application scenarios of: (1) eliminating false positives from the recognition results, and (2) eliminating samples not belonging to the training classes. In the experimental section, we evaluate the performance of our method on these two applications respectively.

11.4.1 The Balance Guaranteed Optimized Tree Method

A hierarchical classifier h_{hier} is designed as a structured node set. Fundamentally, a node is defined as a triple: $\text{Node}_t = \{\text{ID}_t, \tilde{\text{F}}_t, \hat{\text{C}}_t\}$, where ID_t is a unique node number, $\tilde{\text{F}}_t \subset \{\mathbf{f}_1, \ldots, \mathbf{f}_m\}$ is a feature subset chosen by a feature selection procedure that is found to be effective for classifying $\hat{\text{C}}_t$, which is a subset of classes and their groups. We only consider binary splits (until the final layer), so each node has at most two groups. All samples that are classified as the same group will be transmitted into the same child node for later processing. An example with 15 classes is shown in Fig. 11.4, where the ID_t is illustrated in each node and $\hat{\text{C}}_t$ are the local groups. The binary splitting process stops when each group has at most 4 classes (e.g., Node ID 4, 5, 6, 7) in order to limit the maximum depth of the tree and avoid overtraining. All the leaf nodes are multiclass SVMs using the One-versus-One strategy.

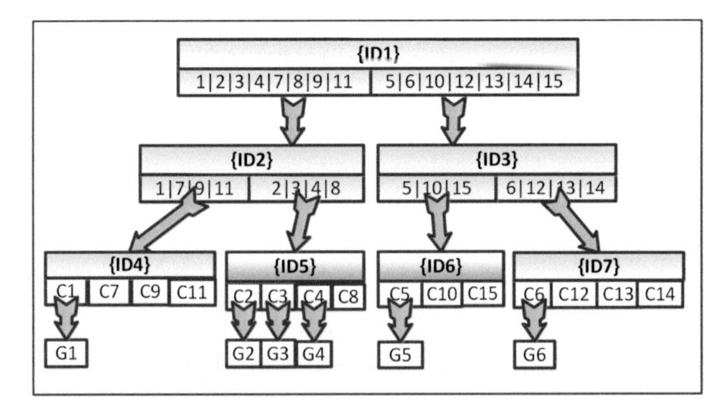

Fig. 11.4 GMM for rejection post-processing for classes C1, ..., C6 in hierarchical classification, integrated with a BGOTR method

This hierarchical classification method is presented as an assembly of individual multiclass classifiers. These classifiers are treated as tree nodes. At each node, there are at least two groups of classes. We use the term "group" to indicate a super-class, which includes several classes as a single item. In the following paragraph, we will introduce our strategy to organize training classes into groups. Every child node corresponds to a choice of group. During classification, every sample starts from the root node at the top, and goes through the hierarchical architecture. At a non-leaf node, the classification decision determines which group the test sample belongs to. The sample is then passed to the corresponding child node for further classification. The procedure continues until the test sample reaches a leaf node whose classification result is a single class, instead of a group of classes.

To construct the hierarchical tree, we first aim at finding an optimal split of the given classes at the current node by minimizing the mean misclassification rate between the two child nodes. We search for all possible splits of the classes into two nearly equal sets of classes. We also select the feature subset that achieves the best accuracy for the given split, using forward sequential feature selection based on grouped subset of features. This process is repeated for each child node. A well-designed hierarchical tree can help improve the accuracy of some confusable classes while suppressing the error accumulation. We propose two heuristics for how to organize a single classifier and construct a hierarchical tree with higher accuracy.

1. Arrange more accurate classifications at a higher level and leave similar classes to deeper layers.
2. Keep the hierarchical tree balanced to minimize the max-depth and control error accumulation. Here we split the tree by equal number of classes, but one could also use other splits, such as by equal a priori fish appearance probabilities, or non-equal numbers of classes to minimizing error.

To help choose a good classifier for each level of the hierarchy, we tried the Random Forests method (Breiman 2001) as an exploration on a small dataset of 7200 fish images of 15 fish species (Table 11.1), when the full dataset of 241,500 images was still in progress. A Random Forest is made of a number of decision trees with binary splits for classification. It predicts responses for new data with the ensemble learned model. In our experiment on 15 species of fish, the Random Forests method was implemented with 50 decision trees. Each tree was constructed

Table 11.1 Fish recognition exploration for choosing the most effective classifier

Method	AR (%)	AP (%)	AC (%)
Random Decision Forests (Ho 1995)	0.772	0.662	0.914
Random Forests (Breiman 2001)	0.625	0.782	0.903
Ada-Boost (Liang et al. 2010)	0.753	0.769	0.923
SVM (Cortes and Vapnik 1995)	0.863	0.858	0.934

Average Recall (AR), Average Precision (AP), Accuracy by Count (AC) are introduced in the experimental Sect. 11.5

using 500 randomly selected features. This Random Forests method and another popular method, Ada-Boost (Liang et al. 2010), were implemented to compare with the multiclass SVM method, as an exploration to choose the appropriate classifier. The experimental results demonstrated that the performance of the multiclass SVM method was better than the Random Forests and Ada-Boost methods.

11.4.2 Trajectory Voting Method

In the view of a traditional fish recognition system, the classifier predicts fish species according to individual images. Some classification errors occur due to varying illumination arising either by the fish orientations or light field. Using the fish recognition results from consecutive frames of the same trajectory helps eliminate these minor errors and improves the overall accuracy. We have applied the image set classification to the live fish recognition scenario in a non-constrained environment. This method uses a set of observations to recognize test samples. The image set is from a video sequence containing multiple images of the same target. In the literature concerning the image set integration, there are mainly two categories of theories regarding the underlying sequence of result integration: the early integration strategy and the late integration strategy. The former method uses the observations to determine the similarity between image sets, before matching. Shakhnarovich et al. (2002) consider the features of multiple observations as a whole, and propose a classification based on their distributions. On the other hand, the late integration strategy uses likelihoods after matching. These likelihoods could be calculated either by product or by maximizing of the individual decisions (Maron and Lozano-Pérez 1998; Zhang and Goldman 2001; Yang et al. 2005).

In our live fish recognition system, we have applied the majority voting algorithm to make use of the temporal information embedded in fish trajectories, and to minimize the environmental influence. This is a late integration strategy. As all fish are freely swimming in a varying illumination environment, the detected fish may have different orientations and appearances. Therefore, the recognition results may vary even for a fish in the same trajectory. A trajectory based winner-take-all voting mechanism is applied after the individual classification. It combines the single frame classification results. The trajectory voting method enhances the fish recognition accuracy by exploiting the consistency in labels expected from tracking each fish individually.

11.4.3 Gaussian Mixture Model For Reject Option

A GMM is employed to represent the hypothetical clusters of density distributions in feature space because individual component Gaussian functions were not sufficient to model the underlying characteristics of the given classes. For example, in fish recognition, some species of fish have specific colors, fin shapes, stripes or texture. It is reasonable to assume that the extracted features represent the domain knowledge and represent them by the density distributions. Each characteristic is expressed both by the mean value μ_i and the covariance matrix Σ_i. The training procedure is unsupervised (after assigning the training class), the GMM captures the prominent density distributions and is not constrained by the label information. There are several variables to be fit in this step, like μ_i, Σ_i. The Expectation Maximization (EM) algorithm (Shental et al. 2003), which is guaranteed to converge to a local maximum by iteratively searching, is applied to optimize the Gaussian mixture model. Figueiredo and Jain (2002) present an unsupervised learning algorithm to learn a proper mixture model from multivariate data. It can automatically select the finite mixture model by using the minimum message length (MML) with advantages compared to other deterministic criteria, e.g., Bayesian Inference criterion (BIC), Minimum Description length (MDL): in particular, it is less sensitive to the initialization, and avoids the boundary of the parameter space.

One difficulty for rejection in a hierarchical method is how to evaluate a probability score based on the intermediate classification results at different layers. Instead of integrating the result score along the path of the hierarchy, here a GMM model is applied after the BGOTR classification to implement the reject option (Fig. 11.4). The GMM model is trained by a subset of features by using the forward sequential selection method. For each BGOTR result, the final $P(C \mid x)$ for that input is estimated according to the GMM likelihood score. More specifically, the rejection uses the posterior probability for the predicted class C_i giving evidence X:

$$p(C_i \mid X) = \frac{p(C_i)p(X \mid C_i)}{p(X)} = \frac{p(C_i)p(X \mid C_i)}{\sum_j p(C_j)p(X \mid C_j)} \tag{11.2}$$

where the prior knowledge $p(C_i)$ is calculated from the training samples. The features used for training the GMM are the same as for BGOTR but a different subset was selected (using the same feature selection criteria). In Chib (1995), Chib and Siddhartha express the marginal density as the prior times the likelihood function over the posterior density. They found comparable performance of the marginal like-

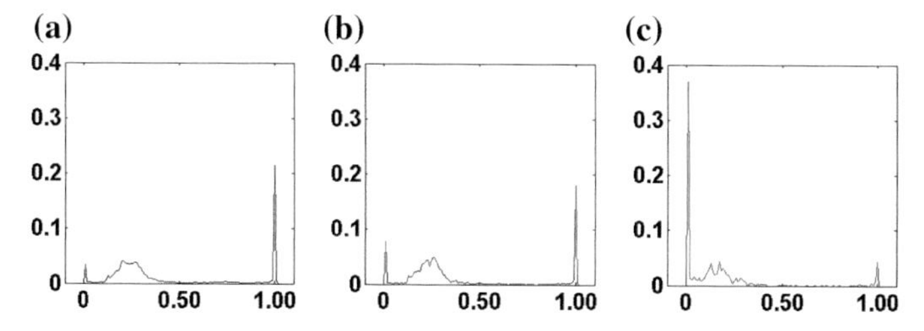

Fig. 11.5 **a** Distribution of posterior probability of the training samples of species *Chromis chrysura*. **b** Distribution of posterior probability of test sample True Positives. **c** Distribution of posterior probability of test sample False Positives. See text for details

lihood with an estimation of the posterior density. Since we address the improvement of rejection in hierarchical classification, we also calculate the posterior density of the testing samples by Bayes rule. For each sample with evidence X and BGOTR prediction C_i, we calculate its posterior probability $P(C_i \mid X)$ from Eq. 11.2 and set a small threshold (i.e., 0.01) to reject all samples whose posterior probabilities are below the threshold. Figure 11.5 illustrates the distribution of the posterior probability $p(C_i \mid X)$ of all samples that are classified as species *Chromis chrysura*. These samples are either correctly classified (True Positives, Fig. 11.5b) or misclassified (False Positives, Fig. 11.5c). The distribution of the posterior probability of False Positives (as shown in Fig. 11.5c) has a peak distribution (about 38 %) around the value of zero while most of the True Positives have higher posterior probability (Fig. 11.5b). The difference between these two distributions is exploited to distinguish False Positives. This algorithm rejects a substantial portion of the misclassified samples with the cost of also rejecting a small proportion of True Positives (see experiment section for details).

11.5 Fish Recognition Experiments

Our data is acquired from a live fish dataset of the 15 different species shown in Fig. 11.6. This figure shows the fish species name and the numbers of observations and trajectories in the ground-truth. The data is very imbalanced, where the most frequent species is about 500 times more common than the least one. Note, the images shown here are ideal images as many of the others in the database are a bit blurred, and have fish at different distances and orientations or are against coral or ocean floor backgrounds.

All fish are manually labeled by following instructions from marine biologists (Boom et al. 2012). The labeling work was supported by a clustering method. Then, three users checked and cleared the clustering results. The final annotation work was confirmed by two marine biologists. In our experiment, the training and testing sets are isolated so fish images from the same trajectory sequence are not used during both training and testing. We use the pre-processing and feature extraction methods presented in the previous section.

11.5.1 Fish Recognition Experiments Using Ground Truth Data

We use the BGOTR method for fish recognition. Both flat SVM and hierarchical methods are explored. Both linear and non-linear kernel methods are tested. Based on the multi-class classifier, we designed four other classifiers:

1. A multiclass 1v1 flat SVM classifier, which classifies all 15 classes simultaneously, is implemented as a baseline classifier. Forward sequential feature selection

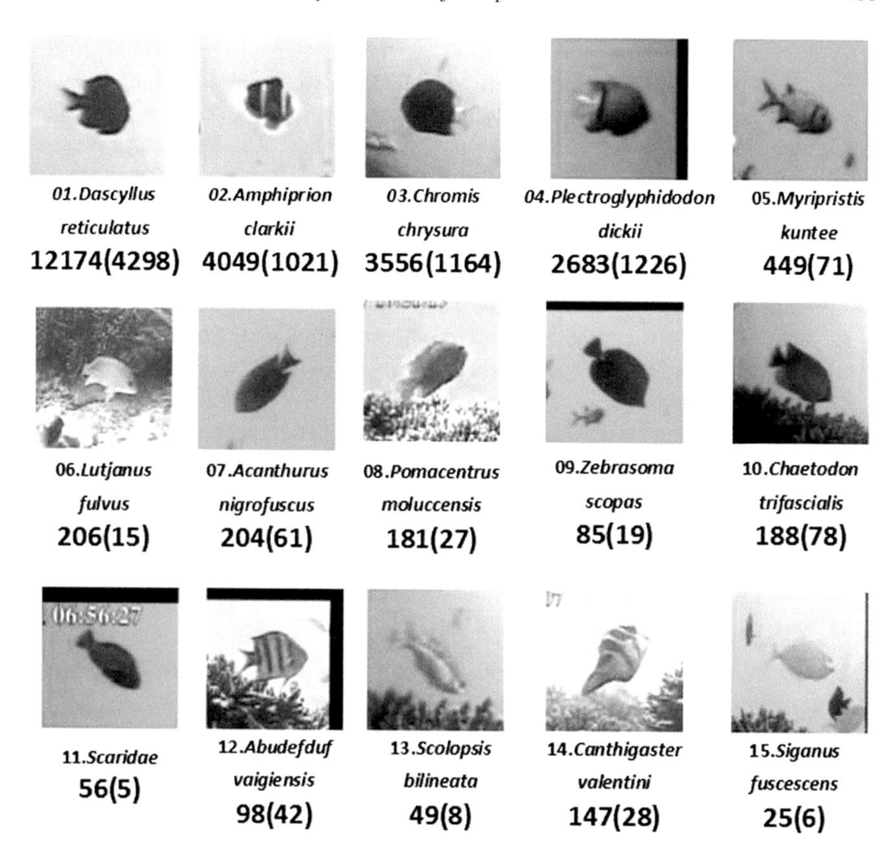

Fig. 11.6 Top 15 species of fish in underwater videos, with the number of observations and trajectories in the ground-truth. All in all, there are 24,150 observations and 8069 trajectories

is applied (named flatSVM-fs) to do greedy selection of the features to maximize the average recall among all classes.

2. The Principal Component Analysis (PCA) algorithm is also implemented as a baseline method for feature selection and classification. It uses singular value decomposition (SVD) to reduce the feature dimensions and we preserve 98 % of the principal component variance (up to 583 dimensions). The processed features are then classified by a 15-class SVM classifier.
3. The Lasso (L1-constrained fitting) algorithm (Tibshirani 1996) is a shrinkage and selection method (Zou and Hastie 2005) for linear regression. It minimizes the usual sum of squared errors, with a bound on the sum of the absolute values of the coefficients. In our experiment, it is implemented as a wrapper procedure using the scoring function of feature subset. We select features such that the MSE is within one standard error of the minimum (up to 763 dimensions). The selected features are then classified by a 15-class SVM classifier.

4. A classical classification and regression tree method (CART (Hastie et al. 2001)) is provided as another automatically generated hierarchical decision tree to be compared with. It starts with a single node, and then looks for the binary distinction which gives the most information about the class. The generating process continues until it reaches the stopping criterion.
5. A taxonomy tree is constructed according to the fish species taxonomy. This tree is pre-defined. It reflects the homologous similarity between species. All the 15 species of fish belong to the *Actinopterygii* class (ray-finned fishes), but in different orders, families and genus. This tree splits all classes into 9 groups at the first level according to their family synapomorphies characteristic and leaves a few similar species to deeper layers where the customized multiclass 1v1 SVM classifier is trained.
6. An automatically generated tree (BGOTR) is designed by recursively choosing a binary split which has the best accuracy over the given classes. Forward sequential feature selection (FSFS) is applied in the BGOTR method to select effective subsets of features at each node of the hierarchical tree and the goal of feature selection is to maximize the average accuracy among all classes, which enhances the weight of less common classes. Feature selection typically selects about 300 of the features at each node.

The experiment is based on 24,150 fish images with a 5-fold cross validation procedure with a leave-$\frac{1}{5}$-out strategy. The training and testing sets are isolated so fish images from the same trajectory sequence are not used during both training and testing. We applied the majority voting algorithm to make use of the temporal information.

Results for the 5 algorithms are listed in Table 11.2 where the AR and AP are recall/precision averaged over all classes rather than over all fish. This is because of the greatly unbalanced class sizes. Three performance metrics are employed to evaluate the accuracy of the proposed system. The first metric is Average Recall (AR, or Macro-Averaged Recall) over all species. It describes on average how many fish are correctly recognized for each species. This score is more important to our experiment because of the imbalance in the classes. The second score is Average Precision (AP, or Macro-Averaged Precision) over all species. It is the probability that the classification results are relevant to the specified species. The third metric is the accuracy over all samples (Accuracy over Count, AC, or Micro-Average Recall), which is defined as the proportion of correct classified samples among the whole dataset.

We compare the hierarchical classification against the linear SVM classifier (AR = 76.9 %). Other non-linear flat SVM methods (polynomial, radial basis function, sigmoid function) are also included but their performances are worse than the linear SVM method. PCA is a popular algorithm to reduce feature dimensions. We apply it before an SVM and achieve almost the same score (AR = 77.7 %). In the seventh row, feature selection before use in a SVM produces slightly better results (AR = 78.4 %) than the flat SVM using all features. The CART algorithm has the lowest AR (53.6 %) among all three hierarchical methods. The taxonomy methodology achieves

Table 11.2 Fish recognition results

Method	AR (%)	AP (%)	AC (%)
SVM (linear)	76.9 ± 4.6	88.5 ± 3.6	95.7 ± 0.5
SVM (polynomial)	61.8 ± 5.0	86.0 ± 7.0	93.0 ± 0.4
SVM (RBF kernel)	70.4 ± 5.6	87.8 ± 6.7	96.0 ± 0.6
SVM (sigmoid)	62.3 ± 5.8	77.1 ± 7.2	85.9 ± 1.0
Lasso	76.6 ± 4.7	85.4 ± 3.3	95.4 ± 0.5
PCA (98 %)	77.7 ± 3.8	88.9 ± 4.1	95.4 ± 0.4
flatSVM-fs	78.4 ± 3.7	88.0 ± 5.5	95.9 ± 0.4
CART (Hastie et al. 2001)	53.6 ± 5.1	52.9 ± 4.6	87.0 ± 0.7
Taxonomy	76.1 ± 5.2	87.2 ± 6.7	95.3 ± 0.4
BGOTR	**84.8*** ± 3.9	**91.4** ± 2.8	**97.5*** ± 0.6

We add the standard deviation of AR/AP/AC over 5-fold cross validation. * means the score is a significant improvement over other methods at 95 % confidence level

a better AR of 76.1 % than CART but is worse than the automatically generated hierarchical tree (84.8 %) which chooses the best splitting by exhaustively searching all possible combinations while remaining balanced. The BGOTR method without node rejection has a lower performance (80.1 % in AR). Most algorithms achieve high AC score, but this is because the classes are very unbalanced. For example, to simply label all fish as class 1 already achieves an AC = 50.4 %. These experimental results demonstrate that reject option has significantly improved the fish recognition performance where comparing to other state-of-the-art techniques, more details are included in Huang et al. (2014).

11.5.2 BGOTR Application to New Real Fish Videos

Our fish recognition system depends on the detection results. Due to the complex environment (e.g., light distortion, fish occlusions and illumination transformation), the fish detection algorithm produces errors that are input to the classification procedure and cause unexpected recognition results. The previous experiments are evaluated on a "clean" dataset where all tested images are valid fish from either known or unknown species. However, in real applications, the acquired data may contain false detections, e.g., blurred images, occlusion by other fish or background objects, non-fish objects (coral, sea flowers, etc.). Some examples of false detections are shown in Fig. 11.7. In this section we experimentally evaluate how many false detections our BGOTR system can reject while preserving the valid ones. We chose 3 underwater videos and have labeled 1000 detections from each video.

The recognition results are shown in Tables 11.3 and 11.4. We use BGOTR to classify the test images and calculate the Average Recall (AR, macro recall) and Averaged Precision (AP, macro precision) among all 15 species. The AR score

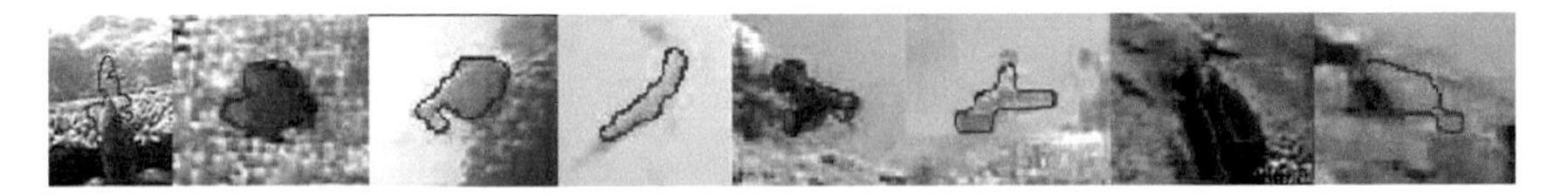

Fig. 11.7 Invalid fish images, chosen from 3 underwater videos. In a normal classifier without a reject option, these images would be classified and cause unexpected results. Our rejection algorithm aims at eliminating them while preserving most valid fish images

Table 11.3 Experiment result for real videos

ID	Average Recall (AR)	Averaged Precision (AP)
Video1	0.815	0.412
Video2	0.804	0.448
Video3	0.725	0.557
Average	0.781	0.472

In each video we select the first 1000 detections and manually label all samples

Table 11.4 Experiment of rejection result in real videos

ID	True detections	False detections	Rejections	TR	FR
Video1	308	692	390	378	12
Video2	148	852	734	705	29
Video3	513	487	380	312	68
Average	323	677	501	465	36

TR = True Rejection, FR = False Rejection

demonstrates that the BGOTR method recognizes about 78 % of the real, untrained valid fish images correctly. The test images include many invalid detections (692, 892, 487, respectively). The BGOTR method filters more than half of these false detections (378, 705, 312, respectively) while it retains most of the valid inputs. Some false detections are not rejected and these inputs lower the average precision score (c. 47 %).

11.6 Conclusion

Live fish recognition in the open sea is fundamentally challenging because of a complex situation where the illumination changes frequently. Prior research is mainly restricted to constrained environments (fish in the tank or on a conveyor system) or dead fish. None of these methods works because of the unconstrained environment and imbalanced dataset. In this chapter, we presented a novel Balance-Guaranteed Optimized Tree (BGOTR) classifier for live fish recognition in a non-constrained environment. Although hierarchical classification is widely applied in machine vision applications, BGOTR improves the normal hierarchical method by two heuristics for how to organize a single classifier and construct a hierarchical tree with higher accuracy. After constructing the tree architecture, a novel trajectory voting method is used to eliminate accumulated errors during hierarchical classification and achieves better performance. The novel rejection system enhances the hierarchical classification algorithm as applied for fish species recognition. We apply a GMM model at the leaves of the hierarchical tree as a reject option. We use feature selection to select a subset of effective features that distinguishes the samples of a given class from others. After learning the mixture models, the reject function is integrated with a BGOTR hierarchical method. It evaluates the posterior probability of the testing samples and reduces the false positive rate, since some misclassification errors in the BGOTR classifier can be overcome at the price of a slightly lower true positive rate due to incorrect rejections. The experimental results demonstrate that the automatically generated hierarchical tree achieves *c.* 6 % improvement of the average recall (AR) and *c.* 3 % improvement of the average precision (AP) compared to the flat SVM and other hierarchical classifiers (Table 11.2). More detailed information is included in Huang et al. (2012, 2014), Huang (2014).

References

Boom, B., P. Huang, J. He, and R.B. Fisher. 2012. Supporting ground-truth annotation of image datasets using clustering. In *Proceedings of 21st international conference on pattern recognition (ICPR)*, 1542–1545. IEEE.

Brehmer, P., T.D. Chi, and D. Mouillot. 2006. Amphidromous fish school migration revealed by combining fixed sonar monitoring (horizontal beaming) with fishing data. *Journal of Experimental Marine Biology and Ecology* 334(1): 139–150.

Breiman, L. 2001. Random forests. *Machine learning* 45(1): 5–32.

Caley, M.J., M.H. Carr, M.A. Hixon, T.P. Hughes, G.P. Jones, and B.A. Menge. 1996. Recruitment and the local dynamics of open marine populations. *Annual Review of Ecology and Systematics* 27: 477–500.

Carlos, S., and F. Alex. 2010. A survey of hierarchical classification across different application domains. *Data Mining and Knowledge Discovery* 22(1–2): 31–72.

Chib, S. 1995. Marginal likelihood from the Gibbs output. *Journal of the American Statistical Association* 90(432): 1313–1321.

Cline, D.E., and D.R. Edgington. 2010. A detection, tracking, and classification system for underwater images. *ICPR Workshop on Visual Observation and Analysis of Animal and Insect Behavior (VAIB)*, Istanbul.

Cortes, C., and V. Vapnik. 1995. Support-vector networks. *Machine Learning* 20(3): 273–297.

Deng, J., A.C. Berg, K. Li, and L. Fei-Fei. 2010. What does classifying more than 10,000 image categories tell us? In *Proceedings of the 11th european conference on computer vision*, 71–84. Berlin: Springer.

Duan, K., D. Parikh, D. Crandall, and K. Grauman. 2012. Discovering localized attributes for fine-grained recognition. In *2012 IEEE conference on computer vision and pattern recognition (CVPR)*, 3474–3481. IEEE.

Edgington, D.R., D.E. Cline, D. Davis, I. Kerkez, and J. Mariette. 2006. Detecting, tracking and classifying animals in underwater video. In *OCEANS*, 1–5.

Figueiredo, M.A.T., and A. Jain. 2002. Unsupervised learning of finite mixture models. *IEEE Transactions on Pattern Analysis and Machine Intelligence* 24(3): 381–396.

Gordon, A.D. 1987. A review of hierarchical classification. *Journal of the Royal Statistical Society* 150(2): 119–137.

Haralick, R., K. Shanmugam, and I. Dinstein. 1973. Textural features for image classification. *IEEE Transactions on Systems, Man and Cybernetics, SMC* 3(6): 610–621.

Hastie, T., R. Tibshirani, and J.J.H. Friedman. 2001. *The elements of statistical learning*, vol. 1. New York: Springer.

He, X.-C., and N.H. Yung. 2004. Curvature scale space corner detector with adaptive threshold and dynamic region of support. In *Proceedings of the 17th international conference on pattern recognition, ICPR*, vol. 2, 791–794. IEEE.

Ho, T.K. 1995. Random decision forests. In *Proceedings of the third international conference on document analysis and recognition*, 278–282.

Huang, X.P. 2014. Balance-Guaranteed Optimized Tree with Reject option for live fish recognition. PhD thesis, University of Edinburgh.

Huang, P. X., B.J. Boom and R.B. Fisher. 2012. Underwater live fish recognition using balance-guaranteed optimized tree. In *Proceedings of the 11th Asian Conference on Computer Vision*, vol. 7724, pages 422–433.

Huang, P.X., B.J. Boom, and R.B. Fisher. 2014. GMM improves the reject option in hierarchical classification for fish recognition. In *Proceedings of Workshop on Applications of Computer Vision 2014*. 371–376.

Larsen, R., H. Ólafsdóttir, and B. Ersbøll. 2009. Shape and texture based classification of fish species. In *Proceedings of the scandinavian conference on image analysis*, 745–749.

Lee, D., R.B. Schoenberger, D. Shiozawa, X.Q. Xu, and P.C. Zhan. 2004. Contour matching for a fish recognition and migration-monitoring system. *Proceedings of SPIE* 5606(1): 37–48.

Liang, Y., J. Li, and B. Zhang. 2010. Learning vocabulary-based hashing with adaboost. In *Proceedings of the 16th international conference on advances in multimedia modeling*, 545–555. Springer.

Maron, O., and T. Lozano-Pérez. 1998. A framework for multiple-instance learning. In *Proceedings of the conference on advances in neural information processing systems*, 570–576.

Mokhtarian, F., and R. Suomela. 1998. Robust image corner detection through curvature scale space. *IEEE Transactions on Pattern Analysis and Machine Intelligence* 20(12): 1376–1381.

Nadarajan, G., Y.-H. Chen-Burger, R.B. Fisher, and C. Spampinato. 2011. A flexible system for automated composition of intelligent video analysis. In *Proceedings of the 7th international symposium on image and signal processing and analysis (ISPA)*, 259–264. IEEE.

Rother, C., V. Kolmogorov, and A. Blake. 2004. "GrabCut": interactive foreground extraction using iterated graph cuts. *ACM Transaction on Graphics* 23(3): 309–314.

Rova, A., G. Mori, and L.M. Dill. 2007. One fish, two fish, butterfish, trumpeter: Recognizing fish in underwater video. In *IAPR conference on machine vision applications*, 404–407.

Ruff, B.P., J.A. Marchant, and A.R. Frost. 1995. Fish sizing and monitoring using a stereo image analysis system applied to fish farming. *Aquacultural Engineering* 14(2): 155–173.

Shakhnarovich, G., J.W. Fisher, and T. Darrell. 2002. Face recognition from long-term observations. In *Proceedings of the 7th European conference on computer vision*, 851–865. Springer.

Shental, N., A. Bar-hillel, T. Hertz, and D. Weinshall. 2003. Computing gaussian mixture models with EM using equivalence constraints. In *Advances in neural information processing systems 16*. MIT Press.

Soh, L.-K., and C. Tsatsoulis. 1999. Texture analysis of SAR sea ice imagery using gray level co-occurrence matrices. *IEEE Transactions on Geoscience and Remote Sensing* 37(2): 780–795.

Spampinato, C., S. Palazzo, B. Boom, J. van Ossenbruggen, I. Kavasidis, R. Di Salvo, F.-P. Lin, D. Giordano, L. Hardman, and R. Fisher. 2014b. Understanding fish behavior during typhoon events in real-life underwater environments. *Multimedia Tools and Applications* 70(1): 199–236.

Strachan, N.J.C. 1993. Recognition of fish species by colour and shape. *Image and Vision Computing* 11: 2–10.

Tibshirani, R. 1996. Regression shrinkage and selection via the lasso. *Journal of the Royal Statistical Society. Series B (Methodological)*, 267–288.

Toh, Y.H., T.M. Ng, and B.K. Liew. 2009. Automated fish counting using image processing. In *International conference on computational intelligence and software engineering*, 1–5.
Wang, Y.-C.F., and D. Casasent. 2009. A support vector hierarchical method for multi-class classification and rejection. In *Proceedings of the international joint conference on neural networks IJCNN*, 3281–3288.
Yang, J., R. Yan, and A.G. Hauptmann. 2005. Multiple instance learning for labeling faces in broadcasting news video. In *Proceedings of the 13th annual ACM international conference on multimedia*, 31–40.
Zhang, Q., and S.A. Goldman. 2001. EM-DD: An improved multiple-instance learning technique. In *Advances in neural information processing systems*, 1073–1080.
Zou, H., and T. Hastie. 2005. Regularization and variable selection via the elastic net. *Journal of the Royal Statistical Society: Series B (Statistical Methodology)* 67(2): 301–320.

Chapter 12
Fish Behavior Analysis

Cigdem Beyan

Abstract In this chapter, we address fish behavior analysis in unconstrained underwater videos. Assessing this is based on unusual fish trajectory detection which tries to detect rare fish behaviors, which can help marine biologists to detect new behaviors and to detect environmental changes observed from the unusual trajectories of fish. Fish trajectories are classified as *normal* and *unusual* which are the common behaviors of fish and the behaviors that are rare respectively. We investigated three different classification methods to detect unusual fish trajectories. The first method is a filtering method to eliminate normal trajectories, the second method is based on labeled and clustered data and the third method constructs a hierarchy using clustered and labeled data based on data similarity. The first two methods can be seen as preliminary works while the results of them are significant considering the challenges of underwater environments and highly imbalanced trajectory data that we used. In this chapter, we briefly summarize these two methods and mainly focus on the third method (hierarchial decomposition) which presented improved results and performed better than the state of art methods.

12.1 Introduction

The study of marine life is important especially for understanding environmental effects such as pollution, climate change, etc. Fish behavior analysis is helpful to detect such environmental effects by detecting the changes in behavior patterns or finding unusual behaviors and detecting the behavior distinctness of different species.

The traditional way to analyze fish behavior is based on visual inspection by marine biologists (Papadakis et al. 2012) such as by diving to observe underwater using photography or acoustic systems (Spampinato et al. 2008). However, this analysis is very time consuming and needs a huge amount of human labor. Moreover,

C. Beyan (✉)
PAVIS - Pattern Analysis and Computer Vision, IIT Istituto Italiano
di Tecnologia, Via Morego 30, 16163 Genova, Italy
e-mail: cigdem.beyan@iit.it

R.B. Fisher et al. (eds.), *Fish4Knowledge: Collecting and Analyzing Massive Coral Reef Fish Video Data*, Intelligent Systems Reference Library 104,
DOI 10.1007/978-3-319-30208-9_12

manually analyzing the data compared to automatic systems implies a decrease in the amount of data that can be analyzed. Therefore, computer vision and machine learning methods could play an important role to analyze the underwater videos.

In computer vision, behavior understanding studies can be classified into two categories: activity recognition and unusual behavior detection (Piciarelli et al. 2008). Activity recognition is very difficult when the variety of behavior models in an uncontrolled and uncooperative real-world data is considered (Piciarelli et al. 2008). On the other hand, unusual behavior detection analysis has become popular in recent years. In this kind of approach, the system does not have any prior knowledge about the behaviors. Unusual behaviors are generally defined as outliers or rare events and are detected with an unsupervised fashion (Anjum and Cavallaro 2008; Jiang et al. 2010).

The aim of our work is to present an unusual fish behavior detection system that uses underwater environment videos. We are making use of detected and tracked fish by the fish detection and tracking components mentioned in previous Chaps. 9 and 10. We have two classes of trajectories: *normal* and *unusual*. Normal fish trajectories are defined as the trajectories which contain frequently observed behaviors while unusual trajectories are defined as outliers or the behaviors not frequently observed. In all of our analysis, we used the trajectories of *Dascyllus reticulatus* since it is the most frequently detected and most accurately recognized fish in the Fish4Knowledge repository. We believe that the methods proposed in this chapter are helpful to understand the unusual behavior of fish species. Furthermore, detecting unusual behaviors can be a preliminary stage to understand specific behaviors of fish species such as feeding, predator-prey, reproduction, etc.

In the rest of this chapter, we first define the problem and give related definitions and challenges (Sect. 12.2). Following this, we present a literature review on fish behavior understanding (Sect. 12.3). In Sect. 12.4, the three methods that we proposed are presented. The first two methods are summarized very briefly as they are preliminary works while the third method is described more deeply. Experiments, data set and results are also discussed in this section. Finally, in Sect. 12.5, we conclude by making a summary of the chapter and by giving possible future directions.

12.2 Problem Description, Definitions and Challenges

In the literature, the definition of unusual behavior is quite ambiguous. Unusual behavior can be used interchangeably with the terms *abnormal*, *rare*, *outlier*, *suspicious*, *subtle*, *interesting*, and *anomaly* depending on the definition of the studies. For instance, Morris and Trivedi (2008) used the words abnormal, anomaly and unusual interchangeably denoting behaviors that do not fit into the typical cluster. In most of the study, the model of normal behavior is learned automatically. Using this learned model, a new (test) behavior is classified as normal or unusual. However, in real life scenarios, it is very difficult to predefine all possible normal and unusual behaviors. Therefore, behaviors are often unusual because there are no previous occurrences of it (Choudhary et al. 2008). Similarly, an event that cannot be classified by the learned models was defined as abnormal in Varadarajan and Odobez (2009). Xu et al. (2010) defined unusual behavior as interesting (not expected) and rare while Varadarajan et al. (2009) assumed that an unusual event is the one that occurs at an unusual location and an unusual time while it is fundamentally different in appearance and/or order. On the other hand, Dickinson and Hunter (2008) defined unusual events as rare events and due to lack of training data, which was detected by the deviation from a model of normal behaviors. Jiang et al. (2010) defined normal events using some rules and classified the events which do not obey the rules as anomalies. In this context, anomalies appear rarely and different from the commonality while the events with large groups are normal.

In the unusual behavior detection area, studies mostly focused on clustering-based methods and did not use labeled data. This is mainly because the labeling is very time consuming given that unusual behaviors are very rare. However, we claim that given a large enough data set, it is possible to find and label some unusual trajectories (although more difficult to find compared to normal trajectories since they are very rare) and apply supervised learning techniques to obtain more normal and unusual trajectories. In our work, we present three different supervised learning methods. For all methods, we consider two classes: normal and unusual.

When we compare fish trajectory data sets from underwater videos with the other unusual behavior detection data sets (for instance traffic surveillance, human abnormal trajectory detection etc.), there are certain challenges:

- Fish are not usually goal-oriented which produces highly complex trajectories in contrast to people or vehicles.
- Fish in the open sea can freely move in three dimensions hence there are no defined rules or roads such as exist in a traffic surveillance scenario.
- Fish usually make erratic movements due to currents in the water which increases the complexity of trajectories and also makes encoding the behavior difficult.

12.3 Literature Review on Fish Behavior Understanding

In the literature, fish behavior monitoring studies which are utilizing computer vision technology are generally for studies on water quality monitoring and toxicity identification such as Thida et al. (2009), Anitei (2011), Nogita et al. (1988), Schalie et al. (2001). Beside this aim, studies focusing on fish stress factor identification (Papadakis et al. 2012) or automatically monitoring abnormal behavior to help the farm operator in aquaculture sea cages (Pinkiewicz et al. 2011) also exist. Some of the research on fish behavior understanding has focused on the behavior of individual fish such as Anitei (2011), Nogita et al. (1988), Schalie et al. (2001) while others have studied fish group behaviors (Thida et al. 2009; Chew et al. 2009). Some studies analyzed only one species like Papadakis et al. (2012), Pinkiewicz et al. (2011), Chew et al. (2009), Kato et al. (2004), Xu et al. (2006). The majority of works in this area analyze the fish trajectories in an aquarium, a tank or a cage which makes the analysis simpler, decreases the number of behavior varieties and also removes the effects of habitat on the behavior of fish. On the other hand, the number of studies using natural habitat underwater environments is very few (Spampinato et al. 2010; Amer et al. 2011).

12.4 Proposed Methods

Our research on detection of unusual fish behaviors covers three methods: (i) A rule based method for filtering normal fish trajectories (Sect. 12.4.1), (ii) A method using clustered and labeled data which is also called the *flat classifier* (Sect. 12.4.2) and (iii) a hierarchical decomposition method (Sect. 12.4.3).

For each method, to obtain the fish trajectories, the tracker in Spampinato et al. (2012) is used and a trajectory is defined by the sequence of centers (x, y) of the fish rectangular bounding boxes which tightly surrounds the detected fish in the image. For any fish i tracked through n frames the trajectory is defined as:

$$T_i = \{(x_1, y_1), (x_2, y_2), \ldots, (x_n, y_n)\} \tag{12.1}$$

12.4.1 A Rule Based Method for Filtering Normal Fish Trajectories

The unusual trajectories are generally defined as outliers in the clusters or rare trajectories. In this scope, the clusters with small numbers of elements are expected to represent rare trajectories and the samples that are different from the main distributions of samples in the same cluster are considered as outliers (Anjum and Cavallaro 2008). Although this approach is reasonable, when the number of trajectories is huge like hundred thousands, millions etc. and/or the number of normal trajectories is much bigger than the number of unusual trajectories, such as 100 times bigger (or more), normal trajectories can dominate unusual trajectories, so extracting small clusters detecting outliers might be inaccurate. This might be even worse if the normal and unusual classes contain sub-varieties even though they are labeled as the same class (Beyan and Fisher 2012).

The aim of the proposed filtering mechanism is to reject all normal trajectories while not rejecting any unusual trajectories. In each step of that method, the trajectories satisfying a rule (filtered) are defined as normal trajectories (such as Normal1, Normal2 in Fig. 12.1). The trajectories which do not satisfy the rule (not filtered) are called the remainders of the corresponding filter (Remainder1, Remainder2 in Fig. 12.1) and are used as inputs to the following filter. First, all fish trajectories are filtered by Filter 1. Then, the remainders of Filter 1 are filtered by Filter 2. This is continued until all the filters are used. In the end, the remainders of last filters are

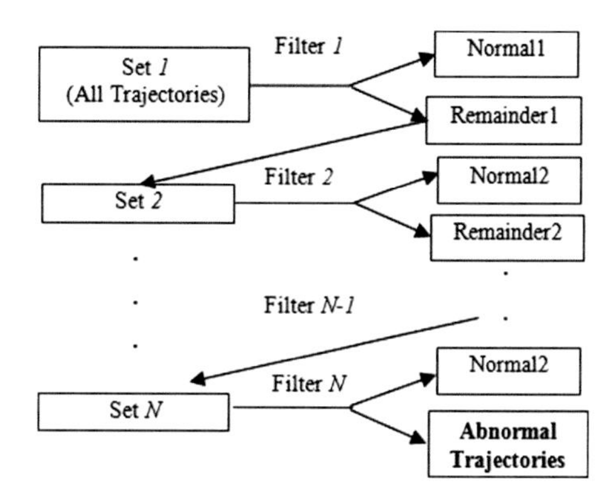

Fig. 12.1 The block diagram of the rule based normal fish trajectory filtering method (Beyan and Fisher 2012)

called unusual trajectories. The filtering order is independent since the rules of filters are independent.

The definitions of the filters are based on fish detection which is in one of two categories: straight and/or cross motions and being stationary. Straight and/or cross motions includes all possible motions in all directions such as left to right, right to left, up to down, down to up and includes a search area as being stationary does. The description of straight and/or cross motions and being stationary can be found in Beyan and Fisher (2012).

Filters are defined as one, two and three length combinations of these motions such as moving right to left (length is one), moving right to left and then being stationary (length is two), moving left to right and then up to down (length is two), being stationary for a while, then moving down to up and then left to right (length is three) etc. Similar trajectories like going left to right and right to left are modeled by same filter. Altogether 21 rules were used (similar trajectories like going left to right and right to left are modeled by same filter method).

In the training phase, for each filter the best parameters: search area for straight and/or cross motions, search area for being stationary and combinations of filters are found. The best parameter for each filter is the one which does not filter out any unusual trajectories. In the case of having more than one parameter which does not filter out any unusual trajectories, the one that filtered the most normal trajectories is selected. If there is no parameter that does not filter out any unusual trajectories, then that filter is not used and the process continues with the following filter. During testing phase, the filters with the best parameters are used to classify new trajectories. Filters that were removed during training are not used during testing.

12.4.1.1 Conclusions for Filtering Normal Fish Trajectories

The proposed rule based filtering method is successful to filter out large amounts of trajectories with a very low time complexity. This method has been used as a preliminary method to collect ground truth data (especially unusual trajectories) thanks to low time complexity and having low false positive and false negative detections (see Sect. 12.4.4). This method can be combined with any unusual fish trajectory detection system which might lead to increase the detection performance since the remaining data set will be more balanced as many normal trajectories will be filtered. It can be applied especially when the number of normal fish trajectories is much bigger than the number of unusual fish trajectories and/or when the number of trajectories is very huge.

12.4.2 Detecting Unusual Fish Trajectories Using Clustered and Labeled Data: Flat Classifier

In this section, we present an approach to detect unusual fish trajectories using multiple features. The presented method is mainly based on clustering. To find the unusual trajectories, an outlier detection method which is based on the sample size of clusters and a distance function is used. Clustered and labeled data are used together to select the best feature set (the feature set that provides the best performance) using a training set (Beyan and Fisher 2013a). This method consists of four steps: (i) feature extraction, (ii) clustering, (iii) outlier detection and (iv) feature selection (Fig. 12.2) and includes the basis of the hierarchical method given in Sect. 12.4.3.

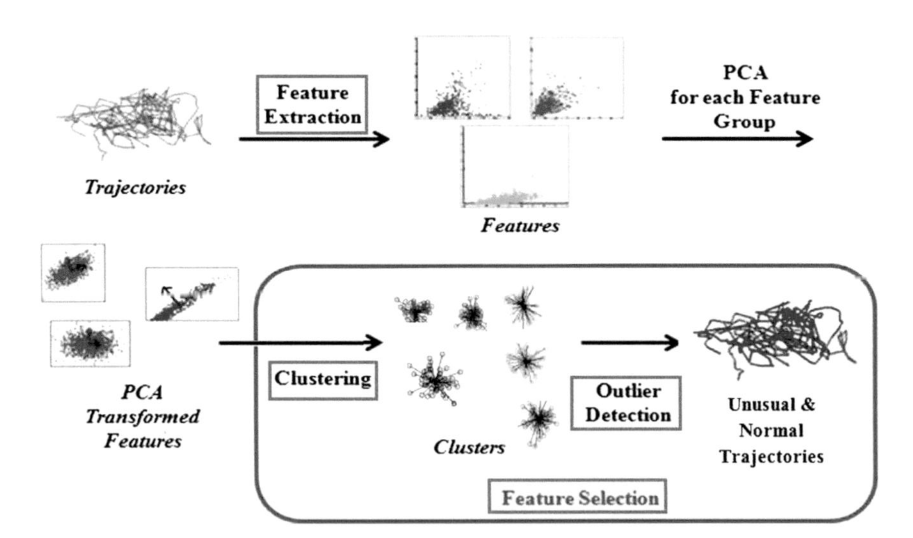

Fig. 12.2 Overview of the flat classifier, see Beyan and Fisher (2013a) for the description of the process

12.4.2.1 Feature Extraction

One challenge of fish detection and tracking in an open and uncontrolled underwater environment is that there may be gaps in the fish trajectory due to occlusions by other fish, etc. To overcome this problem, before extracting features, all trajectories are linearly interpolated. Ten groups of features are extracted and in total, 776 features are obtained in the feature extraction step. These features are generally correlated with each other, therefore to prevent a possible over-training Principal Component Analysis (PCA) is applied to each group of features individually. While applying PCA, to obtain a useful set of components the smallest number of components that represent 90 % of the sum of all eigenvalues is used. As a result of this step, 179 features are obtained as the final feature set. Some of the extracted features are as the following (for all of them please refer to Beyan and Fisher (2013b)):

Curvature Scale Space (CSS) Based Features
Trajectories are first represented using CSS description (Bashir et al. 2006). CSS is calculated using the curvature at every point on the curve by the formula given in Eq. 12.2. This trajectory description is shaped-based and rotation and translation invariant.

$$K_i = \frac{x'_i y''_i - y'_i x''_i}{(x'^2_i + y'^2_i)^{3/2}} \tag{12.2}$$

To find the scale position of the CSS, a Gaussian kernel is used. At each level of space the standard deviation of the Gaussian kernel is increased and the curvature at that level is found. The CSS is represented with a CSS image. As features, statistical properties such as mean and variance of length of curves, number of zero crossings for each standard deviation etc. which are extracted from the CSS image are used. Additionally, for each standard deviation value, statistical features of absolute curvature are extracted. In total 580 features are obtained (Beyan and Fisher 2013a).

Moment Descriptors Based Features
The shape of fish trajectories can be distinguished by using moment descriptors. We utilize affine moment invariants as proposed in Suk and Flusser (2004) in addition to moment, central moment and translation and scale invariant moments. In total 55 features are extracted using those moment descriptors.

Velocity and Acceleration-Based Features
Statistical properties: mean, standard deviation, minimum, maximum, number of zero crossings, number of local minima and maxima etc. of velocity and acceleration are extracted in three dimensions considering the fact that fish can swim in three dimensions in an open sea. However, since the trajectory description in our data repository is in two dimensions, we estimated the third dimension using the width (w) and height (h) of fish detection bounding box ($1/\sqrt{wh}$). In total 42 features are obtained.

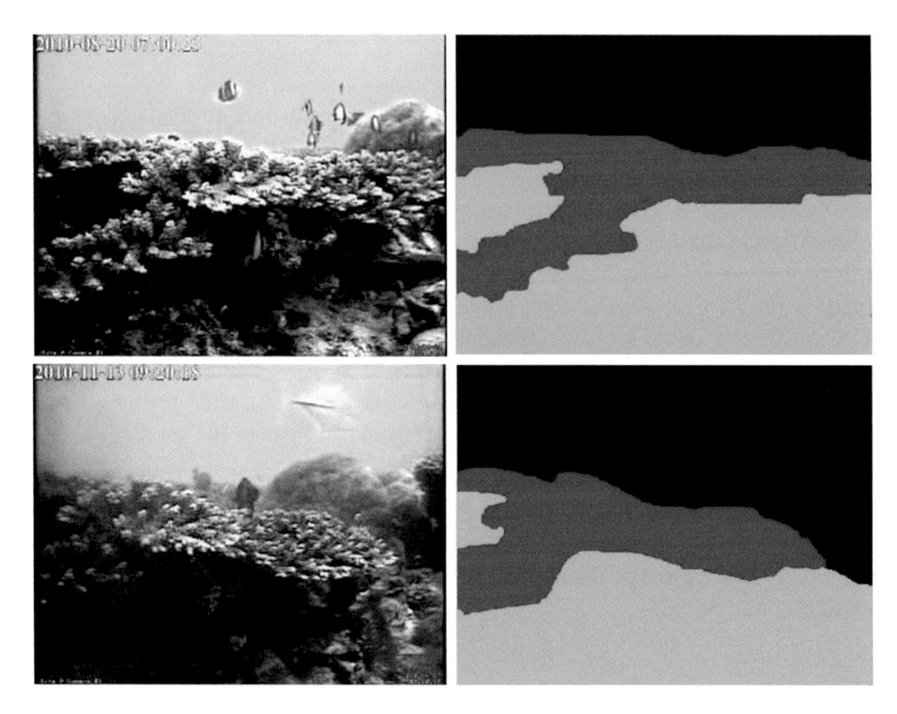

Fig. 12.3 Segmented regions of underwater image; *black* for open sea, *red* for above the coral and *green* for under coral (Beyan and Fisher 2013a)

Fish Pass by Features
Fish trajectories are affected by the geographical properties of the underwater environment and their trajectories can be different in different locations. In this study, we divide the underwater environment into three regions: *open sea*, *under the coral* and *above the coral* (Fig. 12.3). We manually segmented each video scene once and utilize them to obtain the features corresponding to all fish trajectories of a video. As features the frequencies of being in different locations and frequency of crossings from one location to another location are extracted. In total 12 features are obtained.

12.4.2.2 Clustering

We used affinity propagation (AP) (Frey and Dueck 2007) as the clustering method. AP was used by many studies for different purposes including anomaly detections. AP can produce smaller clusters and produce uneven sized clusters which make it compatible with the outlier detection strategy that we use. Furthermore, it is fast, non-parametric, does not depend on sample order and does not need initialization.

12.4.2.3 Outlier Detection

Outlier detection is used to detect unusual trajectories. In this study, we adapted the outlier detection given in Anjum and Cavallaro (2008). Basically, there are two types of unusual trajectories: (i) those located in small clusters, (ii) those in dense clusters but far from cluster centers.

The samples in small clusters are classified as outliers which makes them unusual trajectories. For the samples belonging to dense clusters, an unusual trajectory is detected using the Euclidean distance between the sample and the cluster exemplar. A data sample which is far away compared to threshold $\tau = \mu + w\sigma$ (with mean (μ), weight (w) and standard deviation (σ) of all distances between all samples and the cluster center) is defined as an outlier (unusual trajectory). As can be inferred this threshold is different for each cluster and calculated using the specific cluster.

12.4.2.4 Feature Selection

For feature selection, Sequential Forward Feature Selection (Pudil et al. 1994) is applied together with clustering and outlier detection. Feature selection provides the proper feature sets which also decreases the chance of over-fitting. It eliminates irrelevant and redundant features. Moreover, it helps to eliminate the features which might misguide the clustering. Feature selection is applied as given in Beyan and Fisher (2013a, b).

In the test phase, the new trajectories are classified using outlier detection parameter w and the number of clusters that are found during training. In detail, first clustering is applied to the testing trajectories using the same number of clusters that are found in training and outlier detection is applied with the selected w parameter from the training.

12.4.2.5 Conclusions for Flat Classifier

In this section, we represented fish trajectories with novel descriptors which were never used before for fish behavior analysis. Clustered and labeled data were used together to select the best feature set and classify trajectories as normal or unusual. The flat classifier improved performance of unusual fish detection compared to the filtering mechanism (given in Sect. 12.4.1) where results are given in Sect. 12.4.4. The performance of the flat classifier is successful especially considering the challenges of underwater environments, low video quality, noisy data and erratic movement of fish. Additionally, it is good at detecting normal trajectories as well which is promising to help marine biologist by eliminating many normal trajectories with relatively low error rate.

12.4.3 Detecting Unusual Fish Trajectories Using Hierarchical Decomposition

In this section, we present a novel type of hierarchical decomposition method to detect unusual fish trajectories. The basics of the proposed hierarchical decomposition method are the same as the method presented in Sect. 12.4.2. Clustering of data based on selected features without initially using the known labels is the key to partitioning the data into separable subsets. To automatically generate the hierarchy during training, clustered and labeled trajectories are used together. Different from the traditional way which uses the same feature set for every level of hierarchy or from a flat classifier (Sect. 12.4.2), we use different feature sets at different levels of the hierarchy, which allows selecting more specific features (Beyan and Fisher 2013b). The main contribution of this part is presenting a novel approach for unusual behavior detection which constructs a feature or class taxonomy independent hierarchy.

12.4.3.1 Hierarchy Construction

Training phase of the proposed method includes hierarchy construction. At each level of the hierarchy, data is first clustered using the best feature subset found using feature selection (Sect. 12.4.2.4). After clustering, outlier detection is applied to each cluster and outliers (unusual trajectories) for a specific level of the hierarchy are found. Then, for each cluster, the number of false positives (positive class represents the unusual trajectories) and false negatives (negative class represents the normal trajectories) are found. The clusters which do not have any false positives and false negatives are fixed for that level (shown as classifiable samples which belong to perfectly classified clusters in Fig. 12.4). The hierarchy construction recurses similarly with all samples of clusters that have false negatives or false positives (shown as remaining samples which belong to any misclassified clusters in Fig. 12.4). That tree is extended by repeating the clustering, feature selection and outlier detection until there is no cluster which is perfectly classifiable or all the training samples are perfectly classified (Beyan and Fisher 2013b). The leaf nodes of the hierarchy can be either: perfectly classified clusters (which contain classifiable samples) which can be observed mostly at the upper levels or misclassified clusters (which contain remaining samples). These occur only in the leaf nodes belonging to last level of hierarchy.

Perfectly classified clusters can be either:

- Perfectly classified mixed cluster: Includes unusual and normal trajectories which are correctly classified using the outlier detection.
- Perfectly classified pure normal cluster: Includes only normal trajectories which are correctly classified using the outlier detection threshold.
- Perfectly classified pure unusual cluster: Includes only unusual trajectories which are correctly classified due to being in a small cluster where we assume that samples of small clusters are unusual trajectories.

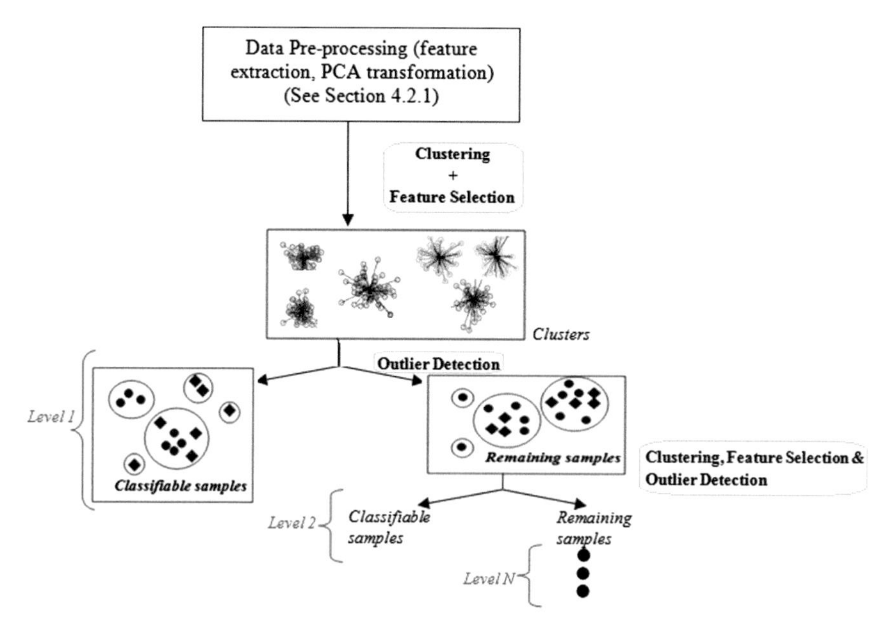

Fig. 12.4 Hierarchy Construction (Beyan and Fisher 2013b)

Misclassified classified clusters can be either:

- Misclassified mixed cluster: Includes both unusual and normal trajectories with at least one sample wrongly classified using the outlier detection threshold.
- Misclassified pure normal cluster: Includes only normal trajectories with at least one trajectory classified as an unusual trajectory using the outlier detection threshold or includes only normal trajectories which are wrongly classified as unusual trajectories due to being in a small cluster.
- Misclassified pure unusual cluster: Includes unusual trajectories where at least one trajectory is classified as a normal trajectory using the outlier detection threshold.

12.4.3.2 New Trajectory Classification Using the Hierarchy

To classify a new trajectory using our method in the test phase, the built hierarchy is used, using all perfectly classified clusters and misclassified clusters at each level, the selected feature subsets for each level and the outlier detection threshold for each cluster are used starting from the top level to the bottom. Testing is based on finding the closest cluster at each level of hierarchy. The closest cluster is found by the Euclidean distance between the new trajectory (in terms of the features selected at the current level) and the cluster centers (including misclassified clusters) at each level of the hierarchy.

At each level in the hierarchy, the closest cluster can be one of the six possible cluster types: (i) perfectly classified pure unusual, (ii) perfectly classified pure normal, (iii) perfectly classified mixed, (iv) misclassified pure normal, (v) misclassified pure unusual, and (vi) misclassified mixed. At each level in the hierarchy, for the new trajectory, three types of class decisions are possible: unusual trajectory, candidate normal trajectory and no effect on the decision.

The decision is based on one of these six cases:

- The closest cluster is a perfectly classified pure unusual cluster which makes the new trajectory an unusual trajectory and classification stops (there is no need to look at any other level of the hierarchy).
- The closest cluster is a perfectly classified mixed cluster and the new trajectory is further than the outlier detection threshold of that cluster which makes the new trajectory an unusual trajectory and classification stops (there is no need to look at any other level of the hierarchy).
- The closest cluster is a perfectly classified pure normal cluster and the new trajectory is further than the outlier detection threshold of that cluster. This makes the new trajectory an unusual trajectory and classification stops (there is no need to look at any other level of the hierarchy).
- The closest cluster is a perfectly classified pure normal cluster and the distance between the new trajectory and the corresponding cluster's center is smaller than the outlier detection threshold of that cluster. This makes the new trajectory a candidate normal trajectory. The new trajectory does to next level of the hierarchy.
- The closest cluster is a perfectly classified mixed cluster and the distance between the new trajectory and cluster center is smaller than the threshold. The new trajectory is a candidate normal trajectory. The new trajectory goes to the next hierarchy level.
- The closest cluster is a misclassified cluster (pure or mixed). The new trajectory proceeds to the next level. This does not have any effect on the classification of the new trajectory unless the closest clusters at each level are misclassified clusters.

Those rules are illustrated in Fig. 12.5.

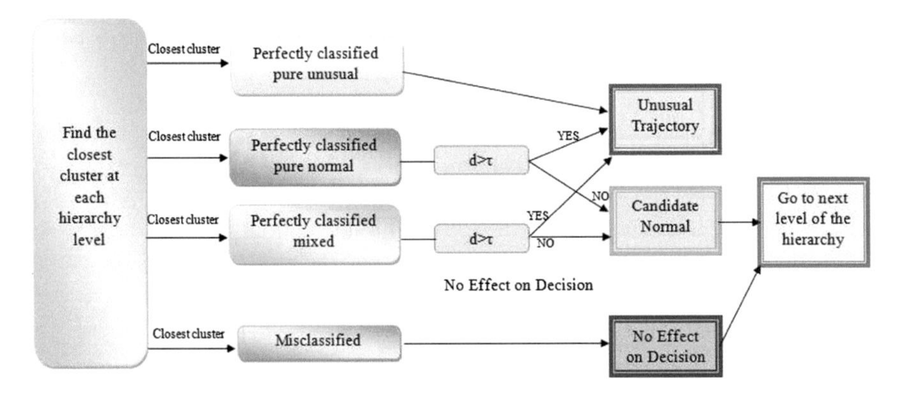

Fig. 12.5 New trajectory classification using the hierarchy

In summary, a single level's decision as unusual trajectory is enough to classify the new trajectory as an unusual trajectory regardless of the level of the hierarchy. On the other hand, if there is no decision as unusual trajectory from any level and if the decision of at least one level is candidate normal then the class of the new trajectory is declared to be normal. However, it is possible that the closest cluster at each level of the hierarchy is a misclassified cluster. In this case, we use the ground-truth labels of the training trajectories and apply the following rules, starting from the top of the hierarchy:

- The closest cluster at the current level contains all normal trajectories by looking at the ground-truth class labels: If the new trajectory is further than the rest of the samples in that cluster this makes it an unusual trajectory and classification stops here. Otherwise the data goes to the next hierarchy level.
- The closest cluster contains all unusual training trajectories by the ground-truth: The new trajectory is classified as an unusual trajectory and classification stops here.
- The closest cluster contains both normal and unusual training trajectories. In this case, we apply the nearest neighbor rule which makes the class of the new trajectory the same as the closest training sample's class. If the class is unusual then classification stops. Otherwise, the data goes to the next level to apply the above rules.
- If the new trajectory reaches the last level and still could not be classified, then it is classified as a normal trajectory.

Those rules are illustrated in Fig. 12.6.

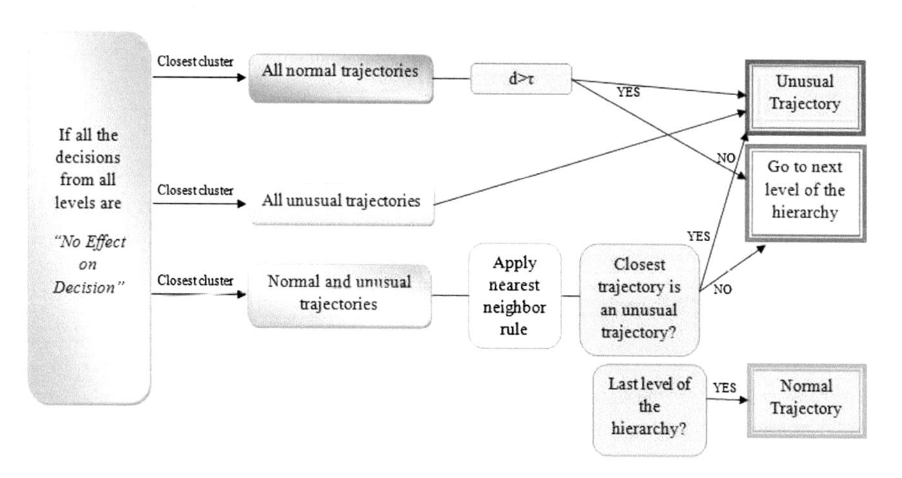

Fig. 12.6 New trajectory classification when the decisions of all levels are "no effect on decision"

12.4.3.3 Conclusions for Hierarchical Decomposition Method

In this section, we presented a hierarchical decomposition method to detect unusual fish trajectories. Considering all three proposed methods in this chapter, hierarchical decomposition method performed the best. Additionally, the comparison between the state of the art methods and the proposed hierarchical method showed that the hierarchical method performs the best in overall (see Sect. 12.4.4.2). Besides, this method is also efficient at classifying new tracks as it is only based on distance calculations between the new trajectory and the cluster centers of each level of the hierarchy. The main contributions of this section are: (i) presenting a novel approach for unusual behavior detection which builds a feature or class taxonomy independent hierarchy, (ii) showing that using different feature spaces in the classification at different levels can improve the performance.

12.4.4 Experiments and Results

In this section, the data set and the state of art classification algorithms to compare their performance with the proposed methods are given. The results are evaluated in terms of *TPrate* (Eq. 12.3), *TNrate* (Eq. 12.4) and geometric mean of *TPrate* and *TNrate* (Eq. 12.5).

$$TPrate = TP/(TP + FN)\ unusual\ trajectory\ class\ accuracy \tag{12.3}$$

$$TNrate = TN/(TN + FP)\ normal\ trajectory\ class\ accuracy \tag{12.4}$$

$$Geometric\ Mean\ of\ TPrate\ and\ TNrate\ (GeoMean) = \sqrt{TPrate \times TNrate} \tag{12.5}$$

where *TP* is the number of correctly classified unusual trajectories, *TN* is the number of correctly classified normal trajectories, *FN* is the number of misclassified unusual trajectories and *FP* is the number of misclassified normal trajectories.

12.4.4.1 Data Set

The proposed methods and all the states of art methods (such as Random Forest (Breiman et al. 1984), Spectral Clustering (Izo and Grimson 2007), and Local Outlier Factor instead of LOF (Janssens 2009)) were applied to 3102 trajectories (3043 normal, 59 unusual trajectories). To the best of our knowledge, this data set is the largest fish trajectory data set in the underwater environment and the largest labeled data set in general. Data includes a single fish species which is *Dascyllus reticulatus* living in the Taiwanese coral reef. Data was collected from 93 different videos having 320 × 240 resolutions, 5 frames per second. Considering that the fish behavior can

Fig. 12.7 **a–b** Normal fish trajectory examples, **c–d** Unusual fish trajectory examples (Beyan and Fisher 2013a, b)

change during the time of the day and *Dascyllus reticulatus* is more active in the morning we used the videos that were captured between 6 and 12.

The normal and unusual behaviors are determined by visual inspection and also examined by the marine biologists. The most usual and frequent behaviors in the data set are hovering over the coral and freely swimming fish in open sea (Fig. 12.7a, b) which represent normal behaviors. On the other hand, unusual trajectories are such as fish suddenly (in one frame) changing direction (predator avoidance, Fig. 12.7c), fish biting at coral (also interaction with plankton, Fig. 12.7d) and so forth. A trajectory that has normal and unusual segments is assumed as unusual.

12.4.4.2 Results

The proposed methods were compared with several state-of-art classification methods and other popular trajectory analysis methods (see Table 12.1). Nine-fold cross validation was performed and training, validation and test sets were constituted randomly and the normal and unusual trajectories are distributed equally in each set.

Table 12.1 The used state of the art classification methods, popular trajectory analysis methods and the proposed methods

Method	Parameters	Abbreviation
k-Nearest Neighbors	$k = \{1, 2, 3, 4, 5, 10, 15, 25\}$ were used as the common parameters. Sequential forward feature selection was applied as given in Sect. 4.2.4	kNN
SVM	As the kernel function, radial basis function with varying kernel parameters was used. Hyperplanes were separated by Sequential Minimal Optimization Sequential forward feature selection was applied as given in Sect. 4.2.4	SVM
Random Forest with Balanced Training	A number of trees *{10, 30, 50, 70, 100, 120, 150, 200, 500, 1000}* were tested and the trees are grown without pruning. For node splitting, the Gini index was used	RFBT
Spectral clustering based method	Normalized cuts special clustering was applied to unusual and normal trajectories individually and each cluster of behavior was modeled as a mixture of Gaussians in the spectral embedding space for normal and unsual classes and based on the likelihood the new track is classified as a normal or unusual trajectory. Different sigma values such as *{1, 10, 20 etc.}* and different cluster size *{10, 15, 20, 30, 40, 50, 60, 80, 90}* for normal and unusual clusters were tested	Spec
Local Outlier Factor	Local Outlier Factor is a density based method which considers a sample to be an outlier if its surrounding space contains few samples. It does not use any clustering technique. Training is performed only using normal classes. During validation normal and usual class trajectories are used and the best feature set is selected using sequential forward feature selection. Neighborhood is defined with a parameter called k. k was taken as *{1, 3, 5, 10, 15, 20 and 25}*	LOF
Filtering method	Pixels *{2, 4, 8, 16, 20}* were taken to define the search area	Proposed M1
Flat classifier	Outlier detection parameter w was taken as *{−1, −0.3, 0, 0.3, 0.6, 0.9, 1, 2, 3, 6}*	Proposed M2
Hierarachical decomposition	Outlier detection parameter w was taken as *{0, 0.3 and 1}*	Proposed M3

Table 12.2 shows the best results in terms of *TPrate*, *TNrate* and average of geometric mean (*GeoMean*) of *TPrate* and *TNrate*. For each evaluation metric the standard deviation (considering cross validation folds) is also given after ± sign. The best results in terms of each evaluation metric are emphasized in bold-face.

The results show that the hierarchical decomposition method has highest unusual fish trajectory detection rate (*TPrate*) and is the best method in overall. On the other

Table 12.2 Results of each method in terms of average of *TPrate*, *TNrate* and *GeoMean*

	TPrate	TNrate	GeoMean
kNN	0.37 ± 0.28	$\mathbf{0.99 \pm 0.01}$	0.55 ± 0.27
SVM	0.81 ± 0.16	0.93 ± 0.03	0.86 ± 0.09
RFBT	0.88 ± 0.01	0.91 ± 0.10	0.89 ± 0.05
Spec	0.57 ± 0.20	0.85 ± 0.11	066 ± 0.04
LOF	0.62 ± 0.17	0.97 ± 0.01	0.77 ± 0.08
Proposed M1	0.80 ± 0.20	0.77 ± 0.04	0.78 ± 0.09
Proposed M2	0.81 ± 0.17	0.76 ± 0.02	0.78 ± 0.09
Proposed M3	$\mathbf{0.94 \pm 0.10}$	0.88 ± 0.02	$\mathbf{0.91 \pm 0.05}$

The best results are emphasized in bold-face

hand, the flat classifier (Sect. 12.4.1) and filtering method (Sect. 12.4.2) are as good as SVM in terms of unusual fish trajectory detection but worse than SVM in terms of normal trajectory detection (*TNrate*). The kNN algorithm has the best *TNrate*, but this is at a considerable miss classification that produces lowest *TNrate* and *GeoMean*.

12.5 Concluding Remarks

In this chapter, we addressed fish behavior with a unusual fish trajectory detection schema using underwater environment videos. We distinguished the fish trajectories as normal and unusual trajectories. All the analysis in this chapter were applied to the trajectories of *Dascyllus reticulatus* from the Taiwanese coral reef during morning time. We presented three different classification methods to detect unusual fish trajectories. The first method (filtering method) is more specific to eliminating normal trajectories. The other methods (flat method and hierarchical decomposition) aimed detecting unusual fish trajectories and performed better than filtering mechanism. The results show that the proposed hierarchical decomposition method is good at detecting unusual fish trajectories while it performed the best compared to the state of art methods. As future work, the proposed methods can be applied to larger fish data sets and other fish species.

References

Amer, M., E. Bilgazyev, S. Todorovic, S. Shah, I. Kakadiaris, and L. Ciannelli. 2011. Fine-grained categorization of fish motion patterns in underwater videos. In *International Conference on Computer Vision Workshops*, 1488–1495.

Anitei, S. 2011. Fishes used against terrorist attacks. http://news.softpedia.com/news/Fishes-Used-Against-Terrorist-Attacks-36818.shtml.

Anjum, N., and A. Cavallaro. 2008. Object trajectory clustering for video analysis. *Transactions on Circuits and Systems for Video Technology* 18(11): 1555–1564.

Bashir, F., A. Khokhar, and D. Schonfeld. 2006. View-invariant motion trajectory based activity classification and recognition. *Multimedia Systems* 12(1): 45–54.
Beyan, C., and R. Fisher. 2012. A filtering mechanism for normal fish trajectories. In *Proceedings of International Conference on Pattern Recognition*, 2286–2289.
Beyan, C., and R. Fisher. 2013a. Detecting abnormal fish trajectories using clustered and labeled data. In *Proceedings of International Conference on Image Processing*, 1476–1480.
Beyan, C., and R. Fisher. 2013b. Detection of abnormal fish trajectories using a clustering based hierarchical classifier. In *Proceedings of British Machine Vision Conference*.
Breiman, L., J. Friedman, R. Olshen, and C. Stone. 1984. *Classification and regression trees*. Monterey: Wadsworth and Brooks.
Chew, B., H. Eng, and M. Thida. 2009. Vision-based real-time monitoring on the behavior of fish school. In *Proceedings of International Association for Pattern Recognition Conference on Machine Vision Applications*, 3–16.
Choudhary, A., M. Pal, S. Banerjee, and S. Chaudhury. 2008. Unusual activity analysis using video epitomes and pLSA. In *Proceedings of Sixth Indian Conference on Computer Vision, Graphics and Image Processing*, 390–397.
Dickinson, P., and A. Hunter. 2008. Using inactivity to detect unusual behavior. *IEEE Workshop on Motion and Video Computing*, 1–6.
Frey, B.J., and D. Dueck. 2007. Clustering by passing messages between data points. *Science* 315(5814): 972–976.
Izo, T., and W. Grimson. 2007. Unsupervised modelling of object tracks for fast anomaly detection. In *Proceedings of International Conference on Image Processing*, 529–532.
Janssens, J. 2009. Outlier detection with one-class classifiers from ML and KDD. In *Proceedings of International Conference on Machine Learning Applications*, 147–153.
Jiang, F., J. Yuan, S. Tsaftaris, and A. Katsaggelous. 2010. Video anomaly detection in spatiotemporal context. In *Proceedings of IEEE International Conference on Image Processing*, 705–708.
Kato, S., T. Nakagawa, M. Ohkawa, K. Muramoto, O. Oyama, A. Watanabe, H. Nakashima, T. Nemoto, and K. Sugitani. 2004. A computer image processing system for quantification of zebrafish behavior. *Journal of Neuroscience Methods* 134: 1–7.
Morris, B., and M. Trivedi. 2008. A survey of vision-based trajectory learning and analysis for surveillance. *Transactions on Circuits and Systems for Video Technology* 18(8): 1114–1127.
Nogita, S., K. Baba, H. Yahagi, S. Watanabe, and S. Mori. 1988. Acute toxicant warning system based on a fish movement analysis by use of AI concept. *International Workshop: Artificial Intelligence for Industrial Applications*, 273–276.
Papadakis, V., I. Papadakis, F. Lamprianidou, A. Glaroulos, and M. Kentouri. 2012. A computer vision system and methodology for the analysis of fish behavior. *Aquacultural Engineering* 46: 53–59.
Piciarelli, C., C. Micheloni, and G. Forestl. 2008. Trajectory based anomalous event detection. *Transactions on Circuits and Systems for Video Technology* 18(11): 1544–1554.
Pinkiewicz, T., G. Purser, and R. Williams. 2011. A computer vision system to analyze the swimming behavior of farmed fish in commercial aquaculture facilities: A case study using cage-held atlantic salmon. *Aquacultural Engineering* 45: 20–27.
Pudil, P., J. Novovicova, and J. Kittler. 1994. Floating search methods in feature selection. *Pattern Recognition Letters* 15(11): 1119–1125.
Schalie, W., T. Shedd, P. Knechtges, and M. Widder. 2001. Using higher organisms in biological early warning systems for real-time toxicity detection. *Biosensors and Bioelectronics* 16(7): 457–465.
Spampinato, C., Y.-H. Chen-Burger, G. Nadarajan, and R. Fisher. 2008. Detecting, tracking and counting fish in low quality unconstrained underwater videos. In *Proceedings of the 3rd International Conference on Computer Vision Theory and Applications (VISAPP)*, vol. 2, 514–519.
Spampinato, C., D. Giordano, R. Di Salvo, Y.-H.J. Chen-Burger, R.B. Fisher, and G. Nadarajan. 2010. Automatic fish classification for underwater species behavior understanding. In *Proceed-*

ings of the First ACM International Workshop on Analysis and Retrieval of Tracked Events and Motion in Imagery Streams, ARTEMIS'10, 45–50. New York: ACM.

Spampinato, C., S. Palazzo, D. Giordano, I. Kavasidis, F. Lin, and Y. Lin. 2012. Covariance based fish tracking in real-life underwater environment. In *Proceedings of International Conference on Computer Vision Theory and Applications*.

Suk, T., and J. Flusser. 2004. Graph method for generating affine moment invariants. In *Proceedings of International Conference on Pattern Recognition*, 192–195.

Thida, M., H. Eng, and C.B. Fong. 2009. Automatic analysis of fish behaviors and abnormality detection. In *Proceedings of International Association for Pattern Recognition Conference on Machine Vision Applications*, 8–18.

Varadarajan, J., and J. Odobez. 2009. Topic models for scene analysis and abnormality detection. In *International Conference on Computer Vision Workshop*, 1338–1345.

Xu, X., J. Tang, X. Liu, and X. Zhang. 2010. Human behavior understanding for video surveillance: Recent advance. In *Proceedings of International Conference on Systems Man and Cybernetics*, 3867–3873.

Xu, J., Y. Liu, S. Cui, and X. Miao. 2006. Behavioral responses of tilapia (oreochromis niloticus) to acute fluctuations in dissolved oxygen levels as monitored by computer vision. *Aquacultural Engineering* 35(3): 207–217.

Chapter 13
Understanding Uncertainty Issues in the Exploration of Fish Counts

Emma Beauxis-Aussalet and Lynda Hardman

Abstract Several data analysis steps are required for understanding computer vision results and drawing conclusions about the actual trends in the fish populations. Particular attention must be drawn to the potential errors that can impact the scientific validity of end-results. This chapter discusses the means for ecologists to investigate the uncertainty in computer vision results. We address a set of uncertainty factors identified by interviewing both ecology and computer vision experts, as discussed in Chap. 2. We investigate state-of-the-art methods to specify these uncertainty factors. We identify issues with conveying the results of ground-truth evaluation methods to end-users who are not familiar with computer vision technology, and we present a novel visualization design addressing these issues. Finally, we discuss the uncertainty factors for which evaluation methods require further research.

13.1 Introduction

As scientists, ecologists have requirements of transparency regarding the data collection process and its potential errors and biases. There are several uncertainty factors that potentially impact computer vision end-results, as discussed in Chap. 2. Each uncertainty factor has specific effects on end-results, hence requiring specific evaluation methods. We interviewed both marine ecology experts and computer vision experts to gain insights on the effects of uncertainty factors, and on the methods for measuring them. In this chapter, we detail the potential effects of each uncertainty factor, the goals of their evaluation, the state-of-the-art evaluation methods, and the uncertainty visualizations developed within the project. Sections 13.2 and 13.3 investigate uncertainty related to computer vision algorithms, while Sects. 13.4 and 13.5

E. Beauxis-Aussalet (✉) · L. Hardman
Centrum Wiskunde & Informatica, Science Park 123,
1098 XG Amsterdam, The Netherlands
e-mail: Emmanuelle.Beauxis-Aussalet@cwi.nl

L. Hardman
e-mail: lynda.hardman@cwi.nl

R.B. Fisher et al. (eds.), *Fish4Knowledge: Collecting and Analyzing Massive Coral Reef Fish Video Data*, Intelligent Systems Reference Library 104,
DOI 10.1007/978-3-319-30208-9_13

investigate uncertainty related to the in-situ deployment of the Fish4Knowledge system. Section 13.6 investigates the impact of both computer vision algorithms and in-situ system deployment uncertainties on end-results. Finally, Sect. 13.7 discusses the uncertainty issues that are not fully addressed by the state-of-the-art evaluation methods.

We show that the Fish4Knowledge project is supported by well-established methods for evaluating the uncertainty factors due to computer vision algorithms. The evaluation of the remaining uncertainty factors requires methods beyond the state-of-the-art. However, the Fish4Knowledge project developed simple evaluation methods for these factors. Directions for future work are suggested with the aim of enabling further scientific rigor in ecology research based on computer vision systems. An overview of the uncertainty factors and their evaluation methods is given by Fig. 13.1 and Table 13.1. The latter refers to the user interface designed to communicate computer vision results and their uncertainty to end-users. The interface organizes information in 5 tabs addressing specific uncertainty issues, and is further discussed in Appendix A.

13.2 Evaluating Uncertainty Due to Computer Vision Algorithms

Computer vision algorithms can introduce errors in end-results by misidentifying fish and non-fish objects, or by misidentifying fish species. To convey this uncertainty to end-users, we consider the two stages of information extraction as two distinct algorithms: *Fish Detection* for identifying fish from other objects (Chaps. 9 and 10), and *Species Recognition* for identifying the fish species (Chap. 11). Besides algorithms themselves, two factors can impact the quality of the output. Algorithms use ground-truth sets of fish examples to learn how to identify fish from each species. The *Ground-Truth Quality* directly impacts the quality of end-results. Further, the *Image Quality* of video recordings can induce errors, e.g., low image quality yields fish appearances that are more difficult to recognize. The interactions between these uncertainty factors are shown in Fig. 13.1 (blue boxes). In this section, we present these uncertainty factors and their evaluation methods.

Fish Detection and Species Recognition Errors—Computer vision algorithms identify the fish appearing in video footage by classifying them into predefined categories. The *Fish Detection* algorithm has two categories, fish or non-fish objects, and *Species Recognition* has one category for each fish species. For both algorithms, objects are assigned to a single category. The fish from the *Fish Detection* results are classified further by *Species Recognition*.

Each fish category is defined by a model constructed from ground-truth sets. Objects are compared to the models, and if similar enough, are classified in the related categories. Similarity between objects and models is usually represented with a *score*. *Score* thresholds are used for selecting the objects to classify, and are

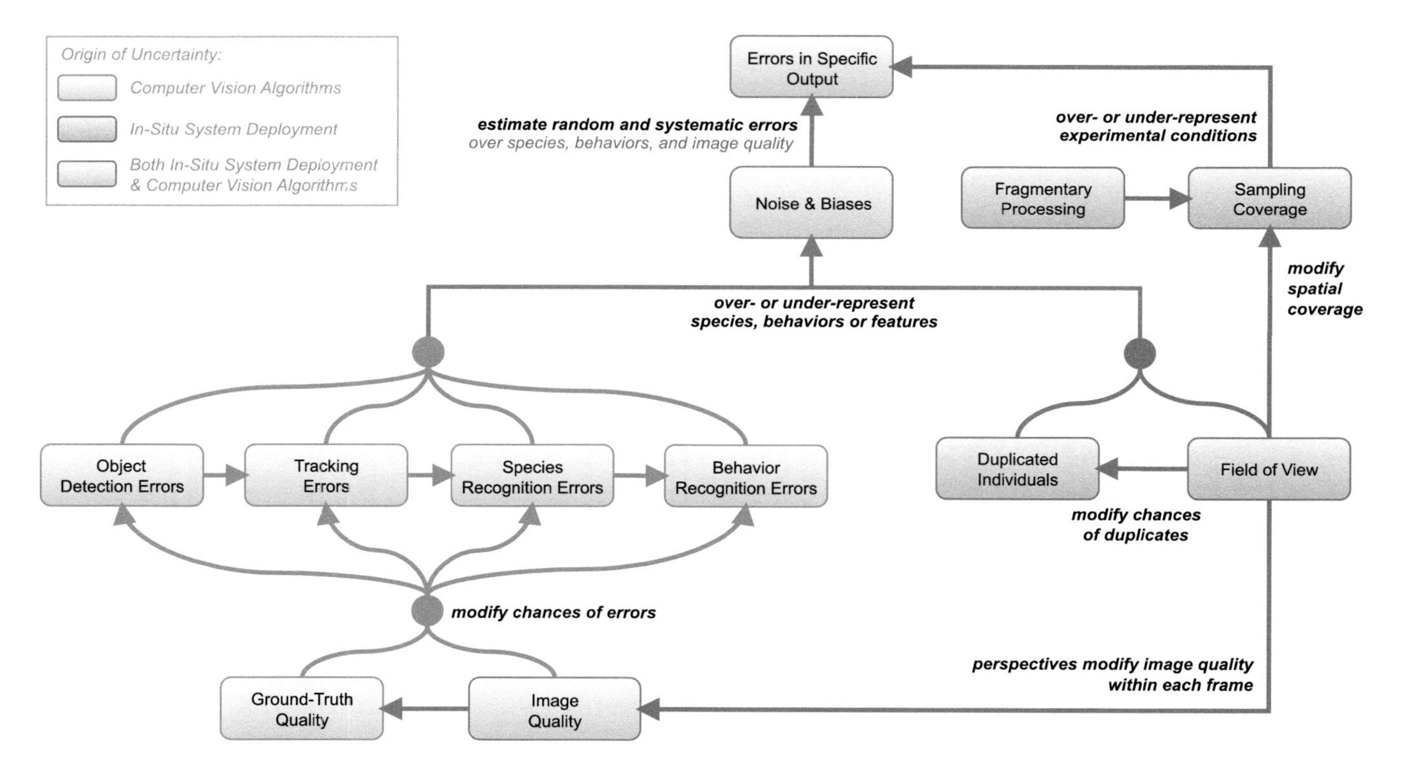

Fig. 13.1 Model of interactions between uncertainty factors. Factors in *blue boxes* are introduced by computer vision algorithms, while factors in *red boxes* are introduced when deploying the system. Factors in *purple boxes* are impacted by both phases of system implementation

Table 13.1 Uncertainty factors and user interface tabs addressing them

Factor	User interface tab and metrics	Figure
Uncertainty due to computer vision algorithms (Sects. 13.2 and 13.3)		
Ground-Truth Quality	**Video Analysis tab**: Number of ground-truth items in test sets (which are proportional to numbers of items in training sets)	Figs. 13.3, 13.4, 13.5, 13.6 and 13.7
Fish Detection Errors	**Video Analysis tab**: Number and proportion of errors per *Image Quality*	Figs. 13.3, 13.4, 13.5 and 13.6
Species Recognition Errors	**Video Analysis tab**: Number and proportion of errors per species	Figs. 13.4, 13.5, 13.6 and 13.7
Uncertainty due to in-situ system deployment (Sects. 13.4 and 13.5)		
Field of View	**Video tab**: Video browsing supports elementary control of Fields of View over time and locations	Fig. 13.8
Duplicated Individuals	**Video tab**: Video browsing supports elementary control of repeated occurrences of fish in groups or coral heads (e.g., over-estimation of *schooling* and *sedentary* species discussed in Chap. 2)	Fig. 13.8
Sampling Coverage	**Video** and **Visualization tabs**: Number of 10-min video samples over time and locations	Figs. 13.8 and 13.9
Fragmentary Processing	**Visualization tab**: Number of processed and unprocessed 10-min video samples. Mean number of fish per video sample. Additional video processing can be requested through the user interface (**Workflow sub-tab**, Appendix A, Fig. A.5 p. 270)	Fig. 13.10
Uncertainty due to computer vision algorithms and deployment conditions (Sect. 13.6)		
Image Quality	**Video** and **Visualization tabs**: Number of videos from each image quality (bottom widget called *Video Quality*) to correlate with *Fish Detection Errors*	Figs. 13.8 and 13.9

(continued)

Table 13.1 (continued)

Factor	User interface tab and metrics	Figure
Noise and Biases	**Video** and **Video Analysis tabs**: Video browser and visualization of computer vision errors, to identify potential biases due to *Field of View, Duplicated Individuals, Image Quality, Fish Detection* and *Species Recognition Errors*	Fig. 13.11
Uncertainty in Specific Output	**Visualization tab**: Measures of dataset characteristics, to correlate with *Noise and Biases* estimates (number of videos over time, location, *Image Quality* and *Field of View*), and *Certainty scores* indicating the similarity of fish with their species model. **Report tab**: Uncertainty can be described by gathering and commenting visualizations	Figs. 13.10 and 13.11

usually set by computer vision experts. Errors occur when objects are not classified into their true category, or when they are not classified at all (i.e., not detected in the videos).

Fish Detection output is impacted by two types of errors. Errors of *Type I*, also called *False Positives* (FP), are non-fish objects classified as fish and contained in the output. Errors of *Type II*, also called *False Negatives* (FN), are undetected fish not contained in the output. These errors are usually measured using ground-truth sets distinct from those used to learn the fish models. Manual fish detections are compared to those of the algorithm, and the numbers of errors are encoded in a table called a *confusion matrix*. Table 13.2 illustrates a typical confusion matrix for *Fish Detection Errors*.

Table 13.2 Example of a confusion matrix for *Fish Detection Errors* (with synthetic data)

		Classification from Ground-Truth	
		Fish	Non-Fish
Classification from Fish Detection Software	Fish	**85 (True Positive TP)**	7 (False Positive FP)
	Non-Fish	**15 (False Negative FN)**	93 (True Negative TN)

The color coding is used in our visualization design to facilitate the identification of type I and II errors

Table 13.3 Example of a confusion matrix with synthetic data for *Species Recognition Errors* (left) and basic metrics for type I and II errors (i.e., FP and FN, respectively)

		Classification from Ground-Truth					**Basic Metrics**			
		A	B	C	D	E	TP	FN	FP	*TN*
Classification from Species Recognition Software	A	**85**	1	4	3	12	**85**	**25**	20	384
	B	■17 ■	**78**	1 ■	7 ■	2 ■	**78**	17	**27**	392
	C	■ 1	2	**90**	6	6	**90**	22	15	387
	D	■ 5	7	2	**77**	1	**77**	18	15	404
	E	■ 2	7	15	2	**81**	**81**	21	26	386

Table 13.4 Advanced metrics commonly used in computer vision

Precision	Recall or TP Rate	FP Rate	Accuracy	F1 Measure
$\frac{TP}{TP+FP}$	$\frac{TP}{TP+FN}$	$\frac{FP}{FP+TN}$	$\frac{TP+TN}{TP+TN+FP+FN}$	$\frac{2TP}{2TP+FP+FN}$

Species Recognition Errors are fish that have been assigned to the wrong species. They are also measured using dedicated ground-truth sets, and encoded in a confusion matrix. Table 13.3 shows an example of a confusion matrix for *Species Recognition Errors*. Type I and II errors also apply to *Species Recognition*. Considering a set of fish assigned to one species, e.g., *Species A*, errors of *Type I* (*False Positives*) are fish from another species erroneously classified as *Species A*. errors of *Type II* (*False Negatives*) are fish not classified as *Species A* but actually belonging to it.

Confusion matrices for *Species Recognition* are more complex to analyze than those of *Fish Detection*. An important concept for understanding them is that False Positives erroneously assigned to one species are False Negatives for their true species. For instance, if *Species A* misses 17 False Negatives erroneously attributed to *Species B*, then *Species B* gains 17 False Positives from *Species A*. Hence counting all the errors for one species requires to sum the False Negatives assigned to all other species (i.e., column-wise sum in Table 13.3), as well as summing the False Positives added by all other species (i.e., row-wise sum in Table 13.3). This examples is illustrated in Table 13.3, e.g., the cell with both red and gray squares indicates: 17 False Negatives (FN) for species A; 17 False Positives (FP) for species B. These 17 errors are counted both in the cell with red background (i.e., summing the cells with red squares) and in the cell with gray background (i.e., summing the cells with gray squares).

In the computer vision domain, classification errors are usually synthesized further using advanced metrics derived from the basic measure of TP, FP, FN and TN. Advanced metrics are rates of correct and incorrect object detection over total numbers of objects belonging to the category (TP and FN) or not (FP and TN). Table 13.4 shows the metrics and formulas commonly used in most of the state-of-the-art evaluations of computer vision errors. Advanced metrics are usually plotted by pairs in Precision/Recall or ROC curves (Receiver Operating Characteristics).

Measurements are usually repeated for several parameter thresholds, e.g., a *score* representing the similarity between fish images and species models (i.e., fish below thresholds are discarded from the species). Figure 13.2 shows examples of such visualization of pairs of advanced metrics.

Image Quality—Varying image quality can be a source of bias. For instance, end-results drawn from one type of image quality can systematically contain different numbers of errors than for another image quality. This biases the comparison of end-results drawn from different types of image quality. Ground-truth evaluations of *Fish Detection* and *Species Recognition Errors* can be used to evaluate this type of bias.

Hence we need to provide ecologists with evaluations of *Fish Detection* and *Species Recognition Errors* detailed for each type of image quality. However, this requires an extensive ground-truth containing sufficient numbers of annotations for all combinations of species and image quality. The considerable cost of such ground-truth collection is likely to be unaffordable, as it was the case for the Fish4Knowledge project. *Species Recognition Errors* could not be fully evaluated for each image quality. Hence we focused on evaluating *Fish Detection Errors* for each type of *Image Quality*.

Image quality is automatically detected prior to *Fish Detection*, and specific parameter tuning is applied for adapting the computer vision algorithm to the characteristics of image quality. To investigate uncertainty due to image quality, ground-truth evaluation of *Fish Detection Errors* were performed for each *Image Quality*. When analyzing fish counts from a set of video samples, users can relate the numbers of videos from each image quality with the errors measured for each image quality.

Ground-Truth Quality—Ground-truth sets contain examples of fish that were manually annotated by ecology experts, or by non-experts recruited from crowd-sourcing (Chap. 14). Computer vision algorithms learn to recognize fish and their species by constructing fish models on the basis of these examples. Hence ground-truth is essential to ensure the quality of information extraction. Issues arise with ground-truth sets that are not representative of the possible fish appearances, and with scarcity of fish examples, e.g., for rare species. To be representative of the fish populations, ground-truth sets need to contain examples of the typical fish appearances. For instance, if a species color can vary between gray and black, the ground-truth must contain examples of both gray and black appearances. Similarly, if cameras often record blurred and low-contrast images, then the ground-truth should contain examples of fish for each image quality. This is usually ensured by selecting ground-truth images through a random sampling among all images collected from all cameras.

Ground-truth can contain outliers such as erroneous annotations, or rare fish appearances (e.g., odd fish poses). With scarce ground-truth, outliers can have a great impact on computer vision errors. For instance, if a small ground-truth set contain an image of seaweed, then the fish model can be distorted so as to be compatible with seaweed appearances. Hence a high number of non-fish objects can be included in end-results. Large ground-truth sets potentially lower the impact of outliers, as outliers' distortion of fish models is likely to be overridden by numerous counter examples.

Hence, to evaluate uncertainty due to ground-truth quality, we need to measure the representativity of ground-truth sets, their annotation errors, and the quantity of ground-truth items. Ground-truth quantity is the number of examples of each type of fish to recognize: examples of fish and non-fish objects for the *Fish Detection* algorithm, and examples of each species for the *Species Recognition* algorithm. Regarding annotation errors, several metrics exist: number of annotators for each image, level of expertise of annotators (e.g., professor, student, or inexperienced), and level of agreement amongst annotators if annotations are contradictory (e.g., Cohen's kappa). They are typically applied for evaluating ground-truth sets collected through crowd-sourcing (Chap. 14). For ground-truth representativity, we need to take into account the *Image Quality* of the recordings. The number of ground-truth items for each image quality indicates potential scarcity for one type of image, which increases uncertainty in end-results drawn from such videos. A randomized selection from a large quantity of ground-truth items ensures a priori that ground-truth sets are representative of the fish appearances. This ground-truth collection method is recommended both for the ground-truth sets used for learning fish models, and the sets used for evaluating *Fish Detection* and *Species Recognition Errors*. However, future work is needed for formally assessing ground-truth representativity, and for assessing that sufficient numbers of items are collected.

During interviews with ecologists, we explained the ground-truth annotation process. Ecologists were interested in the numbers of ground-truth images, and in browsing them. Further metrics, such as numbers of annotators and their level of agreement, were not introduced at first to avoid overwhelming users. We focused on providing the numbers of ground-truth items correctly or incorrectly classified, for each species or image quality. Future work can investigate the benefits of providing further metrics to end-users, e.g., the level of agreement between annotators, to improve user confidence.

13.3 Visualizing Uncertainty Due to Computer Vision Algorithms

End-users who are not familiar with computer vision are likely to encounter difficulties in understanding ground-truth evaluations and their technical concepts (Beauxis-Aussalet et al. 2013a). Some metrics may be misunderstood or may not fully address the uncertainty factors. This section summarizes these issues, and presents a visualization design adapted for end-users who are not necessarily experts in computer vision.

13.3.1 Usability Issues with Computer Vision Evaluations

Confusion matrices need to be read both column- and row-wise, which is tedious and error prone. For instance, considering the cell with both red and gray squares in Table 13.3, if read row-wise it indicates False Positives added to *Species B*. If read column-wise, it indicates False Negatives lost by *Species A*. Memorizing all cell values, and their semantics, is an important cognitive load. Users may forget cell values, or may read only columns or rows.

To limit cognitive efforts, confusion matrices can be synthesized by accumulating errors for each species (i.e., basic metrics in Table 13.3). However, it is no longer possible to distinguish which species are likely to be confused with another. For instance, the cells with red or gray background in Table 13.3 do not indicate the original true species of the misrecognized fish. Users need this information to identify correlations between species populations that are induced by *Species Recognition Errors*, and hence, that are not representative of the actual trends in fish populations. For instance, an important increase of one species implies an increase of its False Negatives. A proportion of its fish are attributed to other species, and this can induce deceiving increases of other species, especially for species of much inferior abundance. Hence, users need to inspect errors between pairs of species, rather than the synthesis of False Positives and False Negatives accumulated for all species.

Finally, advanced metrics are more complex and convey specific types of errors, and thus non-expert users may misinterpret them. For instance, with *Species Recognition*, True Negatives are fish correctly discarded from a species. They are accumulated over all other species, and are usually of a much higher magnitude than True Positives, False Positives and False Negatives, as shown in Table 13.3. High numbers of True Negatives yield low *False Positive Rate* and high *Accuracy* (see formulas in Table 13.4). Hence this may conceal important numbers of False Positives or False Negatives. The visualizations commonly used by computer vision experts use pairs of advanced metrics (e.g., Fig. 13.2). Considering the above-mentioned issues, such visualizations are likely to be overwhelming and misleading for end-users who are not familiar with computer vision. Moreover, advanced metrics no longer indicate the number of items in the ground-truth, and thus possible ground-truth scarcity. Confusion matrices originally provide this information, i.e., the numbers of test items which

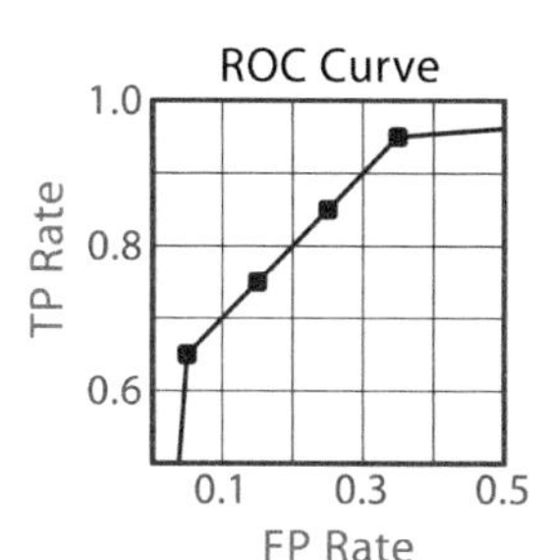

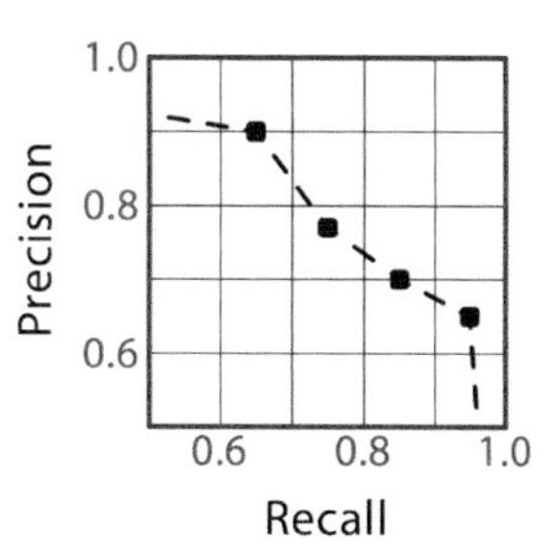

Fig. 13.2 Example of an ROC (*left*) and Precision/Recall curve (*right*). Error rates are given for different parameter settings, e.g., the points on the curves show 4 *score* thresholds discarding objects that are not similar enough to the fish model

are usually proportional to the numbers of training items. Hence we investigated the ways to communicate numbers of test items correctly or incorrectly classified, rather than the ways to communicate complex and potentially misleading error rates.

13.3.2 Preliminary User Study

Ecologists need to understand computer vision errors, but ground-truth evaluations techniques are complex and may overwhelm them. Hence we investigated which level of detail needs to be disclosed to end-users (Beauxis-Aussalet et al. 2013a). We exposed 7 marine ecologists to explanations progressively disclosing the concepts of ground-truth evaluations. Explanations were given in 3 steps. Each step consisted of (i) a visualization of fish counts, as produced by our computer vision system; (ii) a visualization of a ground-truth evaluation of our system, introducing new technical concepts; and (iii) a questionnaire evaluating the impact of the new details introduced. The first step introduced the concepts of ground-truth sets used for training and evaluating the video analysis software. The uncertainty visualization simply compared manual and automatic fish counts. The second step introduced the concepts of True Positive (TP), False Negative (FN) and False Positive (FP). The uncertainty visualization showed manual and automatic fish counts, with details about the amount of TP, FN and FP. True Negative were omitted to avoid overwhelming users. The third step introduced the concepts of fish model and *score thresholds* of classifiers. The *scores* measure how fish images look like the *fish models*, as discussed in Sect. 13.2. The visualization presented sets of fish counts produced by using different *score thresholds*, and ground-truth evaluation of TP, FN, and FP given for each *score threshold*.

At each step, a questionnaire measured (i) *user trust* in the computer vision software's ability to count fish; (ii) *user acceptance* of the software for scientific research; (iii) *user understanding* of the technical concepts; and (iv) the satisfaction of *user information needs* for uncertainty evaluation. The questionnaires were independently analyzed by two experts in Human-Computer Interfaces. A 4-grade scale was used (*Very Low* −−, *Low* −, *High* +, *Very High* ++) to qualify user trust, acceptance understanding and information needs. Table 13.5 summarizes the results of this experiment. It shows that the technical concepts were generally difficult to understand. Extensive time and additional explanations were required for ecologists to familiarize with them. Further, users information needs were not fully satisfied. For instance, users required to watch videos themselves and to inspect other uncertainty factors. Besides these issues, user acceptance remained globally unchanged over explanation steps. User acceptance is relatively high since computer vision can greatly reduce material costs and human efforts. The third step, introducing *score thresholds*, had a slightly positive impact on user trust and acceptance, i.e., in respectively 4 and 2 cases out of the 6 cases that could be improved (User 3 already had maximum score).

Table 13.5 Qualitative analysis of the experiment introducing technical concepts of ground-truth based evaluations

	Trust			Acceptance			Understanding			Info. Needs		
Step:	**1**	**2**	**3**	**1**	**2**	**3**	**1**	**2**	**3**	**1**	**2**	**3**
User 1	+	+	-	+	+	+	-	-	-	--	--	--
User 2	-	-	+	+	+	+	+	-	+	--	-	-
User 3	++	++	++	++	++	++	+	--	--	-	-	--
User 4	-	-	-	-	-	-	++	++	++	-	-	-
User 5	--	--	-	+	+	++	++	++	++	--	--	--
User 6	--	--	-	--	-	-	+	+	-	--	--	--
User 7	+	-	+	+	+	++	--	--	-	--	+	-

The quality of user trust, acceptance, understanding, and satisfaction of information needs is either Very High (++), High (+), Low (−), or Very Low (−−). Green cells indicate a positive effect of the explanation steps, orange cells indicate a negative effect, uncolored cells indicate no significant effect

13.3.3 Visualization Design for Non-expert Users

We designed visualizations intended to limit cognitive load and misunderstandings, while addressing the 4 uncertainty factors related to computer vision algorithms. Our first design choice is to omit the True Negatives. They are not contained, and should not be contained, in end-results as they are not informative from a user viewpoint. Further, Beauxis-Aussalet et al. (2013a) shows that understanding the concepts of True Positive, False Negatives and False Positives is already likely to overwhelm users. Finally, as the magnitude of True Negatives can largely exceed that of errors (False Positives and False Negatives), True Negatives may conceal uncertainty (see Sect. 13.3.1).

To further limit cognitive load and misunderstandings, we avoid advanced metrics (Table 13.4). Our visualizations primarily show the numbers of ground-truth items yielding True Positives, False Negatives or False Positives in computer vision results. Figure 13.3 gives an example of such a display for *Fish Detection Errors*. It shows the numbers of ground-truth items, an important aspect of *Ground-Truth Quality* which is abstracted in traditional ROC or Precision/Recall curves.

The layout of our visualization intends to intuitively convey the concepts of *correct fish* (i.e., True Positives), *missed fish* (i.e., False Negatives) and *added fish* (i.e., False Positives). Stacked charts show the fish contained in end-results above a horizontal line, with *correct fish* below and *added fish* on top. *Missed fish* are displayed below the horizontal line. Colors reinforce the perception of errors. *Correct fish* are shown in blue, a positive or neutral color. *Added fish* are shown in light gray, to express an elusive presence contrasting with the saturated blue of *correct fish*. *Missed fish* are shown in red, a negative color expressing a warning. It aims at creating an intuitive perception that *missed fish* below the line are not included in end-results, and that *added fish* create over-estimations.

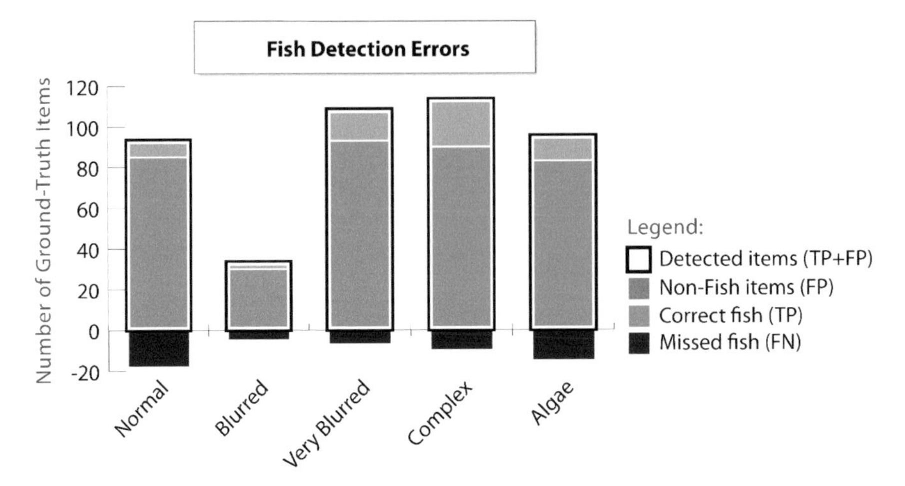

Fig. 13.3 Example of our novel visualization design detailing *Fish Detection Errors* for each type of *Image Quality*

For visualizing *Species Recognition Errors*, the same design principles are applicable. True Positives, False Positives and False Negatives can be displayed for each species. However, ecologists need to investigate which species are often confused with another, so as to identify potential biases with look-alike species. Hence, we need to detail which species add False Positives, or receive False Negatives, and what is the magnitude of errors. Multiple confusions between pairs of species can occur, especially since errors are directional: e.g., fish from Species A misclassified as Species B ($FN_{a \to b}$), and inversely, fish from Species B misclassified as Species A ($FN_{b \to a}$). With N_s species $N_s(N_s - 1)$ pairs of species need to be investigated. This complexity can clutter the visualization and overwhelm users. To address this issues, our visualization displays the most important inter-species confusions, and summarizes the remaining errors. For each species, we select the 2 other species yielding the most FP and FN, and display the related errors in distinct stacked block. The remaining errors from other species are displayed together in one block. Figure 13.4 gives an example of such display. Users can select a species to display errors only for that species, as shown in Fig. 13.5.

The numbers of ground-truth items can greatly vary amongst classes of species and image quality, e.g., scarcity for some classes, or abundance of other classes. In these cases, ground-truth errors can be difficult to visualize. Hence users can switch between visualizing errors either as: (i) numbers of ground-truth items; or (ii) proportional measure of errors (Figs. 13.6 and 13.7). *Fish Detection Errors* are given proportionally to the total number of detected items ($TP + FP$), using Eqs. (13.1) and (13.2). This choice of denominator intends to support the extrapolation of errors in subsets of end-results, for which only the total numbers of detected items are known. For instance, given a set of N_i fish detected in a set of videos with image quality Q_i, a user can extrapolate that it contains $N_i \frac{FP_i}{TP_i+FP_i}$ False Positives, and $N_i \frac{FN_i}{TP_i+FP_i}$ False

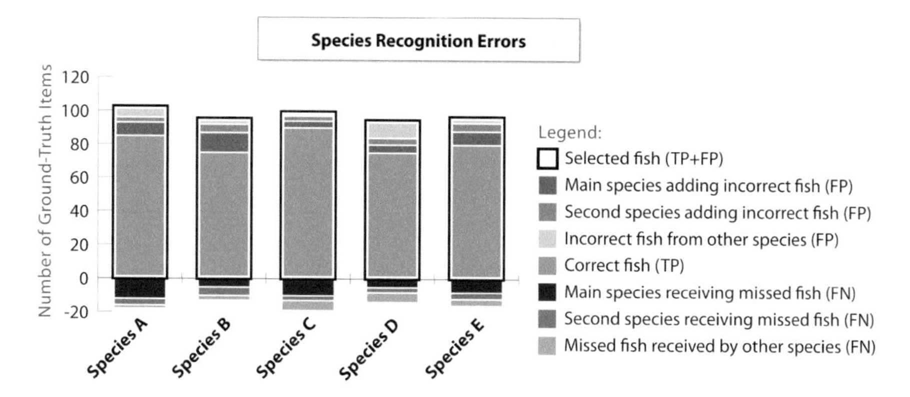

Fig. 13.4 Example of our novel visualization design for *Species Recognition Errors*

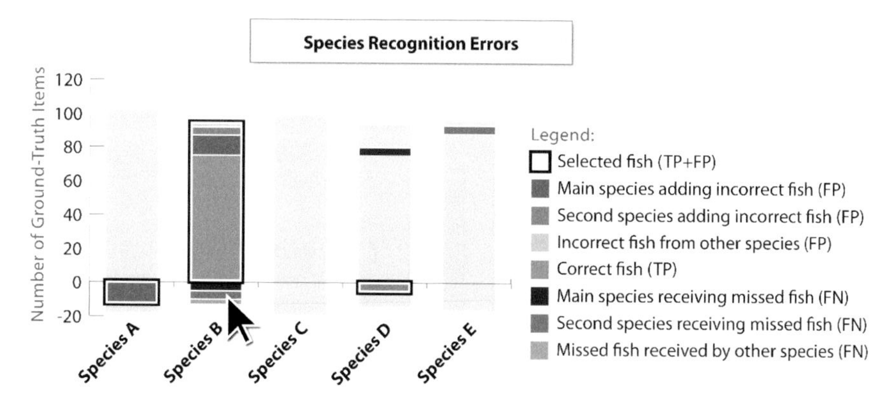

Fig. 13.5 Selecting a species of interest highlights the errors for that species. It shows from which species its False Positives (FP) came from (*gray stacked bars*) and to which species its False Negatives (FN) are attributed (*red stacked bars*)

Negatives (FP_i, FN_i and TP_i being measured from a ground-truth set representative of image quality Q_i).

$$Type\ I\ Error\ Rate\ Q_i = \frac{FP_i}{TP_i + FP_i} \tag{13.1}$$

$$Type\ II\ Error\ Rate\ Q_i = \frac{FN_i}{TP_i + FP_i} \tag{13.2}$$

Equations (13.1) and (13.2): *Type I* and *Type II Error Rates* Q_i are, respectively, the ratio of non-fish objects (FP_i) and undetected fish (FN_i) on the total numbers of detected items ($TP_i + FP_i$), measured in a ground-truth set of image quality Q_i. TP_i

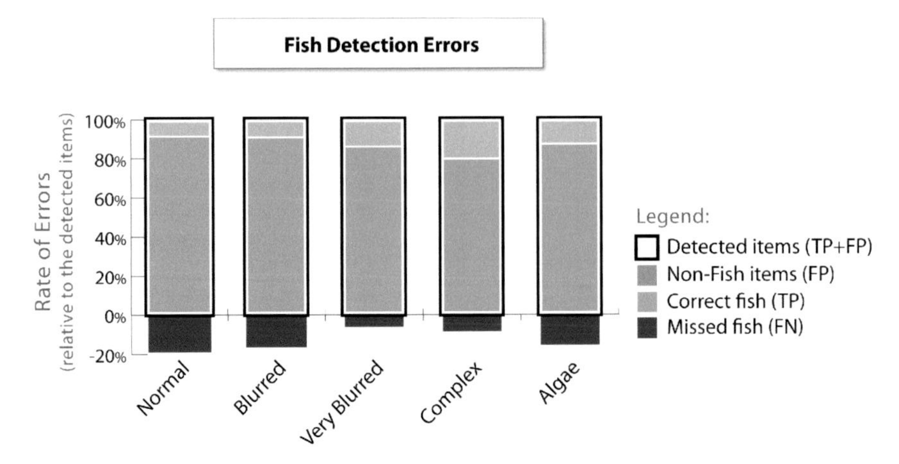

Fig. 13.6 Visualization design from Fig. 13.3 showing rates of errors from Eqs. 13.1 and 13.2

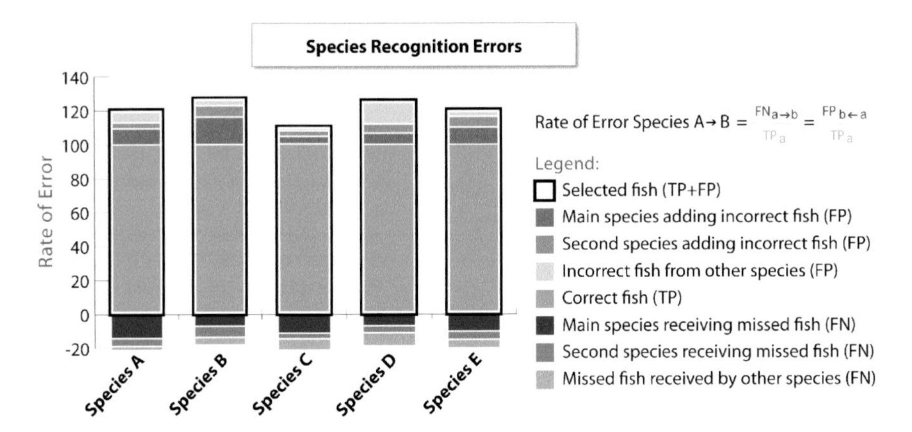

Fig. 13.7 Visualization design from Fig. 13.4 showing rates of error from Eq. 13.3

is the number of fish correctly detected for the ground-truth set. Equation (13.1) is equivalent to *Precision*, and Eq. (13.2) to *False Discovery Rate*.

For *Species Recognition Errors*, the False Negatives transferred from species A to species B ($FN_{a\rightarrow b}$), which are also the False Positives attributed to species B while truly belonging to species A ($FP_{b\leftarrow a}$), are given proportionally to the True Positive for species A using the Eq. (13.3). The choice of a denominator is different from that of error rates for *Fish Detection Errors* because in the case of *Species Recognition Errors*, False Positives and False Negatives are not independent between classes, i.e., between species. False Negatives for one species are False Positives for other species. Hence, in ground-truth evaluations, the number of False Positives observed for one species arbitrarily varies depending on the number of ground-truth items for other species, independently from computer vision algorithms. Furthermore, each subset of end-results may have different proportions of each species, as population dynamics may be different over seasons or locations. For instance one species may be more abundant at specific periods of time than others. Hence, in end-results, the number of False Positives attributed to one species arbitrarily varies depending on the population sizes of other species, independently from computer vision algorithms. Therefore, False Positives are excluded from the denominator of error rate (13.3). The choice of a denominator also intends to support extrapolations of errors in subsets of end-results, as for the error rates (13.1) and (13.2). To do so, the denominator needs to represent the fish counts as observed in subsets of end-results, i.e., total numbers of fish detected for each species. Therefore, False Negatives are excluded from the denominators, as they are not contained in end-results' fish counts for each species. Using only True Positives as the denominator of error rates in Eq. (13.3) is a tradeoff between representing fish counts as observable in sets of end-results (i.e., $TP + FP$), and accounting for numbers of errors that are proportional to the population size of their true species (i.e., excluding FP which are not proportional to the size of their attributed species).

$$Pairwise\ Error\ Ratio\ S_a \rightarrow S_b = \frac{FN_{a\rightarrow b}}{TP_a} \tag{13.3}$$

Equation (13.3): *Pairwise Error Ratio* $S_a \rightarrow S_b$ is the ratio of fish belonging to species A (S_a) erroneously attributed to species B (S_b). $FN_{a\rightarrow b}$ is the number of ground-truth items attributed to S_b while truly belonging to S_a (e.g., the cell with both red and gray squares in Table 13.3). Note that $FN_{a\rightarrow b} = FP_{b\leftarrow a}$, i.e., the number of False Positives attributed to species B while truly belonging to species A. TP_a is the total number of TP for S_a. Note that $FN_{a\rightarrow b}$ is different from $FN_{b\rightarrow a}$ and *Pairwise Error Ratio* $S_a \rightarrow S_b$ is different from *Pairwise Error Ratio* $S_b \rightarrow S_a$.

To conclude, our visualization supports the evaluation of 4 uncertainty factors due to computer vision algorithms by showing a simple but complete representation of ground-truth evaluation results. *Fish Detection* and *Species Recognition Errors* are evaluated by visualizing absolute and relative numbers of errors in ground-truth sets (Figs. 13.3, 13.4, 13.5, 13.6 and 13.7). The uncertainty due to *Ground-Truth Quality* is evaluated by visualizing absolute numbers of ground-truth items, and the uncertainty due to *Image Quality* is evaluated by visualizing *Fish Detection Errors* for each type of image (Fig. 13.3).

13.4 Evaluating Uncertainty Due to In-Situ System Deployment

During our study of user requirements (Chap. 2), we identified uncertainty factors that are not related to computer vision algorithms but to the deployment of the system. The deployment of cameras over marine ecosystems can greatly impact end-results, independently from potential computer vision errors. The cameras' field of view can increase or decrease the chances to observe specific species, hence creating biases. Some fields of view increase the chances of counting repeatedly the same individuals swimming back and forth in front of the camera, hence creating further biases. These types of biases typically concern *benthic* (i.e., living on the seabed), *sedentary* (i.e., living in coral heads), *schooling* (i.e., living in groups), and *herbivorous or carnivorous* species.

Furthermore, camera deployment over geographical locations may not provide a sufficient sampling of the ecosystem. Ecologists usually need redundant measurements to ensure the statistical validity of their observations (Cochran 1977). Hence sufficient numbers of cameras need to be deployed. Additionally, the extent of the sampling coverage can be reduced if all collected videos are not processed due to technical issues. In this section, we discuss evaluation methods for these uncertainty factors related to the in-situ deployment of the Fish4Knowledge system.

Field of View and Duplicated Individuals—The different types of coral are populated by specific species feeding on its organisms, or hiding in its structure. Thus the placement of cameras needs to reflect the different habitats of interest. If some habitats are not observed, their species are under-represented, and end-results are biased. For instance, observations of *benthic, sedentary* and *carnivorous and herbivorous* species are biased if fields of view do not cover the specific habitats where these species are living.

Some species swimming behaviors (e.g., coming in and out of coral cavities) yield repeated occurrences of the same fish in the cameras' field of view. For instance, *schooling* and *sedentary* species (e.g., living in groups, or coral head cavities) are likely to yield *Duplicated Individuals* in end-results. Fields of view contribute to biases due to multiple re-identification of the same fish. For instance, close-ups on specific coral heads increase the chances of observing duplicated fish from *sedentary* species. Similarly, groups of *schooling* fish may not be consistently observed between close-ups and open sea views. Further, the depth of field of view modifies the sampling coverage of the area. For instance, compared to open sea views, close-up views cover a smaller area of the ecosystem.

The state-of-the-art does not offer well-established methods for handling these uncertainty factors. Future work needs to develop measures of rates of *Duplicated Individuals* depending on fish species and *Fields of View*. For example, a measure of such potential bias can indicate that *schooling* species S observed from field of view V are over-estimated by $FP_{s,v}\ Rate = \frac{FP_{s,v}}{TP_{s,v}+FP_{s,v}}$, similarly to error rate (13.1).

Finally, the Fish4Knowledge system relies on fixed cameras which fields of view are expected to remain the same over time. However, fields of view may vary over time. Small shifts can occur during maintenance and lens cleaning operations, and larger shifts can occur with environmental events such as typhoons. Hence, accidental changes of field of view need to be controlled and monitored over time.

Sampling Coverage and Fragmentary Processing—The Fish4Knowledge system stores continuous video footage into 10-min excerpts. Ecologists need to take into account the number of 10-min video samples from which computer vision results are drawn, since it influences the statistical representativity of the patterns observed in fish populations. For instance, fish counts observed from a few videos may not be representative of the actual populations of the ecosystem. Further, if different fish counts are drawn from video sets of different size, the more videos the more fish, hence comparison is biased. Therefore users need evaluations of sampling size (e.g., the numbers of videos over time periods and locations), and a comparable measure of fish abundance for end-results drawn from different sampling sizes.

The primary metric for sampling size is the number of video samples from which end-results are extracted. Additionally, the number of unprocessed videos, i.e., still in the workflow processing queue (Chap. 8), indicates that further video processing could complement the end-results. The Fish4Knowledge system offers functionalities for manually requesting that specific videos of interest are processed with high priority (Beauxis-Aussalet et al. 2013b).

To analyze sets of end-results extracted from varying numbers of video samples, averaging fish count per video as in (13.4) offers a comparable metric of fish abundance. Further, measuring variance over samples as in (13.5) complements the estimation of uncertainty. Such measure of variance over samples is often used as a basis for statistical analysis (Cochran 1977).

$$Mean\ Fish\ Count\ per\ Video = \frac{Number\ of\ Fish}{Number\ of\ Videos} \tag{13.4}$$

Equation (13.4): Measure of fish abundance for comparing fish counts drawn from varying numbers of videos.

$$Variance\ over\ Videos = \frac{1}{N_v} \sum_{i=1}^{N_v} (Mean\ Fish\ Count\ per\ Video - N_i)^2 \tag{13.5}$$

Equation (13.5): Measure of variance in fish abundance. N_v is the number of 10-min video sample, N_i is the number of fish in the ith video sample.

However, the measures of mean and variance of fish counts per video in (13.4) and (13.5) must be used with care as they face three problems:

(1) *Video duration* must be identical over samples. Video samples of longer duration are likely to contain more fish that samples of shorter durations, thus biasing the mean fish count per video as in (13.4). For video samples of unequal duration, fish abundance can be assessed by averaging fish counts over a time unit, e.g., $Mean\ Fish\ Count\ per\ Minute = \frac{1}{N_v} \sum_{i=1}^{N_v} \frac{Number\ of\ Fish\ in\ Video\ Sample\ i}{Duration\ of\ Video\ Sample\ i\ (in\ \mathrm{min})}$. We recommend the use of video samples of equal duration. Using videos of different durations would considerably complicate the measurement of uncertainty, particularly for analyzing the variance of fish abundance over different cameras while taking into account missing videos, as explained below.

$$Mean\ Abundance\ per\ \text{10-min} = \sum_{j=1}^{N_c} Fish/Video\ at\ Camera\ C_j \tag{13.6}$$

Equation (13.6): Measure of fish abundance for analyzing fish counts drawn from several cameras, with varying numbers of video per camera. N_c is the number of cameras. $Fish/Video\ at\ Camera\ C_j = \frac{Number\ of\ Fish\ at\ Camera\ C_j}{Number\ of\ \text{10-min}\ Videos\ at\ Camera\ C_j}$, i.e., the result of Eq. (13.4) for one camera.

(2) *Fish abundance over different cameras*, and for the same time period, must be measured by summing the results of Eq. (13.4) for each camera separately, as in Eq. (13.6). It would be conceptually inaccurate to measure fish abundance as the result of Eq. (13.4) for all cameras globally, i.e., $\frac{Number\ of\ Fish\ for\ All\ Cameras}{Number\ of\ Videos\ for\ All\ Cameras}$. This is because the cameras observe the same time periods. For instance, if cameras 1 and 2 observe the same 10-min time period, yielding 2 video samples with respectively N_1 and N_2 fish occurrences, then the overall fish abundance is $N_1 + N_2$ rather than $\frac{N_1+N_2}{2}$. To clarify what the metric represent, we recommend to use the label *Mean Abundance per* 10-min rather *Mean Fish Count per Video*. Note that if the cameras' field of view observe the same overlapping areas, the overall fish abundance cannot be assessed as in (13.6).

(3) *The variance of fish abundance over different cameras* (i.e., the variance of Eq. (13.6) results) is equivalent to the variance of a sum of random variables (the variables being the results of Eq. (13.4) for each camera). Such variance is measured by summing the covariances of Eq. (13.4) results over cameras, as in Eq. (13.7). Measuring such covariance assumes that video samples are available for all cameras, and for all the 10-min time periods. For instance, if *Mean Fish Counts per Video* at C_1 include the 10-min time period t_1 (e.g., 16:00–16:10 on Jan. 1st 2011), then video samples must be available at all cameras for the same time period t_1. Then covariances can be measured using Eq. (13.8). If video samples are missing, i.e., if a 10-min time period is covered by at least 1 camera but not by all cameras, then it is not possible to measure covariances using Eq. (13.8). However, statistical methods can address this problem (Little and Rubin 2014). They consists of discarding incomplete sample subsets or using replacement values for missing samples (imputation). None of them can provide perfect results, and the choice of a method depend on each use case constraints.

$$Variance\ over\ Cameras = \sum_{j=1}^{N_c}\sum_{k=1}^{N_c} Cov(Fish/Video\ at\ C_j,\ Fish/Video\ at\ C_k) \tag{13.7a}$$

$$= \sum_{j=1}^{N_c} Var(Fish/Video\ at\ C_j) + 2\sum_{j=1}^{N_c}\sum_{k>j}^{N_c} Cov(Fish/Video\ at\ C_j,\ Fish/Video\ at\ C_k) \tag{13.7b}$$

Equation (13.7): Measure of variance in fish abundance observed from several cameras, i.e., the variance of Eq. (13.6) results. $Fish/Video\ at\ C_j$ is the *Mean Fish Count per Video* for camera j, i.e., the result of Eq. (13.4) for one camera. $Var(Fish/Video\ at\ C_j)$ is the variance of $Fish/Video\ at\ C_j$, i.e., the result of Eq. (13.5) for one camera. $Cov(Fish/Video\ at\ C_j, Fish/Video\ at\ C_k)$ is the covariance of the results of Eq. (13.5) for cameras j and k. N_c is the number of cameras.

$$Cov(Fish/Video\ at\ C_j,\ Fish/Video\ at\ C_k) = \sum_{t=1}^{N_{10\,\mathrm{min}}} \frac{(N_{t,j} - Fish/Video\ at\ C_j)(N_{t,k} - Fish/Video\ at\ C_k)}{N_{10\,\mathrm{min}}} \tag{13.8}$$

Equation (13.8): Measure of covariance as used in Eq. (13.7). $Fish/Video\ at\ C_j$ is the *Mean Fish Count per Video* for camera j, i.e., the result of Eq. (13.4) for one camera. $N_{10\,\mathrm{min}}$ is the number of 10-min time periods covered by the video samples. $N_{t,j}$ is the number of fish observed at Camera j during the 10-min time period t.

To conclude, Eqs. (13.4) and (13.5) provide relatively simple measures of fish abundance which overcome the issues of *Sampling Coverage* and *Fragmentary Processing*. However, these are applicable to analyze fish counts drawn from one single camera. For analyzing the overall fish abundance for several cameras, the applicable measure of fish abundance is given by Eq. (13.6). However, the related measure of variance in fish abundance, i.e., Eq. (13.7), is not directly applicable in the case of missing samples due to *Sampling Coverage* and *Fragmentary Processing*. In such cases alternative methods exists (Little and Rubin 2014) and can be chosen depending on each use case.

13.5 Visualizing Uncertainty Due to In-Situ System Deployment

Although the state-of-the-art does not offer well-established methods for quantifying the effect of *Fields of View* on fish counts, we provide users with elementary means to investigate their impact. A tab of the user interface is dedicated to the browsing of video samples, and is shown in Fig. 13.8. Ecologists can inspect the different *Fields of View* over cameras and time periods. They can estimate which ecosystem is observed, which species are likely to be over- or under-estimated, and the potential *Duplicated Individuals*. Users can also investigate potential changes of field of view over time.

The lower part of the interface contains filtering widgets for selecting the videos of interest. Users can specify the characteristics of the videos of interest (e.g., time, location, *Image Quality*, species observed), in widgets that can be opened and closed on-demand. The widgets also offer an overview of the numbers of video samples for each characteristics. For instance, in Fig. 13.8 the histograms represent numbers of

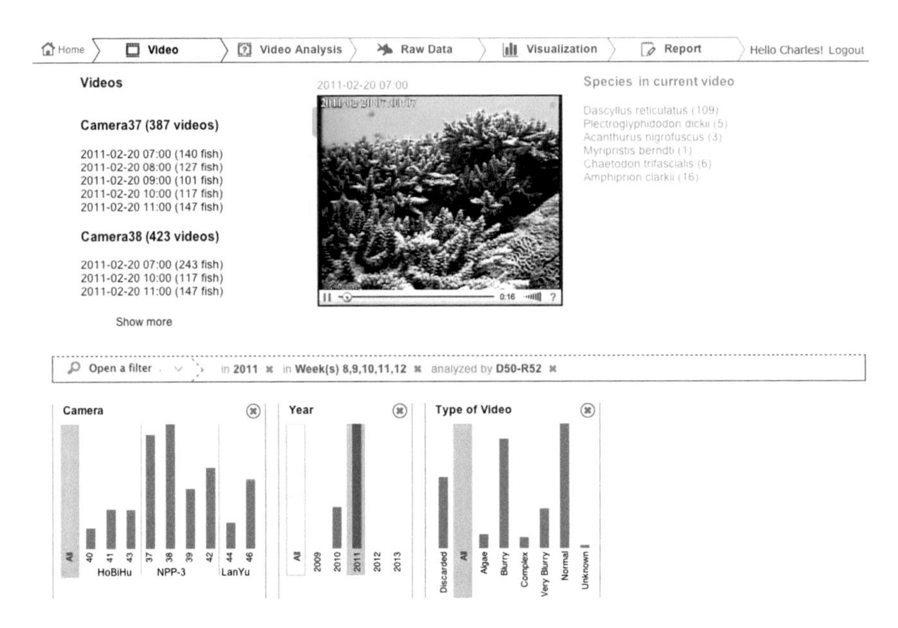

Fig. 13.8 **Video tab**: Video browser and visualizations for estimating uncertainty due to *Field of View*, *Duplicated Individuals*, *Image Quality*, *Sampling Coverage* and *Fragmentary Processing*. The *bottom* histograms show numbers of 10-min video samples, and their distribution over locations (e.g., cameras), time (e.g., year) and *Image Quality*

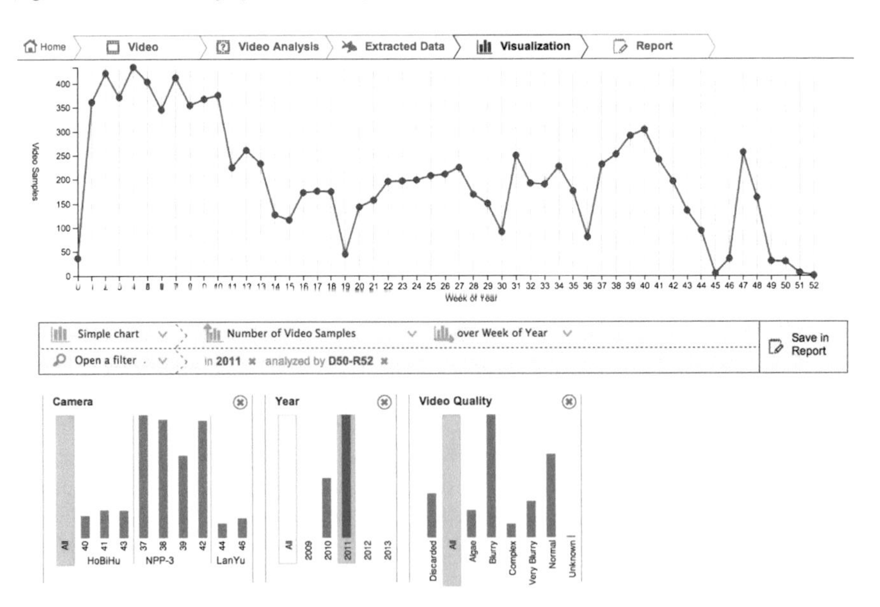

Fig. 13.9 **Visualization tab**: Visualizations for estimating uncertainty due to *Sampling Coverage*, *Fragmentary Processing* and *Image Quality*. The *bottom* histograms are the same as Fig. 13.8, and the main line graph above details the distribution of 10-min video samples over one year (2011)

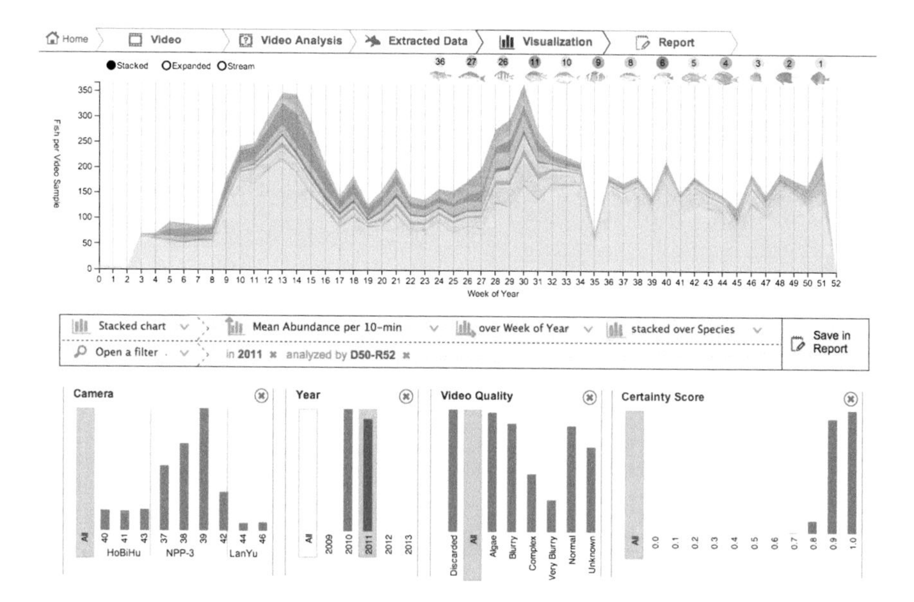

Fig. 13.10 **Visualization tab**: Visualizations for estimating *Uncertainty in Specific Output* due to *Sampling Coverage*, *Fragmentary Processing*, *Image Quality*, *Fish Detection Errors* and *Species Recognition Errors*. The *bottom* histograms and the main graph above show average fish counts per video, a balanced metric of fish abundance addressing *Fragmentary Processing* issues. The Video Quality widget shows fish abundance for each *Image Quality*. It indicates potential biases due to *Fish Detection Errors* that can arbitrarily vary depending on *Image Quality*, rather than natural phenomena. The Certainty Score widget shows the distribution of fish *scores*, which are used by computer vision algorithms to represent the similarity between each fish and their species models (Sect. 13.2). These indicate potential biases due to *Species Recognition Errors* since errors are more likely to occur for fish with low score

videos over locations, year, and *Image Quality*. This offers basic means to investigate uncertainty due to *Sampling Coverage, Fragmentary Processing* and *Image Quality*.

Uncertainty due to these factors can be further detailed in another tab of the interface, shown in Fig. 13.9. This tab offers the same widgets, and overview of numbers of video samples. Numbers of videos can be detailed in the main graph, on the upper part of the interface. Further, the main graph and the widgets' histograms can also display absolute numbers of fish, and mean abundance per 10-min as in Eq. (13.6). Figures 13.9 and 13.10 show visualizations of these metrics. The main graph can also display boxplots for visualizing the variance of fish abundance over sets of samples.

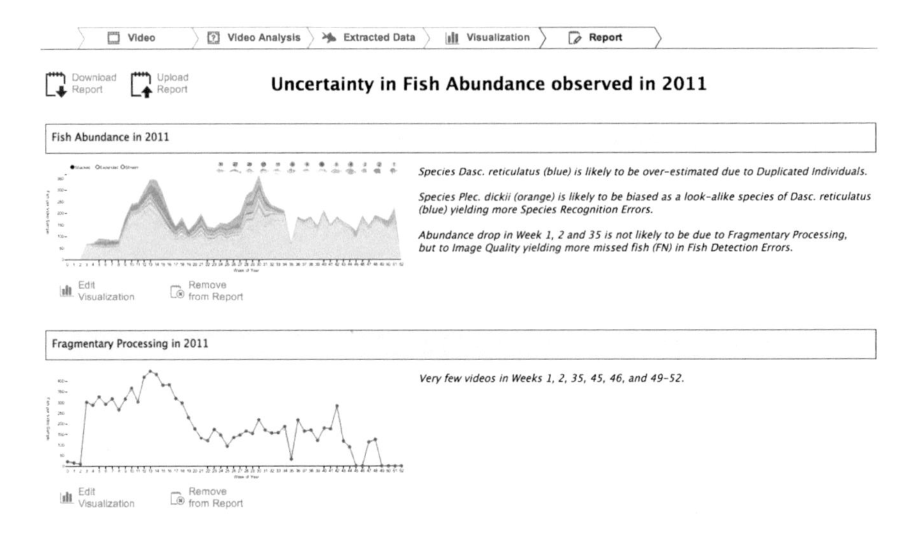

Fig. 13.11 **Report tab**: Example of visualizations gathered and annotated for describing *Uncertainty in Specific Output* due to *Noise and Biases*

13.6 Uncertainty Due to both Computer Vision Algorithms and In-Situ Deployment

Ecologists need to evaluate the uncertainty in end-results. These are impacted by uncertainty factors due to both computer vision algorithms and system deployment. Uncertainty factors interact with each other, as summarized in Fig. 13.1. Although there is a variety of factors and interactions between them, their overall impact can be synthesized as two types of effect: noise, i.e., random errors yielding measurement variance, and biases, i.e., systematic errors yield under- or over-estimated measurements. Biases occur when measurements are systematically different under conditions that are independent from natural phenomena, such as *Image Quality*. This section discusses the means to measure the level of noise in the data, and to identify systematic differences of measurements. Levels of errors are usually measured under controlled conditions, e.g., using ground-truth datasets with specific characteristics. This section also presents the means to investigate errors in specific sets of end-results, which characteristics can be different than those of ground-truth sets.

Noise and Biases—Random errors, i.e., noise in end-results, is commonly measured using metrics of mean and variance such as (13.4) and (13.5). These metrics are a well-established basis for statistics investigating of all sorts of populations (Cochran 1977). Significant differences of means and variances can be observed under conditions that are independent of natural phenomena. An example can be considered with a same fish population observed with different image qualities, or fields of view. If computer vision yields significantly different means and variances of the fish counts, then the different observation conditions can potentially bias the end-results.

As mentioned in Sect. 13.4, no well-established methods are available for evaluating biases due to *Field of View* and *Duplicated Individuals*. Hence the rest of the discussion focuses on identifying biases due computer vision algorithms. As discussed in Sect. 13.2, the Fish4Knowledge project was able to measure *Fish Detection Errors* for each *Image Quality*, and *Species Recognition Errors* for each species. Such measurements can support the evaluation of potential biases due to *Image Quality* and *look-alike* species.

If measurements of *Fish Detection Errors* (e.g., Eqs. (13.1) and (13.2) and Fig. 13.3) vary significantly with different *Image Quality*, then they indicate potential biases in end-results. Sets of end-results from a specific image quality can be artificially over- or under-estimated, compared to end-results from another image quality. If error rates are of the same magnitude for all image qualities, then they do not indicate potential biases. They rather indicate a general level of noise, even if error rates are high. End-results drawn from image qualities having similar levels of uncertainty are potentially over- or under-estimated in the same way, and hence, are comparable. Contrarily, high error rates for *Species Recognition Errors* indicate potential biases between look-alike species, even if error rates are of the same magnitude for all species.

The visualizations of *Fish Detection* and *Species Recognition Errors*, presented in Sect. 13.3.3 and Figs. 13.3, 13.4 and 13.5, support the identification of significant difference in error rates indicating potential biases in end-results. We assume that the significance of error magnitude depends on the study at hand, and their specific requirements with uncertainty issues. For instance, a descriptive survey of fish population may tolerate higher uncertainty than a survey intended to demonstrate causal effects of specific environmental conditions.

Uncertainty in Specific Output—Measurements of *Fish Detection* and *Species Recognition Errors* in ground-truth sets potentially support extrapolations of errors in other sets of computer vision results. Error rates in Eqs. (13.1)–(13.3) can be used to extrapolate errors in end-results, by multiplying them with the numbers of fish in the output. For instance, given a set of fish detected in video samples of image quality Q_i, the potential number of False Positives in end-results could be computed using Eq. (13.9).

$$\textit{Non-Fish in Samples } Q_i = N_i * \textit{Type I Error Rate } Q_i = N_i \frac{FP_i}{TP_i + FP_i} \tag{13.9}$$

Equation (13.9): *Non-Fish in Samples* Q_i is the extrapolated number of False Positives in a set of end-results extracted from videos of image quality Q_i. N_i is the number of fish in the end-results. TP_i and FP_i are the numbers of True Positives and False Positives measured for a ground-truth set of image quality Q_i.

However, the validity of such extrapolation relies on the assumption that errors measured in ground-truth evaluations are representative of errors occurring in computer vision outputs. Further research is needed to control this assumption. For instance, the proportions of fish and non-fish objects may vary across videos of the same image quality, and this would bias the results of (13.9). Alternative methods exist for the case of varying class proportions (Beauxis-Aussalet and Hardman 2015), and can potentially provide more accurate counts of individuals. However, future work is needed to assess their reliability. Hence the Fish4Knowledge user interface did not retain uncertainty methods such as (13.9). Metrics such as (13.1)–(13.3), complemented with numbers of fish and videos samples in sets of end-results, were retained for simple indications of uncertainty in end-results, without extrapolating the numbers of errors.

To complement the evaluation of uncertainty in end-results, the user interface can display the *certainty scores* measuring the resemblance of each fish with the model of its species. Figure 13.10 shows the widget conveying the distribution of fish over *certainty scores*. Species models are constructed using ground-truth images dedicated to the learning of fish appearances. *Scores* are used by the *Species Recognition* algorithm for selecting which fish to classify in each species. The higher the *score*, the more likely the fish truly belongs to the species, and the lower the chances of errors. The *score* is not a measure of error probability, but a measure of visual similarity between fish occurrences and fish models. Measures of error probability can be developed on the basis of this *score*, and such probability can be used to improve the computation of fish abundance (see Chap. 15).

We investigated the impact of such *scores* on user understanding of uncertainty (Beauxis-Aussalet et al. 2013a). As shown in Sect. 13.3.2, user trust and acceptance was slightly improved by providing *score* thresholds to select fish to retain in end-results. Hence we retained the use of such *score* in the user interface. A filter widget displays the distribution of fish over *scores*, and allows the manual selection of a threshold (see Fig. 13.10).

13.7 Future Work

Ground-truth evaluations are well-established methods for evaluating uncertainty due to computer vision algorithms. However, future work is needed to enable the extrapolation of errors in end-results. The representativity of ground-truth needs to be assessed. Large numbers of ground-truth items are randomly sampled from the entire collection of images. This ensures a priori that the ground-truth is representative of the entire video collection. However, this method does not demonstrate that the magnitude of errors measured for the ground-truth sets is similar to that of computer vision performed on other video sets. An approach to estimate how the ground-truth is generalizable could consist of repeating ground-truth measurements, and computing the mean and variance of numbers and rates of error. This method can support extrapolation of errors in end-results, and the measure of confidence intervals for extrapolated errors. But it may require an extensive ground-truth collection. Another approach can make use of error probabilities estimated from *certainty scores*, as discussed in Chap. 15. The accuracy and the costs of these approaches can be compared.

During user interviews, ecologists often asked how to evaluate if the ground-truth sets contain enough fish examples. This aims at estimating the cost of ground-truth collection implied for integrating the detection of a new species. It also aims at deciding on collecting further ground-truth items for the species that are difficult to recognize. Future work is needed for establishing methods to estimate optimal ground-truth size. An approach could consist of repeating ground-truth evaluation for the same computer vision algorithm, but trained using ground-truth sets of different sizes. If the numbers of errors are relatively stable although ground-truth size is increased, then users can consider that the number of ground-truth items is sufficient.

Uncertainty due to in-situ deployment of the system requires important future work. Metrics for *Duplicated Individuals* need to be researched, and to take into account the species and *Fields of View* at stake. Such metrics can be of the same form as *Type I Error Rates* in (13.1), and *Duplicated Individuals* can be considered as False Positives. To extrapolate *Duplicated Individuals* in end-results, the measurements can be repeated over ground-truth sets to compute the mean and variance. Similarly to extrapolations of computer vision errors in end-results, this supports the estimation of confidence intervals for the numbers of *Duplicated Individuals* extrapolated in end-results. Finally, future work needs to address the challenge of extrapolating potential biases and errors in end-results by taking into account the different uncertainty factors. To do so, a unified framework of compatible uncertainty metrics needs to be researched. It needs to integrate metrics of biases and errors from *Fish Detection* and *Species Recognition Errors*, *Image Quality, Fields of View*, and *Duplicated Individuals*, and metrics of species abundance accounting for *Fragmentary Processing* and geo-temporal *Sampling Coverage*, e.g., average fish counts per unit of time or area.

References

Beauxis-Aussalet, E., E. Arslanova, L. Hardman, and J. Van Ossenbruggen. 2013a. A case study of trust issues in scientific video collections. In *Proceedings of 2nd ACM international workshop on multimedia analysis for ecological data*, 41–46.

Beauxis-Aussalet, E., S. Palazzo, G. Nadarajan, E. Arslanova, C. Spampinato, and L. Hardman. 2013b. A video processing and data retrieval framework for fish population monitoring. In *Proceedings of 2nd ACM international workshop on multimedia analysis for ecological data*, 15–20.

Beauxis-Aussalet, E., and L. Hardman. 2015. Multifactorial uncertainty assessment for monitoring population abundance using computer vision. In *Proceedings of the IEEE conference on data science and advanced analytics (DSAA)*.

Cochran, W.G. 1977. *Sampling techniques*, 3rd ed. New York: Wiley.

Little, R.J., and D.B. Rubin. 2014. *Statistical analysis with missing data*. New York: Wiley.

Chapter 14
Data Groundtruthing and Crowdsourcing

Jiyin He, Concetto Spampinato, Bastiaan J. Boom and Isaak Kavasidis

Abstract Human annotated data is a prerequisite for the training and evaluation of computer vision algorithms. Such data is referred to as "ground truth" data. In this chapter, we describe the strategies and systems we have devised in order to obtain the ground truth data required by each of the computer vision components within the Fish4Knowledge system, including fish detection, tracking, and recognition.

14.1 Introduction

Labeled data is a prerequisite for successfully applying and evaluating machine learning techniques to a wide range of problems. Within the Fish4Knowledge project, each of the video analysis components needs labeled video footage, i.e., ground truth data, to train and validate our computer vision algorithms, including algorithms for fish detection, tracking, and recognition.

Creating ground truth data for computer vision research is often a time consuming task done by humans using dedicated tools such as those presented in Spampinato et al. (2012a). Recently, crowdsourcing as a collaborative problem solving strategy has received much attention. In particular, within the computer vision community,

J. He (✉)
Centrum Wiskunde & Informatica, Science Park 123,
1098XG Amsterdam, The Netherlands
e-mail: he@cwi.nl

C. Spampinato · I. Kavasidis
Dipartimento di Ingegneria Elettrica,
Elettronica e Informatica Universita' di Catania,
Viale A. Doria 6, 95125 Catania, Italy
e-mail: cspampin@gmail.com

B.J. Boom
Cyclomedia, Van Voordenpark 1b, 5301 Zaltbommel, The Netherlands
e-mail: bas.boom12@gmail.com

I. Kavasidis
e-mail: ikavasidis@gmail.com

R.B. Fisher et al. (eds.), *Fish4Knowledge: Collecting and Analyzing Massive Coral Reef Fish Video Data*, Intelligent Systems Reference Library 104,
DOI 10.1007/978-3-319-30208-9_14

the wisdom of crowd was shown to provide effective solutions in a wide range of problems where *large scale* ground truth data are needed, e.g., Russell et al. (2008), Yuen et al. (2009), von Ahn and Dabbish (2004), von Ahn et al. (2006), Chen et al. (2011). Within the Fish4Knowledge project, a number of crowdsourcing techniques have been studied to obtain the ground truth data needed by various components of our video analysis system (Kavasidis et al. 2013a; Boom et al. 2012; He et al. 2013a, b).

Each of our video analysis components required a different strategy to create its ground truth data. For instance, the fish detection and tracking component required labels indicating whether or not a fish appears in a video clip, which can be readily casted as a crowdsourcing task manageable by the majority of the crowd-workers. However, recognizing the species of a fish—in order to obtain ground truth data needed by the fish recognition component—requires highly specialized domain knowledge, hence not manageable for most crowd-workers. In this chapter, we discuss the different crowdsourcing strategies and systems we have devised to obtain the ground truth data that were used in the development and evaluation of the F4K system.

The rest of this chapter is organized as follows. In Sect. 14.2 we describe a collective labeling tool for collecting ground truth for fish detection and tracking. In Sects. 14.3 and 14.4 we present a clustering-based approach to label fish species. In particular, we discuss how we convert a labeling task requiring expert knowledge to a task that can be accomplished by laymen.

14.2 Ground Truth for Fish Detection and Tracking

Generating annotations for object detection and tracking in videos is a tedious and error-prone task. While no domain knowledge is required (e.g., unlike fish species recognition), it requires annotators to be very precise. Therefore, such annotation tools must be user-oriented, providing simple visual interfaces and methods that are able to guide and speed-up the annotation process. Following this need, within the F4K project we developed a number of annotation tools (Kavasidis et al. 2012, 2013b) for fish detection and tracking. Below, we describe PerLA, a collaborative tool employed to create ground truth for evaluating our fish detection and tracking approaches. We also demonstrated how involving more people (even non-experts) in the labeling process results in more accurate annotations.

14.2.1 Generating High Quality Annotations Using Collaborative Efforts

Performance evaluation and Labeling Annotation (PerLA) tool[1] is a web-base collaborative environment which allows users to share their own annotations with

[1]http://f4k.ing.unict.it/perla.dev.

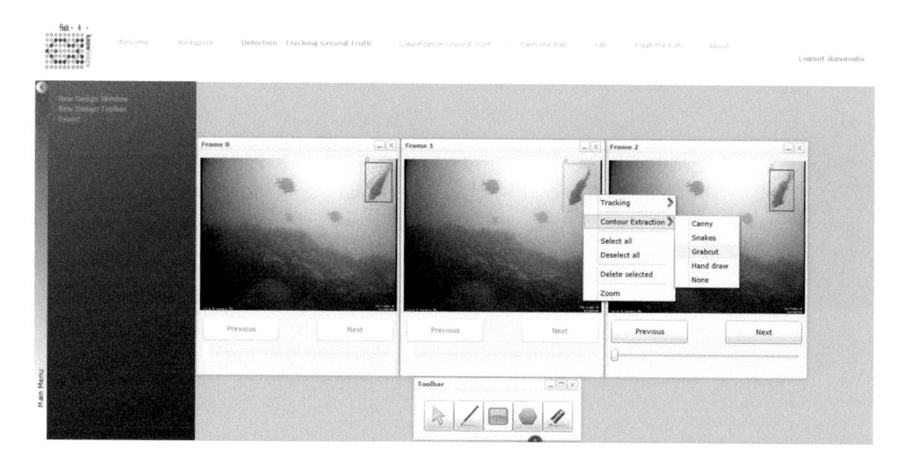

Fig. 14.1 A labeling window when a semi-automatic contour extraction method is used

others (Kavasidis et al. 2013a). It integrates multiple annotations in an inherent user supervision manner to achieve highly effective and efficient annotations. Given an input video stream, PerLA extracts video frames and provides a set of basic tools (polygon and pencil) to annotate each video frame and to follow objects across frames. However, manual annotation is often discouraging, especially in long videos where the number of annotations might be overwhelming: e.g., in the F4K repository videos where more than 20,000 fish for a ten-minute videoclip can be found. Under these circumstances, it is necessary to assist users in drawing object contours as efficiently as possible.

To this end, PerLA implements several automatic contour extraction methods, e.g., Grabcut (Rother et al. 2004), Snakes (Kass et al. 1988), etc., which were chosen because they offer a good trade-off between needed resources and quality of the results. The automatic contour extraction can be applied by drawing the bounding box containing the whole interesting object, right clicking on it and selecting from the "Contour Extraction" sub menu one of the available methods (Fig. 14.1). This is a trial-and-error process that does not always yield the desired result, because the success of the automatic image contour extraction algorithm depends on the type of image used on (image color patterns, contrasts etc.). In case of automatic contour extraction failure, the user can resort to the manual drawing tools.

For object tracking annotations, PerLA exploits the capabilities of multiple-window applications to allow users to follow objects across consecutive frames in an intuitive way. In particular, to be able to annotate multiple instances of the same object in consecutive frames, the user must arrange side-by-side multiple drawing windows. When the user places two windows with their boarders in direct contact, they become, what we call, a "drawing chain". When an adequate, for the user's

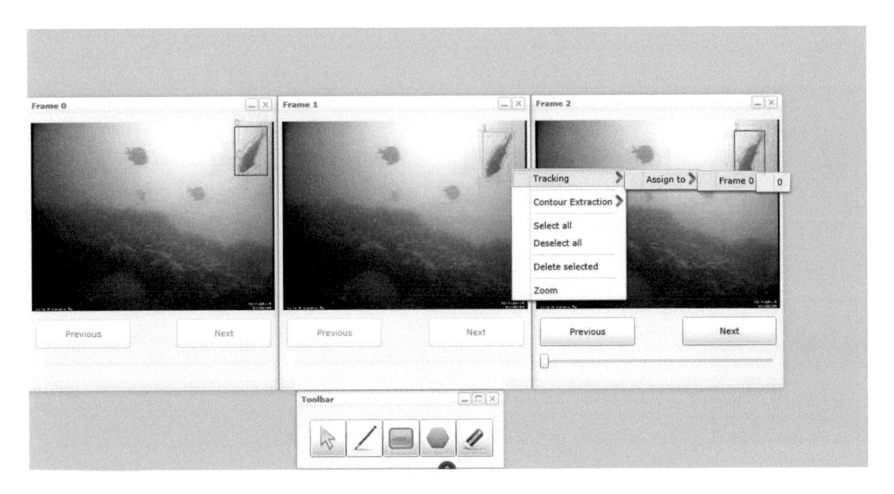

Fig. 14.2 A three-window chain for tracking ground truth generation

needs, chain is formed the user can draw an object and bring up the context menu by right clicking on it, then select the voice "Tracking" and select an object from the previous frames he/she wants to assign the clicked object to (see Fig. 14.2).

14.2.1.1 Combining Multiple Annotations

The collaborative nature of PerLA implies that there may exist multiple annotations for the same object. Such multiple annotations are combined in order to produce a much more accurate object representation since combined opinions are generally more objective than single ones (Howe 2006; Brabham 2008). Specifically, each videoclip receives more than one annotation from the same or different users, which are integrated through a voting policy in order to generate a "best ground truth" (BGT). The BGT building process involves two basic steps: (i) add newly annotated objects to the BGT, (ii) integrating objects' contours.

Let us suppose that the BGT has already been built for a given video V and a user annotates V again. For each newly annotated object A, two cases may occur:

New object instance: The object A has never been previously annotated and it is added directly to the BGT. This exploratory strategy avoids limiting the number of objects on each ground truth; however, to prevent noisy ground truth, each object instance in the BGT comes with a count of the number of annotators that have labeled it over the total number of annotators, thus allowing us to filter out the object instances which have received few annotations.

Existing object instance: I.e., there is already an instance (hereto referred to as GT) of object A in the BGT. In this case we assess a matching score between object A and object GT and if this score is over a threshold (in our case 0.75) the contours of A will be combined to those of the GT. The matching score is computed as the weighted mean of the two following measures:

(1) **Overlap Score**: Given the object A and the corresponding object GT of the best ground truth BGT, the overlap score, O_{score}, is given by:

$$O_{score} = \frac{area(A \cap GT)}{area(A \cup GT)}. \tag{14.1}$$

(2) **Euclidean Distance Score**: Pairwise Euclidean distance between A points (X, Y), with $(X_i, Y_i) \in A$, and GT points (x,y), with $(x_{i'}, y_{i'}) \in GT$, computed as:

$$E_{score} = 1 - \frac{\sum_i^n \sqrt{(X_i - x_{i'})^2 + (Y_i - y_{i'})^2}}{max(\sum_i^n \sqrt{(X_i - x_{i'})^2 + (Y_i - y_{i'})^2})}. \tag{14.2}$$

Usually, a resampling procedure is applied, in order to normalize the number of points in the two contours.

The objects' contour combination is based on the assumption that the BGT contours aggregated over multiple annotators are more accurate than the new ones. In detail, once a new object is considered for being integrated into the BGT its contours C_A are combined with the contours C_{GT} of the corresponding BGT object to form the new object contours C_{NGT}, where each point is computed as:

$$C_{NGT}(i, j) = \frac{1}{2^{N-1}} \sum_{n=1}^{N} (w_A \times C_A(i, j) + C_{GT}(i, j)), \tag{14.3}$$

where $w_A \in [T, 1]$ (where $T = 0.75$ is the threshold described above) is the matching score between A and GT and N is the number of different annotations for that given object. Figure 14.3 shows the result of a combination of four annotations (one belongs to the already existing best ground truth) on the same object, whereas Fig. 14.4 shows how object contours evolve as the number of annotators increases.

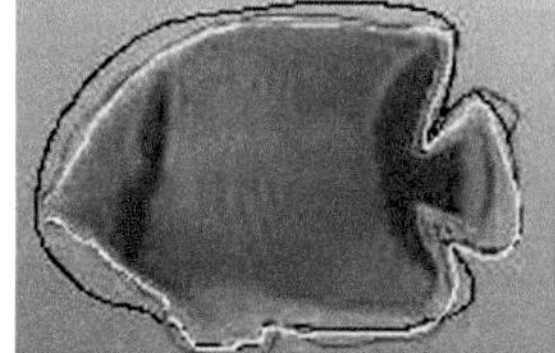

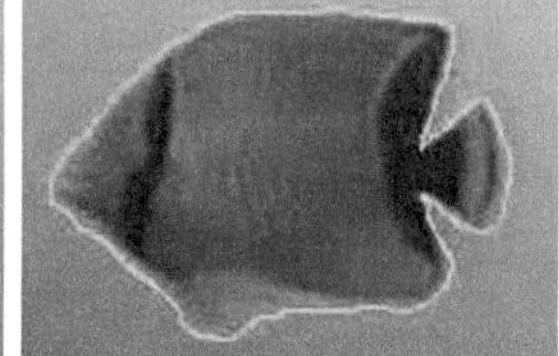

Fig. 14.3 Combination of multiple annotations. On the *left* there are four annotations: three (*black*, *yellow*, *red*) from different users and one (*blue*) belonging to the already existing best ground truth. On the *right*, the "integrated" annotation which is included into the new "best ground truth"

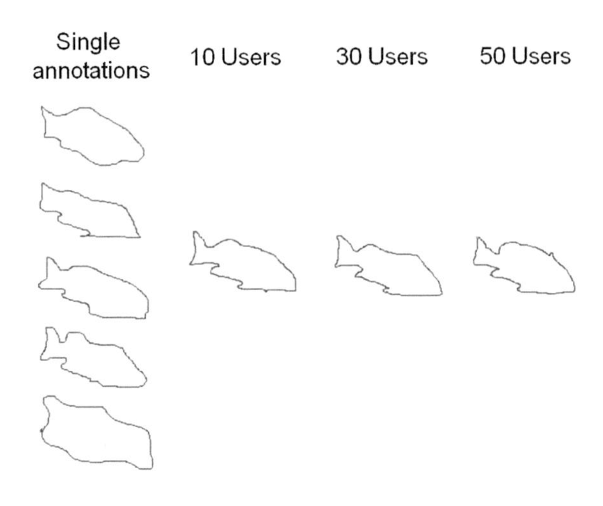

Fig. 14.4 "Integrated" object contours as the number of annotators increases

Finally, each user is assigned a quality score U_{qs} indicating his/her annotation ability, in order to prevent bad annotators from affecting the integrated ground truth:

$$U_{qs} = \frac{1}{N} \sum_{i}^{N_{GT}} q_i n_i, \tag{14.4}$$

where N is the total number of objects that the user has drawn, N_{GT} is the number of the created ground truths, q_i is the quality, computed as the average of the matching scores between all of its annotations and the best ground truth ones, of the ith ground truth and n_i is the number of objects belonging to that ground truth.

14.2.2 Experimental Results

We first evaluate the system performance in terms of the time needed to perform annotation and the accuracy of collected annotations. We asked 50 computer science undergraduates to annotate fish in 200 consecutive frames of 5 different videos (320×240, 5 fps, 10 min long), with PerLA, the GTTool (Kavasidis et al. 2012) and ViPER-GT. The users were asked to complete the task within 2 weeks. The time spent on PerLA was measured automatically. For the GTTool and ViPER-GT, users were asked to accurately take note of the time spent during the whole process. Table 14.1 shows the results. The accuracy of the contours was compared against gold standard ground-truths available for those five videos by calculating the average of the PASCAL score and the Euclidean distance. We evaluated only the performance in annotating contours leaving out the trajectory labeling process (i.e., matching fish in consecutive frames) as it was much easier and users with high quality scores obtained a match of 100 % on the drawn trajectories.

Table 14.1 Comparison between PerLA, the GTTool and ViPER-GT

	PerLA	GTTool	ViPER-GT
Total drawn objects	34,131	43,124	31,409
Manually drawn objects	16,832	14,563	31,409
Automatically drawn objects	17,299	28,561	–
Average time per object	7, 4 s	4, 2 s	11, 2 s
Contour accuracy	89 % (95 %)	90 %	79 %
Learnability	9.1	8.2	3.4
Satisfaction	7.3	7.3	4.3

The number in parenthesis is the accuracy obtained by using the contour integration module (see Fig. 14.5)

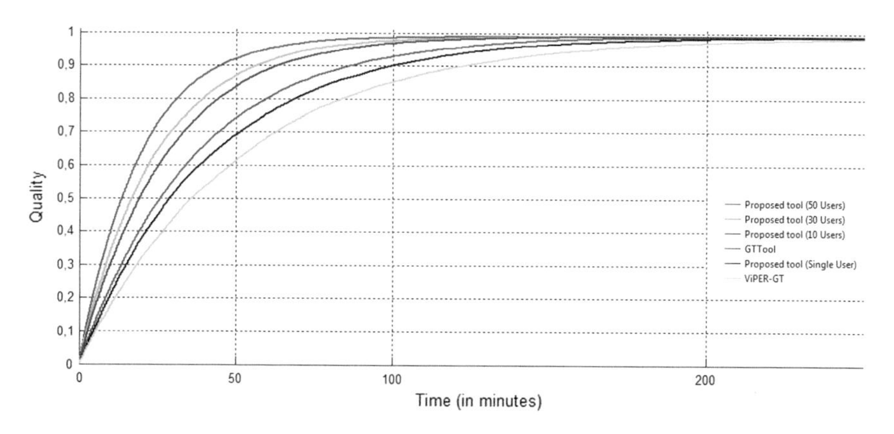

Fig. 14.5 The time needed and the obtained quality, for annotating a video containing 100 objects. In single user cases the graphic represents the time needed by the best performing user. For PerLA, when annotation integration takes place, it represents the time needed for the users to achieve the corresponding quality score, working in parallel (i.e., the time that each user spent)

On average, GTTool needed least time to annotate the videos. This is because GTTools employs automatic object detection (in addition to automatic contour extraction, which is the same as PerLA) to support user annotation, resulting in a large number of automatically drawn objects, as shown in Table 14.1. For the same reason the accuracy of the annotations drawn with GTTool was slightly better than that achieved with the proposed tool, PerLA. ViPER-GT ranked lowest due to its complete manual nature. Note, these results refer to a single user setting and do not include possible advantages that can be exploited by PerLA's multi-user nature.

In order to compare the effort needed to generate high quality ground truth using the aforementioned tools, we measured the time needed to annotate a video containing 100 fish objects. In single user applications, such as ViPER-GT and GTTool, which do not offer any annotation integration method, the time necessary to create a ground truth increases with respect to its quality. Considering that PerLA is devised for collaboration, integrating their annotations gives a significant boost to the quality/effort ratio. Figure 14.5 shows the time needed in order to achieve different quality scores. In the single-user case the best performance was obtained with GTTool: in the best case, it needed 61 min to achieve a ground truth quality of 0.8. When the annotation integration module was used, the same quality was achieved in about 30 min (with 50 users).

Upon annotation completion, a usability questionnaire (Chin et al. 1988) was compiled by the users, in order to obtain feedbacks about the user experience. Users graded the tools in terms of learnability (how easy it is to learn to use the tool) and satisfaction (users' general feeling about the tool usage) in a 1 (worst) to 10 (best) scale. Table 14.1 lists the results. In terms of learnability, PerLA achieves a score of 9.1/10, followed by the GTTool (8.2/10). ViPER-GT ranked third with a very low score (3.4/10) mainly because of its complex interface and the time needed to achieve a satisfactory level of knowledge about its usage. In terms of user satisfaction, GTTool and PerLA arrived at a tie (7.3/10). Users commented that: (1) the GTTool's object detection and tracking algorithms alleviated a large part of the work; while (2) PerLA was easier to use, better organized, and visually more appealing. The worst performer was, again, ViPER-GT because of the total lack of automated tools and its steep learning curve.

14.2.3 Discussion

We have presented a web-based collaborative video annotation tool, which deals with different aspects of the ground truth generation process for object detection and tracking. It allows users to speed-up the generation of high quality annotations due to both the distribution of the workload to multiple users and the reliable integration of multiple annotations. Although the proposed system has proved to be more effective than previous solutions, the process of annotating images/videos at large scale still remains tedious and time-consuming for human operators.

To further support visual data annotation, in a later work (Kavasidis et al. 2013b) we developed an online game relying on user amusement to build large-scale annotations which are then combined by clustering techniques. Evaluation of the annotations generated by gamers and comparison to a hand drawn ground truth (the same used for testing PerLA) confirmed that reliable visual annotations is not necessarily a cumbersome task both in terms of effort and time needed.

14.3 A Cluster-Based Approach to Fish Recognition

14.3.1 Introduction

Typically, the annotation/labeling tasks the annotators are asked to perform in a crowdsourcing setup are simple, that is, little or no expert knowledge is required. However, correctly identifying fish species from a footage requires highly specialized expertise from marine biologists: biologists specialized on the Australian reefs perform not as well as those specialized on the Taiwanese coral reef fish species. Further, since experts are a scarce and expensive resource, it is unlikely that they would provide the amount of image labels needed for the purpose of training and evaluating the fish classification models.

The question is then, can we create a ground truth set of sufficient *quantity* with sufficient *quality* by taking advantage of the collaborative problem solving ability of the crowd, while solving the problem that the crowd generally lacks the domain knowledge required by the task?

A smart way of presenting a problem or decomposing a complicated problem into simpler sub-problems may greatly reduce its difficulty and makes an infeasible task feasible. Typical examples include Foldit (Cooper et al. 2010) that uses a puzzle solving game for protein structure prediction. Another example is Galaxy zoo (Zooniverseteam 2014) that uses "citizens' wisdom" to contribute to morphological classification of galaxies. In this section, we present a clustering-based approach to support the annotation of fish species. The effectiveness of our approach is two-fold. First, it converts the difficult recognition task into a cluster validation task, which reduces the amount of expert knowledge necessary. Second, the annotation process becomes more efficient by using the automatic clustering to select groups of similar images.

14.3.2 Ground-Truth Annotation Using Automatic Clustering

14.3.2.1 Fish Clustering Methods

To support the groundtruth annotation with automatic clustering, a method for clustering, in our case fish images, is necessary. In this research, several methods have been developed: The first method uses sets of color, texture and contour features (respectively the HSV (Hue Saturation Value) space, a Canny edge detector and the Curvature Scale Space representation). For all the individual feature values, we can compute Gaussian Mixture Model (GMM) of, for instance, all color values of a single fish. Using the Kullback–Liebler divergence (*KL divergence*), first introduced by Goldberger et al. (2006) for clustering, a distance between two images (represented by a GMM) is computed. In the second method, we experimented with

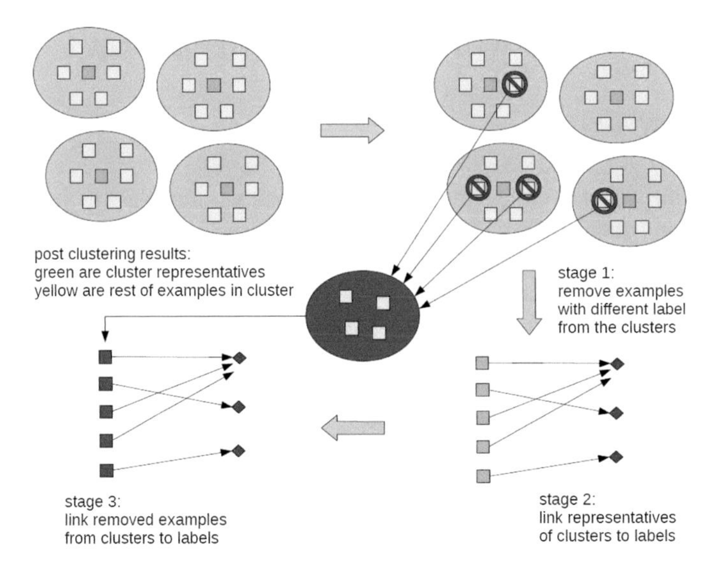

Fig. 14.6 A pictorial representation of our annotation framework, which is supported by a clustering method. In the first stage, images that do not belong to the same class as the representative image are removed from the cluster. In the second stage, the representative images are linked to class labels. A possible third stage is to also link removed images to class labels

pyramid histogram of visual words (dense SIFT features with color information), where the Euclidean distance between normalized histograms is taken as the distance measure. The clusters are computed given all distances between all examples, using Affinity Propagation (Frey and Dueck 2007), which also provides us with a representative image for a cluster. The representative image is important because it allows us to represent the class label of the cluster with a single image. In Boom et al. (2013), instead of using Affinity Propagation (Frey and Dueck 2007) for clustering, we use nearest neighbor methods similar to Locality Sensitive Hashing (Gionis et al. 1999) for large scale database in combination with the GMM and histogram representations.

14.3.2.2 A Cluster-Based Annotation Framework

The clustering results are then used to support our annotation task: annotating images with fish species names. Specifically, an annotator is asked to verify whether each image in a cluster belongs to the same class as the cluster representative. This way, the annotation task is converted from identifying fish species names to judging image similarity, which is relatively easy and requires less domain knowledge. The framework proposed here consists of three stages (shown in Fig. 14.6):

1. Cleaning the clusters (blue ovals in Fig. 14.6): images that are not similar to the representative image (green square) are removed.
2. Merging the clusters: the representative images are linked to labels (shown as purple diamonds).
3. Linking removed ones (shown as red squares): although these images are often the more interesting ones, depending on the size of the dataset relinking them to clusters again can also work.

Here, "cluster" refers to a group of images that are similar according to an automatic clustering algorithm; and "label" refers to a group of images that are judged to belong to the same cluster given by a human annotator (where this group should contain all images in that category). In the Fish4Knowledge project, a label corresponds to all images of fish that belong to a certain species, although the labelers need not know the species name—all that is required is the ability to recognize the visual similarity.

14.3.2.3 Interfaces

The first stage is shown in Fig. 14.7a, where images that do not belong to the same label as the representative cluster image—shown at the top of the screen—are removed. That is, images that are not correctly clustered are removed (including partial and non-fish). After cleaning the clusters, we obtain three types of images: (1) The representative cluster images, (2) the images that belong to the clusters and (3) images that are not part of a cluster.

The second stage is shown in Fig. 14.7b, where the representative images are linked to a label, as well as all other images within the cluster. Further, the automatic clustering methods tend to overcluster (e.g., 156 clusters for 32 labels), and in this stage there is a need to combine some of the clusters into a label. Since it is often

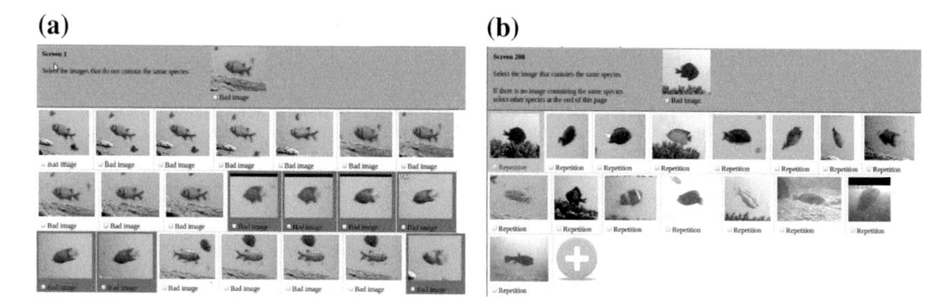

Fig. 14.7 Interfaces. **a** First interface: annotators remove images (by clicking on them) that do not belong to the same label as the representative image shown on the top of the screen. This cleans the clusters. **b** Second interface: annotators link the images in the top row to a label by clicking on one of the gallery images which represent a label. New labels can also be added with a click on the green "plus" button

not known how many labels there are in advance, we allow the annotators to create a new label by pressing the green plus button.

We defined a third stage, where the set of images that were not part of a cluster are further analyzed. In order to compare our work by just labeling the entire dataset one by one, we annotated all the removed images from the clusters using the interface shown in Fig. 14.7b. There are other possibilities, for instance to recluster the images to speedup this stage even more.

14.3.2.4 Combining Multiple Annotators

Annotating images is time-consuming, it is often performed by multiple annotators where these annotators make different errors and/or have different opinions. Following Whitehill et al. (2009), we apply a probabilistic framework to combine annotations from multiple annotators. We have an observed label L_{ij} for each image j of the M images given by each annotator i of the N annotators. For each annotator i, the expertise of this annotator is modeled by the parameter α_i, which gives their accuracy in the labeling task. For each image j, the difficulty of the image is given by the parameter β_j. The groundtruth image label is denoted by Z_j. Using Expectation-Maximization both α_i, β_j and Z_j can be inferred given the observed labels L_{ij}. Further work by Khattak and Salleb-Aouissi (2011) show how to in-cooperate labels of experts, which allows us to compute first estimates of α_i and β_j on a small number of images annotated by experts and interpolate this later on the entire dataset. We extended this work to support also the multi-class problem, allowing us to determine groundtruth labels for all images given multi-class annotations.

14.3.3 Experiment

The proposed framework is used to annotate a dataset of fish images for 32 different fish species, obtained by the Fish4Knowledge project. In this experiment, 6 annotators annotated the dataset using the KL divergence for clustering and 2 of the annotators annotated the dataset again using the Pyramid histogram. A subset of this data is also annotated by marine biologists, which allows us to infer the "gold standard" by using Khattak and Salleb-Aouissi (2011).

In Fig. 14.8a the accuracy at each stage of the annotation framework is shown. Note that the annotations in the first stage are linked to the performance of the clustering methods. In this stage, we focus on the number of images that are correctly and incorrectly clustered. Annotators make more mistakes with incorrectly clustered images compared to the correctly clustered ones. It may be caused by two factors: (i) the images are not scanned very comprehensively, which leads to avoidable mistakes; and (ii) these images are more difficult to be recognized and told apart from the representative images, which may be the reason why the (automatic) clustering

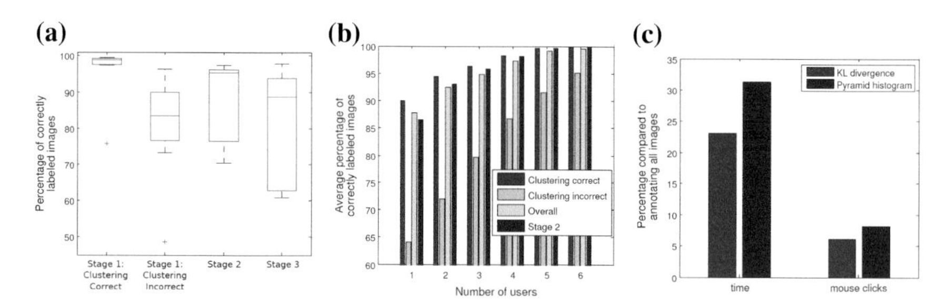

Fig. 14.8 Evaluation. **a** Box plot of the performance in labeling of all annotators for the individual stages. **b** Histogram of the average combined performance of a certain number of annotators. **c** The improvement in both time and mouse clicks over annotating all images

algorithms have made the errors in the first place. In our opinion, the first factor is the more dominant cause for the behavior observed in this figure, although there are some cases where the second factor also plays a role. The stage 2 performance indicates the labeling performance without the support of clustering, as here annotators had to link a single image to a pictorial "label" for each image. From the performance of stage 3, we observed linking the images excluded from stage 1 is more difficult than linking the representative images. This also shows that factor (ii) plays a small role while factor (i) is the dominant factor, because more mistakes are being made on these images, but it is on average not a large difference from stage 2, which shows that in most cases these images can easily be distinguished.

To obtain a groundtruth dataset, we combined the annotations of all the annotators that were annotated with the support of the KL divergence method. Figure 14.8b gives the average performance by combining certain numbers of annotators compare to this groundtruth. This figure clearly shows that there is almost no loss in accuracy due to the clustering, when we compare the "Overall" results with the "Stage 2" results which is a good estimation of the accuracy without clustering. Figure 14.8b also shows the difference between the accuracy on correctly clustered images and that of the incorrectly clustered images. The incorrectly clustered images have a small influence in this case on the performance, because the percentage of incorrectly clustered images for both our clustering methods is not that large (for KL divergence and Pyramid histograms, it is 9.8 % and 16.9 % of the entire dataset, respectively).

Another important measure is the time it takes to annotate the images, where we measured one of our annotators performing non-stop annotation. This give us an estimation of the amount of time it took to complete a screen of the first as well as the second interface, which is 19.7 s and 7.3 s respectively. Figure 14.8c shows a significant improvement in terms of time-based measures based on extrapolation of these values if we compare annotation with the support of clustering to that without the support of clustering (i.e., second interface). In crowdsourcing, annotators are often paid by the number of mouse clicks, where Fig. 14.8c also gives a comparison here. By annotating all M images with the second interface, we would need $2M$ clicks. Using our proposed interfaces, there is only one click if an image needs to be removed from the cluster and an extra to confirm these annotations for each cluster. In the second stage, again two clicks are needed for every screen (one to select the image and on to confirm the decision). Finally, when using KL divergence, we achieve 93 % reduction in mouse clicks.

14.3.4 Discussion

In this section, we presented a framework for annotating images supported by a clustering method. In later work this same framework is also used in combination with a nearest neighbor method (Boom et al. 2013). The quality of the annotation is slightly affected by the interface used and the performance of the automatic clustering method (e.g., 5.1 % error rate compared to 4.2 % error rate when combining 3 annotators). Further, with this method we are able to effectively reduce the time/clicks needed for annotation by two thirds. This framework has been actively used without stage 3 to label a large set of images (91894 fish images) in combination with the nearest neighbor method, making it even more efficient. A lot of images are labeled as "bad images", because of occlusions, low resolution or false positives from the fish detection methods. We have obtained species labels for 28,264 images, which are used as the final ground truth dataset for fish recognition.

14.4 Do You Need Experts in the Crowd? A Case Study in Fish Species Verification

The annotations obtained from the previous sections result in clusters of images, where each cluster corresponds to a fish species. To associate species names to these image clusters, domain knowledge from marine biologists is needed. However, as already discussed, experts are a scarce and expensive resource and are not likely to provide labels in large quantity. To address our problem, we used the expertise of marine biologists to transform the fish identification task into a game based on a visual similarity comparison task that can be performed by a large number of non-experts. We then conducted a user study seeking the answers to the following research questions:

Q1. Can non-expert players of this game achieve acceptable performance evaluated with the labels provided by the experts?
Q2. Can players learn and improve their performance during the game?

We asked experts to label only a small subset of our data as ground truth and developed a cluster-based interface shown in Fig. 14.9a to facilitate their labeling process.

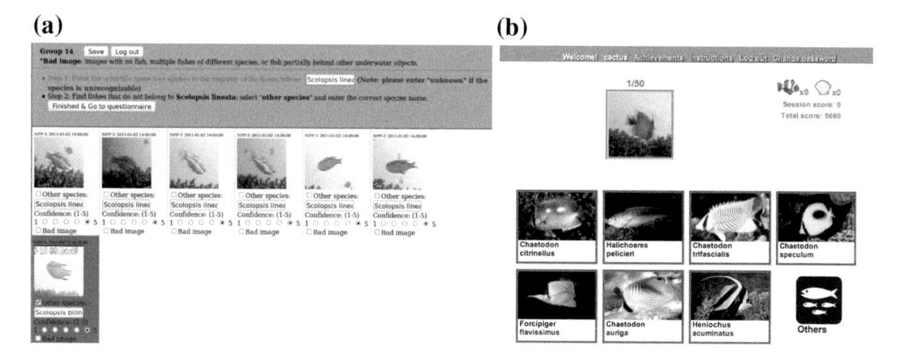

Fig. 14.9 Expert and game interfaces for labeling fish species. **a** Species recognition interface for experts. **b** Game interface for non-experts

To obtain clusters with relatively good quality, two students manually clustered 3,000 images randomly sampled from our video data into 28 clusters. To limit the amount of effort experts need to examine the clusters, at most 30 images are randomly selected from each cluster and shown to the experts. As the size of the clusters is unevenly distributed, we obtained a total of 190 labeled images. Three marine biologists, having a research experience of 30, 10 and 25 years in Taiwanse coral reef fish respectively, were invited to create the ground truth labels.

We make the following observations about the obtained ground truth. (i) Biologists are not always certain about the species of a fish: (a) one of the experts assigned labels such as “A or B” to 3 images, and (b) in 45 cases,[2] a family or higher level label is assigned. In the former case, we consider both labels mentioned; in the latter case, we consider all species under a higher level label as possible target labels. Thus, it is possible that an image has multiple labels assigned by a single expert. In total 288 species and 20 families were mentioned as labels for the 190 images. (ii) Biologists do not always agree. Table 14.2 shows the agreement between biologists in terms of Cohen’s κ,[3] assuming the complete category set consists of all unique species mentioned in the labels provided by the experts. No perfect agreement was achieved, neither at species nor family level. This result suggests that our labeling task is not trivial even for experts.

With the labeling interface presented in Fig. 14.9a, it is very hard, if not impossible, for those who do not have knowledge about coral reef fish species to effectively provide labels. Therefore for non-experts a game-based interface as shown in Fig. 14.9b is developed. Essentially, it follows the idea of converting a fish recognition task to a image similarity comparison task, as discussed in Sect. 14.3.

[2]A case is a {*image, expert label*} pair, thus 190 $\times$ 3 cases in total.

[3]When there exist multiple labels for an image assigned by one expert, we randomly draw one of them to be evaluated; we repeat this process 100 times and report the averaged κ and its standard deviation over the 100 runs. Agreement calculated in this way is rather conservative.

Table 14.2 Cohen's κ for measuring expert agreement

Comparison	Species level		Family level	
	Avg. κ	Sdv.	Avg. κ	Stv.
E1 versus E2	0.55	0.008	0.85	0.004
E1 versus E3	0.48	0.008	0.75	0.000
E2 versus E3	0.67	0.006	0.76	0.0001

The players are asked to compare a *query image*, i.e., the image to be labeled, to a set of *candidate labels*, i.e., textbook images of candidate species. They click a candidate label if they believe that the fish on it and the query image belong to the same species, or "others", if none of the candidates is similar enough to be considered as the correct answer. A feedback score for the chosen label is provided. Ideally, players can learn from the feedback and improve their performance.

To encourage players to achieve higher scores and to play more sessions, we show the top 10 scorers on a leading board.

14.4.1 Experiments

During the manual clustering stage, we found that 53 out of the 190 images were assigned to "wrong" clusters. That is, there exist many fish that look similar but belong to different species. Thus, RQ 1 boils down to

Can non-experts distinguish between similar species when examples of these species are displayed next to each other? To answer this question, we conduct two experiments that simulate two situations.

Experiment 1. We first assume an ideal situation, where the *target label(s)* (labels suggested by the biologists) of the query image is always among the candidates. The primary goal of the experiment is to investigate whether the players can identify the target label when there exist very similar species.

We select candidates that are similar to the target labels as follows. Let $c = \{i_n\}_{n=1}^N$ be a cluster containing N query images, and $f(i)$ maps an image to one of the species $S = \{s_m\}_{m=1}^M$. We compute a relevance score between an image $i \in c$ and a species as $\text{score}(i, s) = \text{count}(\{f(j) = s, j \in c\})/N$. All species with a non-zero score are the ones that were clustered together, which means that they are visually similar. To avoid overloading the players with too many candidates, we limit the number of candidates to 7. Random images are chosen when less than 7 candidates are available, and we make sure that the target labels are in the candidates.

Experiment 2. We then considered a more realistic situation when some target labels are not in the candidates. In practice, we do not have information about the target labels of the query images. We need to select candidates based on certain similarity measures computed with automatic methods, which are most likely imperfect. It is then important to know whether the non-expert players can still make the right choice, that is, select “others”, when similar species are displayed as candidates. We use the same setting as in Expr.1 to select candidates and deliberately remove the target labels from the candidates for a set of randomly selected query images. Note, if too many target labels are removed, users may expect that “others” is always the safe bet when they are not sure. With a few trial runs, we decided to remove the target labels for 25 % of the query images.

Settings of system feedback. Players may be able to improve their performance for different reasons: learning from the system feedback, getting used to the quality of the images, etc. Here we only consider the simplest and ideal system feedback, i.e., we assign scores to each click on an candidate label based on the biologists’ voting. Since experts do not always agree, a click on an option can receive 0, 1, 2, or 3 points. In practice when expert labels are not available, other types of feedback should be used, e.g., peer-agreement, automatic similarity measures.

Aggregation of obtained labels. We apply majority voting to aggregate the labels from multiple players into a single label for evaluation. Since experts may have assigned multiple labels to an image, we do not simply take the winner of the majority voting as the chosen label, but rank the candidates in descending order of their votes. In Expr. 2 we ignore the cases where target labels are not displayed and “others” are *correctly* chosen, as they do not provide information about which label should be assigned to the image, hence neither hurt or help the performance.

14.4.1.1 Evaluation

Quality of non-expert labels. We use Cohen’s κ to measure the agreement between the aggregated non-expert labels and each of the three experts. We compare these to the pairwise agreement among the experts. The top 1 candidate is chosen as a result of majority voting. Further, Normalized Discounted Cumulative Gain (NDCG) (Järvelin and Kekäläinen 2002) is used as it provides an intuitive interpretation of the correctness of the labels. For a query image, given the biologists’ voting, each candidate can be rated as 0, 1, 2, or 3. The ranked list of candidates as a result of majority voting is then evaluated using these graded expert judgements.

Learning behavior of non-expert users. We study users’ performance over time in terms of (1) memorization: when an image is shown again; and (2) generalization: when an unseen image that belongs to a seen species is shown.

We measure the performance of a single label as follows. Let $L = \{l_k\}_{k=1}^{K}$ be the candidate labels for an image, $J(l) = \{0, 1\}$ be a player’s judgement, and $E(l) = \{0, 1, 2, 3\}$ be the expert votes of label l for the image. The performance of a judgement is defined as experts’ votes for the chosen candidate normalized by the maximum votes one can achieve for the set of candidates: $s = \frac{\sum_{l \in L} J(l) \cdot E(l)}{\max_{l \in L} E(l)}$.

Since scores achieved at a certain time point can be sensitive to players’ random errors, we smooth the score at each time point with the scores achieved so far:

Table 14.3 Agreement between experts and non-experts

		E1		E2		E3	
		Avg. κ	Sdv.	Avg. κ	Sdv.	Avg. κ	Sdv.
Expr.1	Species	0.62	0.01	0.65	0.006	0.55	0.009
	Family	0.83	0.008	0.81	0.01	0.72	0.009
Expr.2 (New)	Species	0.65	0.009	0.50	0.008	0.45	0.009
	Family	0.73	0.01	0.73	0.01	0.68	0.01
Expr.2 (Old)	Species	0.53	0.01	0.68	0.01	0.64	0.02
	Family	0.80	0.02	0.78	0.02	0.74	0.01

$s_t = \sum_{i=1}^{t} s_i / t$. t refers to the tth time a player labels the same image (memorization), or a different image in the same species (generalization).

In a session, the first 12 images are randomly selected without repetition. After that, with a probability of 0.5 an image is selected from those that were already labeled in the current session. As images are selected randomly, the repetition of images (memorization) or species (generalization) do not happen the same number of times. In order to conduct reliable statistical testing for comparison (see Sect. 14.4.2), we consider repetitions of images/species that have more than 30 cases.[4] Specifically, we consider $\leq$ 4 repetitions of images for both experiments; $\leq$ 25 repetitions of species for Expr. 1, and $\leq$ 10 for Expr. 2. As fewer sessions were played in Expr. 2, less repetitions are available.

14.4.2 Results and Discussion

Data obtained. We collect labels for the 190 images labeled by the experts. Twenty two players contributed 72 sessions in Expr. 1 and 32 players contributed 49 sessions in Expr. 2. On average each image received 19 and 13 labels, respectively. Our users have a diverse background and age groups, including school age children and university students and researchers.

Performance of non-experts. Table 14.3 shows the result of label agreement at both species and family level. In terms of Expr.1, we see that the agreement between expert and non-expert labels are rather similar to that among the experts themselves (see Table 14.2). In terms of Expr.2, we see that the "new" players (those who only participated in Expr.2) achieve relatively lower agreements with experts compared to players in Expr.1 (compared to 2 of the 3 experts – however, they achieved higher agreement with Expert 1 who has a lower agreement with the other two experts as shown in Table 14.2.) while the performance of "old" players is comparable to that

[4] A case is a $\{image(species), user\}$ pair.

Table 14.4 Non-experts' performance evaluated by NDCG.

Method	Species		Family	
	NDCG@1	NDCG@5	NDCG@1	NDCG@5
Expr.1	0.84	0.88	0.93	0.94
Expr.2.new	0.72▼	0.77▼	0.86▼	0.94
Expr.2.old	0.88	0.86	0.91	0.94

Table 14.5 Comparing the performance in the first sessions under Expr. 1 and 2. Only "new" players are considered

Method	Species		Family	
	NDCG@1	NDCG@5	NDCG@1	NDCG@5
Expr.1	0.84	0.88	0.93	0.94
Expr.2	0.72▼	0.77▼	0.86▼	0.94

of Expr.1. This to some extent suggests that although the experimental condition has changed, the "training" the players received during Expr.1 has an influence on their performance in Expr.2.

Further, Table 14.4 shows the performance of non-expert labels in terms of NDCG. In practice, when using the collected labels as training data, often only the label(s) with the highest scores are considered as target labels. Thus it is important that the top ranked labels are correct according to experts' labels. We list the results of NDCG@1 and 5. The new players in Expr.2 have a significantly lower performance compared to Expr.1, while the performance of "old" users do not show a significant difference compared to that achieved in Expr.1. (▲▼) indicates a significant difference between player performance in the two experiments (p-value$<$ 0.01) tested using Wilcoxon signed-rank test.) We consider two potential explanations: (1) the set up of Expr.2 makes a more difficult task for novice players; or (2) since most of the new players did only one session, the general quality of the labels is not as good as that of Expr.1, where many played more than one session. To distinguish the two cases, we verify if the results from only the first session of each player in Expr.1 still outperforms that of Expr.2. In Table 14.5 we see that indeed, a significant difference exists between the performance of the first session labels in the two experiments. That is, when target labels are absent while similar non-target labels are present, novice players are more likely to be confused. This suggests that selecting a good set of candidate labels is important.

Do non-experts learn? Table 14.6 shows the comparison of the averaged scores achieved at the first label for an image to that of the nth labels. These numbers confirm that there is a significant difference between the scores achieved with the first label and those achieved over time, in both experiments non-experts can learn and improve their labels over time. They do not only learn to provide more accurate

Table 14.6 The impact of learning over time

Memorizing					Generalization					
Labels	1	2	3	4	1	5	10	15	20	25
Expr.1	0.30	0.38▲	0.46▲	0.51▲	0.42	0.51▲	0.59▲	0.63▲	0.67▲	0.70▲
Expr.2.new	0.30	0.40▲	0.44▲	0.52▲	0.37	0.58▲	0.62▲	–	–	–

Wilcoxon rank-sum test is used for significance testing. All comparisons are between the first label and the nth label

labels for images that they have seen before, but also for similar images, i.e., different images that contain species that they have seen before.

14.5 Conclusion

In this chapter, we discussed the methods and tools we have developed to obtain ground truth data that are used for training and evaluating various components of the F4K system. For fish detection and tracking, we have developed PerLA, a collaborative object detection and tracking annotation platform, and demonstrated how aggregating multiple labels improves annotation accuracy. For fish recognition, we developed a clustering-based approach that allows non-expert annotators to perform the annotation task that requires highly specialized knowledge. We have demonstrated that, with such a transformation, the agreement between lay annotators and expert annotators are comparable to that between expert annotators. Further, novice lay annotators can learn and improve their performance over time.

References

Boom, B.J., P.X. Huang, and R.B. Fisher. 2013. Approximate nearest neighbor search to support manual image annotation of large domain-specific datasets. In *Proceedings of the international workshop on video and image ground truth in computer vision applications, VIGTA '13*, 4:1–4:8. New York: ACM.

Boom, B., P. Huang, J. He, and R.B. Fisher. 2012. Supporting ground-truth annotation of image datasets using clustering. In *Proceedings of the 21st International Conference on Pattern Recognition (ICPR)*, 1542–1545. IEEE.

Brabham, D.C. 2008. Crowdsourcing as a model for problem solving an introduction and cases. *Convergence: the international journal of research into new media technologies* 14(1): 75–90.

Chen, Y.-Y., W.H. Hsu, and H.-Y.M. Liao. 2011. Learning facial attributes by crowdsourcing in social media. In *WWW'11*, 25–26.

Chin, J.P., V.A. Diehl, and K.L. Norman. 1988. Development of an instrument measuring user satisfaction of the human-computer interface. In *Proceedings of the SIGCHI conference on Human factors in computing systems*, 213–218. ACM.

Cooper, S., F. Khatib, A. Treuille, J. Barbero, J. Lee, M. Beene, A. Leaver-Fay, D. Baker, and Z.P. Foldit players. 2010. Predicting protein structures with a multiplayer online game. *Nature* 756–760.

Frey, B.J., and D. Dueck. 2007. Clustering by passing messages between data points. *Science* 315(5814): 972–976.

Gionis, A., P. Indyk, R. Motwani, et al. 1999. Similarity search in high dimensions via hashing. In *In VLDB*, vol. 99, 518–529.

Goldberger, J., S. Gordon, and H. Greenspan. 2006. Unsupervised image-set clustering using an information theoretic framework. *IEEE Transactions on Image Processing* 15(2): 449–458.

He, J., J. van Ossenbruggen, and A.P. de Vries. 2013a. Do you need experts in the crowd?: A case study in image annotation for marine biology. In *OAIR'13*, 57–60.

He, J., J. van Ossenbruggen, and A.P. de Vries. 2013b. Fish4label: accomplishing an expert task without expert knowledge. In *OAIR'13*, 211–212.

Howe, J. 2006. The rise of crowdsourcing. *Wired magazine* 14(6): 1–4.

Järvelin, K., and J. Kekäläinen. 2002. Cumulated gain-based evaluation of IR techniques. *ACM Transactions on Information Systems* 20(4): 422–446.

Kass, M., A. Witkin, and D. Terzopoulos. 1988. Snakes: Active contour models. *International journal of computer vision* 1(4): 321–331.

Kavasidis, I., S. Palazzo, R. Di Salvo, D. Giordano, and C. Spampinato. 2012. A semi-automatic tool for detection and tracking ground truth generation in videos. *Proceedings of the 1st international workshop on visual interfaces for ground truth collection in computer vision applications, VIGTA '12*, 6:1–6:5.

Kavasidis, I., S. Palazzo, R. Di Salvo, D. Giordano, and C. Spampinato. 2013a. An innovative web-based collaborative platform for video annotation. *Multimedia Tools and Applications* 1–20.

Kavasidis, I., C. Spampinato, and D. Giordano. 2013b. Generation of ground truth for object detection while playing an online game: Productive gaming or recreational working? In *IEEE conference on computer vision and pattern recognition workshops (CVPRW)*, 694–699. IEEE.

Khattak, F.K., and A. Salleb-Aouissi. 2011. Quality control of crowd labeling through expert evaluation. In *Second workshop on computational social science and the wisdom of crowds (NIPS 2011)*, 1–5.

Rother, C., V. Kolmogorov, and A. Blake. 2004. GrabCut: interactive foreground extraction using iterated graph cuts. *ACM Transactions on Graphics* 23(3): 309–314.

Russell, B.C., A. Torralba, K.P. Murphy, and W.T. Freeman. 2008. Labelme: A database and web-based tool for image annotation. *International Journal of Computer Vision* 77(1–3): 157–173.

Spampinato, C., B. Boom, and J. He (eds.). 2012a. *VIGTA*.

von Ahn, L., R. Liu, and M. Blum. 2006. Peekaboom: a game for locating objects in images. In *CHI '06*, 55–64.

von Ahn, L., and L. Dabbish. 2004. Labeling images with a computer game. In *CHI'04*, 319–326.

Whitehill, J., P. Ruvolo, T. Wu, J. Bergsma, and J.R. Movellan. 2009. Whose vote should count more: Optimal integration of labels from labelers of unknown expertise. In *NIPS*, 2035–2043.

Yuen, J., B.C. Russell, C. Liu, and A. Torralba. 2009. Labelme video: Building a video database with human annotations. In *ICCV*, 1451.

Zooniverse_team (2014). Galaxy zoo. http://www.galaxyzoo.org/. Accessed 16 Nov 2014.

Chapter 15
Counting on Uncertainty: Obtaining Fish Counts from Machine Learning Decisions

Bastiaan J. Boom

Abstract Most questions in the Fish4Knowledge project are related to the ability to count fish, fish species or events using video analysis Chap. 2. Automatic video analysis however brings uncertainty due to False Positive/False Negative classifications, which makes determining the counts based on automatic video analysis difficult. Automatic video analysis software is often able to express a measure of uncertainty for a single detection/recognition using a similarity score, indicating how certain the software is about a single decision. Logistic Regression allows us to combine these similarity scores to compute an estimated count, outperforming the estimates of the underlying machine learning methods used for the original video analysis. We show this works both for the two-class and multi-class problem. Furthermore, we identify potential pitfalls where ensuring a correct sampling procedure is essential. The error in the estimated species counts using Logistic Regression was on average around 15 on a set of 11,585 fish images, while the machine methods had an average error of 800. The key to understanding this huge improvement in accuracy is that we use Logistic Regression over the individual classifications to estimate the size of the whole population, rather than the species classification of any individual observation.

15.1 Introduction

A method is developed to better estimate the abundance of different fish species based on Machine Learning Decisions that contain uncertainty. Machine Learning methods often provide a measure of uncertainty using a similarity score. This similarity score indicates how close the analyzed data example is to the object it would like to recognize. This similarity score could be a probability, percentage or distance in some feature space. Our question is whether given a subset of samples (in our case images) for which we know the ground truth labels, obtain an accurate count of the instances in the entire set (without necessarily knowing precisely which individual

B.J. Boom (✉)
Cyclomedia, Van Voordenpark 1b, 5301 KP Zaltbommel, The Netherlands
e-mail: bas.boom12@gmail.com

R.B. Fisher et al. (eds.), *Fish4Knowledge: Collecting and Analyzing Massive Coral Reef Fish Video Data*, Intelligent Systems Reference Library 104,
DOI 10.1007/978-3-319-30208-9_15

images are included in the set). For marine ecologists who rely on automated methods for their observations, this is an important question, because the automatic video analysis software uses machine learning methods together with similarity scores to decide whether or not a detected object is a fish and the species the fish belongs to. Automatic video analysis software is likely to bias the final counts, except in the unlikely event that the number of False Positives (FP) equals that of the False Negatives (FN). Here we present a method to produce more accurate estimates of counts. Given the large number of fish (11,585 image), it is very difficult to count the fish manually, so an automatic estimate has to be computed. Instead of using a hard decision based on a threshold which is normally done to make a decision, a soft decision is used based on the similarity score. This will give more accurate counts and improve the overall estimate based on Machine Learning methods.

15.2 Related Work

Several methods have been developed to automatically estimate counts of objects, mostly from image data. These methods work on three different levels in the classification process: feature, score and decision level (Lip and Ramli 2012).

To automatically estimate land cover categories in satellite data, improvements based on the confusion matrix are suggested by Hay (1988) and Jupp (1989). Confusion matrices already contain the final decisions of the machine learning methods, where these methods are designed to correct the under- and over-estimates on satellite data.

In microscopic imaging, counting cells can be important for research. Also, in video surveillance, counting people in crowds can be challenging. Both Lempitsky and Zisserman (2010) and Chan et al. (2008) present methods to extract image features and perform regression using those features to achieve highly accurate counts at the feature level, without know anything about individual cells or persons.

Scientists can be interested in both finding individual cells or persons, which requires a separate approach if you use feature level counting methods. Although there are machine learning methods that can accurately find cells or persons, feature level approaches can not use the output of these kind of methods for the estimation.

Work that discusses the problem of estimating the bias of a Machine Learning method is Vucetic and Obradovic (2001), which performs corrections based on estimates of classifier decisions. Saerens et al. (2002) focused on the same problem, computing a new a priori distribution of a dataset based on the features. Unlike all these decision- and feature-level approaches, we propose a simple method that uses similarity scores to count items. This has the benefit of being able to use existing Machine Learning methods, while they should in theory be able to give better estimate than hard decision by using more information.

15.3 Method for Estimation of Counts Based on Similarity Scores of a Classifier

15.3.1 Sampling Strategy

To estimate the counts of the entire dataset based on a subset of the data, the sampling strategy needs to be random thus creating a true representation of the entire set. In this work, we use the following sampling strategy: For the entire dataset, we select two representative subsets (for instance using random sampling) and obtain ground truth information for these subsets by manual annotation. This gives a *training set*, a *validation set* and the remainder of the dataset which we denote as the *test set*. The *training set* is used to train a classifier (using any Machine Learning method) that gives a label based on low level feature data. This subset does not have necessarily have to be a representative subset of the entire dataset. The *validation set* is used to verify the performance of the classifier on untrained examples. Based on the performance of the classifier given by the scores and the ground truth of the *validation set*, our method estimates the counts in the remaining dataset. The *validation set* therefore needs to be representative of the entire dataset (i.e., random sampling). The *test set* is the remaining unknown dataset of which we would like to estimate the number of class labels. If the ground truth labels or statistics of the *test set* are known, we can use this set to verify our method.

15.3.2 Normal Classification Process

The normal process of determining the counts of the entire dataset is by training a classifier on the *training set*. Afterwards the *validation set* is used to determine an optimal threshold t based on the similarity score $s_{i,c}$ of every example i of each class label c. The threshold often already depends on the classifier used, where for instance for the likelihood ratio and Adaboost (Machine Learning method) the threshold $t = 0$ is normal, while in other cases, such as with a distance, this choice is less obvious. A Receiver Operator Characteristic (ROC) or Precision/Recall curves can be used to decide the optimal threshold t to separate the different classes given type I and II errors (false positives and false negatives). The effects of thresholding on the final counts is ambiguous, where an optimal threshold for the *validation set* can bias the counts in the *test set*. Using the similarity score instead of a decision based on a threshold will give better results because a soft threshold can be used instead of a hard threshold.

For a multiclass problem, the choice of the threshold is even less obvious, given that we obtained a similarity score $s_{i,c}$ for every class. Using the threshold t for a certain item i either multiple class-scores are above the threshold or none of the scores are above this threshold. The most common fix in this case is to use the maximum score for a certain item i to provide the final decision. However, by using the maximum score, valuable information can be lost, for instance, if two class-scores of an item were very high compared with the rest of the class-scores, this indicates that the item belongs to one of those classes. By making a decision, this information is lost, while with the scores, this information can be used later in the decision making process which is the obvious advantage of a soft decision.

15.3.3 Estimating Counts Based on Logistic Regression

Estimating the counts based on the similarity scores of a classifier (a Machine Learning method such as Adaboost or SVM) can be done using Logistic Regression. Logistic Regression computes a probability of correct or incorrect classification based on an increasing function, which is given by the Machine Learning scores. This probabilistic function can be fitted based on the similarity scores and the ground truth class-labels of the validation set. This technique is very similar to Platt scaling (Platt 1999) except that we assume similarity scores as input instead of adding the classifiers function into the equations. The advantage of our method is that we can learn on a different validation set instead of assuming that the training set is representative in class distribution to the entire underlying database. Often, to achieve better recognition performance, having a balance training set is better which might not be the case for the underlying problem. First, we explain the two-class problem and afterwards we introduce the multi-class problem.

15.3.3.1 Two Class Problem

The probability $P(y_i|s_i)$ that the item i of N belongs to a given label $y_{i,c} = \{0, 1\}$ depends on the similarity score s_i. In this case, the expected count is given by $E_y = \sum_{i=1}^{N} P(y_i|s_i)$, where N is the total number of items. Logistic Regression allows us to calculate a relation between the score and the probabilities given a validation set. This relation (Logistic Regression) is given by this equation:

$$P(y_i|s_i) = \frac{1}{1 + e^{-(\beta_0 + \beta_1 s_i)}} \tag{15.1}$$

The parameters β_0, β_1 are calculated based on the validation set. This calculation is available in most statistics packages, where we use matlab function β = *glmfit(S, Y, 'binomial', 'link', 'logit')*, in this function S is a matrix/vector of score, Y is the vector of labels and β is the parameter vector. In Eq. (15.1), we assume that the *validation set* is a good representative set for the *test set*, because we selected the *validation set* randomly for the *test set*. Equation (15.1) with the parameters β_0, β_1 from the validation set allow to transform the unknown score in the test set into probabilities which by summing give an expected count of the items belonging to that class.

15.3.3.2 Multiclass Problem

For the multiclass problem, the equations are very similar to Eq. (15.1), but instead of having a single label, we have multiple labels. Instead of using $y = \{0, 1, 2, 3, 4, M\}$, we change this problem to a two-class problem, where we have for each class and each item a binary label $y_{i,c} = \{0, 1\}$, $y = 3$ now becomes $y_{i,3} = 1$. In the multiclass problem, instead of having a single score s_i, we have for each class a score $s_{i,c}$. At first this might be counterintuitive that for a two class problem we have a single score, while for a three class problem we have three scores. The meaning of the similarity score is to indicate the similarity towards the choice of a certain class in the dataset. For a two class problem, having two scores makes no sense because the second score will be the inverse of the first score. Given that we have the ground truth labels $y_{i,c} = \{0, 1\}$ and score $s_{i,c}$, as discussed before we can see this as a two-class problem and use only the score for that class in Eq. (15.1). We argue however that using the scores of all classes to obtain this probability will give more information and thus better results:

$$P(y_{i,c}|s_{i,\zeta}) = \frac{1}{1 + e^{-(\beta_{0,c}+\beta_{1,c}s_{i,c_1}+...\beta_{M,c}s_{i,c_M})}} \tag{15.2}$$

In this case, the probability given the scores of all classes in the set $\zeta = \{c_1, ..., c_M\}$ is given by this formula, where for each class, given the ground truth, we need to compute the parameter β_c with Logistic Regression. Thus for each class, we compute using the validation set the logistic regression parameters based on all the similarity scores, which will give us for each item i thus M probabilities (for each class a single probability giving the chance it belongs to that class). A problem with this approach is that we cannot guarantee that the probabilities for a single item sum to 1. It is also possible to use multinomial logistic regression to alleviate this. However given the large amount of data, we found that this approach would often not give an estimate of the parameters β for the multinomial case in a reasonable time frame (under a week to estimate β). In our case, to obtain the final count of a class, we sum the probabilities of that class: $E_{y,c} = \sum_{i=1}^{N} P(y_{i,c}|s_{i,c})$. In our explanation, we mentioned the Logistic Regression which uses the "logit" kernel.[1]

15.3.4 Limitation in Estimations

There are several limitations in the use of this approach which relate to the **Classifier** and the **Dataset**:

- **Classifier**: The better the classifier score represents how certain the classifier is about the decision, the better the performance of the Logistic Regression estimation

[1] In our experiments, we observed that both the logit kernel, described here, and the probit kernel (Bliss 1934) obtain almost similar results.

of counts. In this case, we expect a increasing function where if the similarity score is higher, the probability that an item indeed belongs to that class becomes also higher. Problems will arise if all positive decisions get a single score 1 and negative decisions get a single score 0 (or if the ranges given by the classifier are very limited), because in that case Logistic Regression will just give positive decisions a single probability $P(y_i = 1|s_i = 1)$ and for negative decisions another single probability, which depends on the validation set. This can however still outperform the guess made by the classifier, but will be less accurate than similarity scores based on an continuously increasing function. On the other hand, a classifier should perform better than random guessing and the better it performs the better the estimates of Logistic Regression are. If a classifier performs almost perfectly, do not expect much better results from Logistic Regression, since that is just a statistical sampling method which is linked to the classifier's performance.
- **Dataset**: The sampling procedure described in Sect. 15.3.1 is important because if the *validation set* is not a representative sample of the *test set* this procedure has no guarantee to work. If there is any doubt on this issue, a possible way to investigate this is using a statistical significance test for Logistic Regression, where for instance the Hosmer–Lemeshow test (Lemeshow and Hosmer 1982) can be used. Another problem arises if the *validation set* is too small to represent the test set. For the multiclass problem, an interesting issue is missing classes that do not exist in the *training set* and the *validation set*, but are present in the *test set*. In this case, it might be possible to make a classifier for unknown classes.

In theory, there are multiple problems that need to be addressed in an implementation of this method. In practice, we found that the estimate counts are often close to the true counts although the experimental setup was not entirely ideal. For instance in the fish count experiment in the coming section, only for the top 3 classes score are calculated, but still good estimates are obtained.

15.4 Counting Fish with Logistic Regression

15.4.1 Experimental Datasets for Counting Fish

In the Fish4Knowledge dataset, there are often large imbalances between the different classes. This is still a significant problem for Machine Learning methods because it often influences the performance of these methods. Our work in Chap. 11 discusses machine learning methods (hierarchical SVM classifier) that are able to deal with unbalanced datasets. In this chapter, we build on that work to count fish species using the hierarchical SVM classifier described in Huang et al. (2014).

For this experiment, new ground truth images are selected and annotated randomly from the subset of videos. These images are different for the set of images used for training and testing the original recognition methods. The main reason is that for the images in the training and test set, we focused on finding more images of rare species, which improves the recognition performance but does not give a true representation

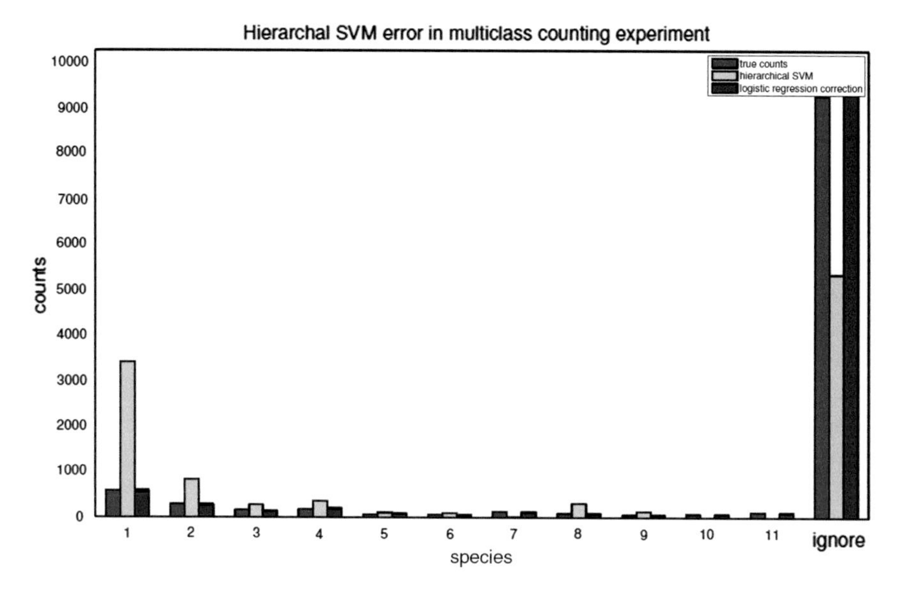

Fig. 15.1 This histogram shows the fish species count, for the ground truth and the estimates based on the classifier and logistic regression. The differences between the true counts and the correction based on Logistic Regression are very small, where for some classes the hierarchical classifier is very biased

of the species in the complete dataset. In this experiment, a set of 11,585 images of 11 species randomly sampled from the Fish4Knowledge datasest was labeled and split into a *validation set* and *test set*, where the set described in Huang et al. (2014) was used for training this classifier, which consisted of 23 species. The histogram in Fig. 15.1 gives the distribution of the fish species in this dataset. We randomly split the set of 11,585 images 20 times based on the videos where the fish were visible, which allows us to observe the statistical parameters like the standard deviation. On the validation set, we computed the logistic regression parameters β, where for each class c, we obtain $c + 1$ parameters β, one for each score together with β_0. For each item i, we obtain c probabilities $P(y_{i,c}|s_{i,\zeta})$ using Logistic Regression based on all the scores. The sum of the probabilities of all items for a single class c gives the final count estimated with logistic regression. In our experiment, we compare the estimated numbers given by the hierarchical SVM classifier (described in Huang et al. 2014) with the estimate number given by Logistic Regression in *test set*.

15.4.2 Results of Counting Fish with and Without Logistic Regression

In Fig. 15.1, the true counts are shown together with estimates given by the Hierarchical SVM classifier and Logistic Regression, with the scores of the Hierarchical SVM classifier. It is clearly visible that for some classes (1, 2, 3, 4, 8), the hierarchical SVM classifier has a large bias, where Logistic Regression is able almost completely correct for this bias. Notice that there is a class "ignore", which are the false positives from the fish detection stages. The classifier developed for the fish recognition is already designed to filter out part of the false positives, however this method shows that further improvement can be achieved. Another problem is that false positives did not get a similarity score how likely the classifier found it to be a false positive or a fish but gave a direct decision. We replaced this decision with the score 0 or 1, hoping Logistic Regression could use the other similarity scores of other classes to determine false positives. Logistic Regression is able to make a more educated decision based on these similarity scores, which while not the ideal in this situation still outperforms the existing solution. Figure 15.2 shows the relative error in fish counts w.r.t. the ground truth, where the axes between the Hierarchical SVM classifier and Logistic Regression are completely different. Logistic Regression does not perform perfectly, but is able to reduce the error significantly.

While Logistic Regression gives better counts, it does not have to improve the performance of the individual classifications, but gives more realistic estimates for

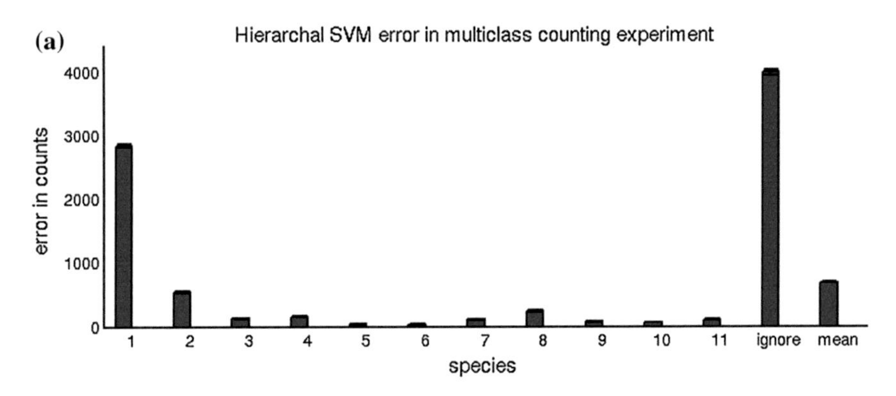

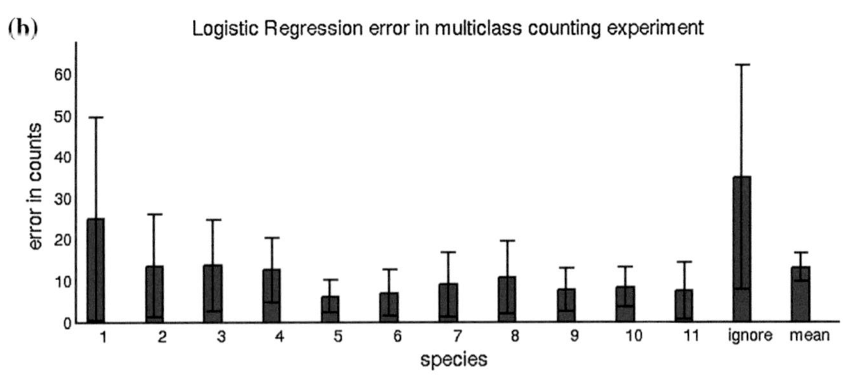

Fig. 15.2 This histogram shows the error between the true counts and the estimates for **a** Hierarchical SVM classifier and **b** Logistic regression. A different scale for the y-axes is necessary to show the error and standard deviation of the Logistic regression estimation

the entire set by removing biases towards certain classes. We have observed similar behavior with different datasets and different classifiers. If the purpose is correct estimation of the quantity of objects present, this seems to be more reliable than relying only on a classifier for this analysis.

15.5 Conclusion

One of the most important questions of marine biologists and ecologists to answer for the Fish4Knowledge system is the abundance of fish species at any given time. We observed that Machine Learning methods are able to determine with an high accuracy the correct species, however often are biased in the final count due to False Positive/False Negative which are unevenly distributed. This method is able to correct for these effects by using logistic regression, allowing an more accurate answer to one of the main questions the marine biologists and ecologists had for this system (What is the abundance of fish species?).

15.6 Discussion

Because this method was developed in the latter stages of the project. We observed that not all classifiers give a nice similarity score, where, for instance, in hierarchical classifiers this is often difficult. This makes it also difficult to use the full potential of an approach like Logistic Regression to correct for the uncertainty, which can be used for both Fish Detection and Fish Recognition.

It is challenging to convince end-users that logistic regression can give better counts than the software used to analyze the data. In the analysis software, users can often directly look at the mistakes made by the software, while Logistic Regression works on a higher level, where users have to verify large counts of items, which is often not possible. In Marine Biology/Ecology, statistical tools are becoming more important, so end-users have worked with Regression or even Logistic Regression methods before, which might improve their confidence and trust in this methodology.

More research can be performed on Logistic Regression in combination with Machine Learning methods. Questions that should be answered in the future are: How many examples are necessary for the validation dataset to get sufficient estimates per class? What is the influence of the Machine Learning methods on the Logistic Regression estimates? How to combine multiple decisions of Machine Learning methods, like the detection and recognition of fish which obviously depend on each other? Is there a theoretical limit to the improvements?

References

Bliss, C. 1934. The method of probits. *Science* 79(2037): 38–39.

Chan, A. B., Z.-S. Liang., and N. Vasconcelos. 2008. Privacy preserving crowd monitoring: Counting people without people models or tracking. In *IEEE conference on computer vision and pattern recognition, 2008. CVPR 2008*, 1–7. New York: IEEE.

Hay, A. 1988. The derivation of global estimates from a confusion matrix. *International Journal of Remote Sensing* 9(8): 1395–1398.

Huang, P.X., B.J. Boom, and R.B. Fisher. 2014. GMM improves the reject option in hierarchical classification for fish recognition. In *Proceedings of the workshop on applications of computer vision* 2014: 371–376.

Jupp, D.L.B. 1989. The stability of global estimates from confusion matrices. *International Journal of Remote Sensing* 10(9): 1563–1569.

Lemeshow, S., and D.W. Hosmer. 1982. A review of goodness of fit statistics for use in the development of logistic regression models. *American Journal of Epidemiology* 115(1): 92–106.

Lempitsky, V. S., and A. Zisserman. 2010. Learning to count objects in images. In *NIPS*, vol. 1, p. 2.

Lip, C., and D. Ramli. 2012. Comparative study on feature, score and decision level fusion schemes for robust multibiometric systems. In *Frontiers in computer education*, vol. 133, ed. S. Sambath, and E. Zhu, 941–948., Advances in Intelligent and Soft Computing Springer: Berlin.

Platt, J.C. 1999. Probabilistic outputs for support vector machines and comparisons to regularized likelihood methods. In *Advances in large margin classifiers*. Citeseer.

Saerens, M., P. Latinne, and C. Decaestecker. 2002. Adjusting the outputs of a classifier to new a priori probabilities: a simple procedure. *Neural Computation* 14(1): 21–41.

Vucetic, S., and Z. Obradovic, 2001. Classification on data with biased class distribution. In *Machine learning: ECML 2001*, 527–538. Heidelberg: Springer.

Chapter 16
Experiments with the Full Fish4Knowledge Dataset

Robert B. Fisher

Abstract This chapter presents some of the experiments possible when using a very large ecological observation dataset, such as the Fish4Knowledge dataset. The dataset was acquired over 1000 days, observing 12 h a day using 9 undersea cameras at 3 locations. 23 different species were recognized. Each day's observations vary considerably, but analysis of the large dataset allows trends to be observed. Key results are (1) that there is only little variation in fish observation through the daylight hours, (2) that typhoons only temporarily disrupt the abundance measures, and (3) different habitats show different ratios of species.

16.1 Introduction

Previous chapters presented much of the technology used in the Fish4Knowledge project. This chapter presents some of the questions that can be answered with the data that was acquired, demonstrates how they might be answered and even gives some of the answers. To answer these questions, the main Fish4Knowledge database is used, which contains processed results from 455,993 videos, each of 10 min duration, leading to 145 million tracked fish.

The analysis presented here is interesting for several reasons: (1) it demonstrates a style of analysis possible once really large ecological datasets are available, and

R.B. Fisher (✉)
School of Informatics, University of Edinburgh,
10 Crichton St, Edinburgh EH8 9AB, UK
e-mail: rbf@inf.ed.ac.uk

R.B. Fisher et al. (eds.), *Fish4Knowledge: Collecting and Analyzing Massive Coral Reef Fish Video Data*, Intelligent Systems Reference Library 104,
DOI 10.1007/978-3-319-30208-9_16

(2) it shows the variability in datasets such as these, which arises because of both natural day-to-day fluctuations, but also all of the opportunities for acquiring faulty data in a complex (and relatively low-cost) big data project such as this.

In spite of the noisy nature of the data, the data shows some trends, and also unexpected non-trends. For example, the data shows that typhoons are only temporarily disruptive in terms of the abundance (or at least number of observations). Secondly, there does not seem to be much variation in the number of fish observed per hour over the daylight hours. And there might be an increase in abundance in February–March.

The main Fish4Knowledge dataset was used for most of the analyses, containing results selected from the 455,993 processed 10 min videos; however, there are also 2 specialized smaller datasets: (1) data from the 08:00–08:10 for every available camera over all recording days (1 Oct 2010–26 June 2013) and (2) all data (06:00–18:00) from one particular day (April 22, 2011). These two subsets were chosen to allow an exploration of how the fish abundances vary over the 1000 observed days at an active time of day (first subset), and how the abundances vary throughout the day (second subset).

The following sections give the details of the acquired data, the methodology of the analysis and discuss issues that arose during the analysis.

16.2 Data

The data analysis presented in this chapter is based on the full Fish4Knowledge dataset. The video system recorded 524,086 10 min video clips (i.e. 6 per hour of video) using 9 cameras at 3 locations (although only 455,993 were analyzed, as described below). Four cameras were in the bay outside the Taiwan Hengchun Nuclear Power Plant on the south coast, three were in HoBiHu harbor, in Kenting, and two were on the coast of LanYu Island, which was about 50 Km southeast of Taiwan, as illustrated here:

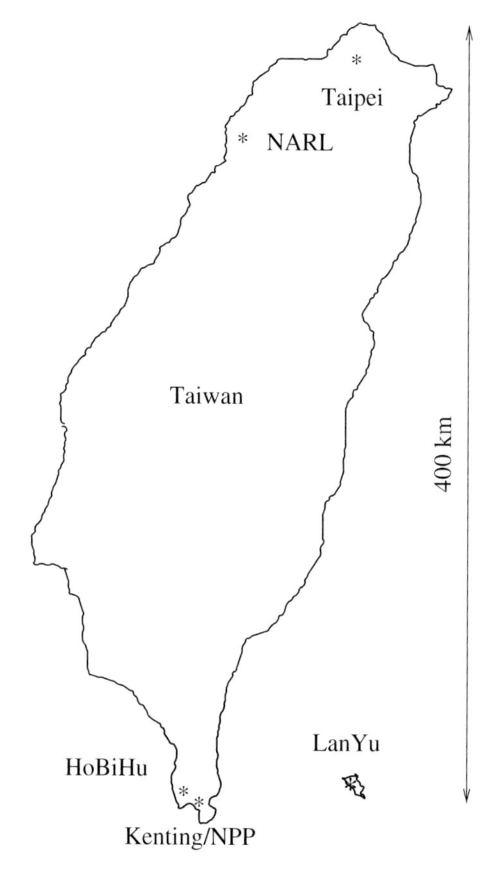

The raw resolution of many of the videos was 320 × 240, at about 5–8 frames per second. However, over half of the videos were recorded at 640 × 480, and some at 24 fps. This upsizing was created by interpolation and not from higher quality video data. The lower spatial and temporal resolution was beneficial for the storage of the approximately 90 thousand hours of video in about 91 terabytes of compressed disk storage. The processed results were stored in a SQL database which required about 400 Gb to store the details of the detected and recognized fish. The main data storage and processing was done at the Taiwan National Center for High-performance Computing. See Chap. 3 for more details about the hardware and Chaps. 4 and 5 for more details about the video and result database.

The major (negative) consequence of the capture schema was the slower frame rate, which meant that a fish could sometimes translate a considerable distance between frames. This could therefore introduce tracking errors (see Chap. 10) when multiple fish were in the field of view. The lower spatial resolution probably reduced the performance of the recognition algorithms (see Chap. 11) somewhat as well.

Of the 455,993 processed videos, all were processed by the fish detection and tracking algorithms (see Chaps. 9 and 10). The first part of the detection process is an analysis of the quality of the videos. A machine-learning algorithm (see Chap. 9)

Table 16.1 Summary of processed videos

Type	Total captured	Processed videos	Percent (%) of captured processed	Percent (%) of processed by type
Algae	49,370	49,165	99.58	11
Blurred	181,965	181,757	99.89	40
ComplexScenes	37,404	37,401	99.99	8
EncodingProblem	108,140	39,920	36.92	9
HighlyBlurred	65,024	65,025	100.00	14
Clear	75,806	76,465	100.87	17
Unknown	6171	6176	100.08	1
Total	524,086	455,993	87.01	100

was used to assess the quality of each video, and was also used to adjust the parameters of the fish detection algorithms and to prioritize the processing of the videos.

Columns 2–6 from Table 16.1 show the number of videos classified (see Chap. 9) as being in each quality category, and also the number (and percentage) from each category that were processed by the recognition algorithm. For a variety of operational reasons, a few videos were processed more than once, leading to some processed percentages being slightly more than 100 % and others being less than 100 %. Some of the operational reasons were the upgrading of algorithms as the project proceeded and storage of videos at different resolutions. In theory, one could have rerun the analysis using the final algorithms once the project development was finished, but in fact the data analysis used on the order of 400 CPU (2.2 Ghz) years of processing (see Chap. 3 for the hardware description). This implied there is a small amount of error and inconsistency in the recorded database.

Of particular interest was the unexpectedly low percentage of videos that were in the Clear/Normal category (14 %) and the unexpectedly high percentage of videos that were in the EncodingProblem category (21 %), which produced random artifacts in the video, many of which could become false fish detections. This categorization was based on a classifier trained on a manually labeled subset of videos (see Chap. 9). The other categories that were classified were Algae (algae on the camera lens, 9 %), slightly Blurred water (35 %), HighlyBlurred water e.g., after a storm (12 %), ComplexScenes where there was much plant and illumination activity in the background (7 %), Unknown (1 %) and NotSet (0.01 %).

The feature extraction stage of the recognition algorithm was considerably slower (implemented in Matlab) and so not all videos were processed by the recognition algorithms (see Chap. 11). In the end, 455,993 (87 %) of the videos were processed. Only about 37 % of the videos that were assessed to have EncodingProblems were processed, because they were introducing too many false positive fish detections (some of which are included in the dataset, unfortunately).

By the end of the project, the detection process (see Chap. 9) detected approximately 1.44 billion individual fish instances of a sufficiently large size (at least 50 by 50 pixels) that recognition was attempted. The detected fish were then linked across video frames (see Chap. 10) to produce approximately 145 million tracked fish. The pixels inside the extracted contours were the inputs into the recognition process (see Chap. 11).

The detection process was greatly affected by the illumination conditions on the background, and also by any movement of background material, typically seaweed. A consequence of this is a high false detection rate over the full video set. Some experiments (Chap. 11) on randomly selected video clips show that false detections can be on the order of 68 %. The recognition algorithm has an 'unknown or bad detection' rejection mechanism that eliminates many of these false detections, which rejected 74 % of the false detections in one experiment. (Although there are also other species besides the 23 that were recognized, we estimate that these account for less than 1 % of the observed fish.) The recognition algorithm reduced the number of tracked and recognized valid fish to approximately 81 million. Furthermore, when the duplication of videos at different spatial resolutions is taken into account, this reduced the number of videos and trajectories to 282,048 videos and 57.4 million trajectories. Videos classified as having encoding errors were eliminated, as were videos from 18:00–19:00 (as the recording for these times was highly incomplete because of the variations in the lighting conditions). Fish classified by the recognition algorithm as being not one of the 23 trained species or non-fish were rejected, leaving 261,751 10 min video clips (43,625 h of video) and 27.4 million trajectories, where each trajectory is normally the same fish tracked over multiple video frames. All results presented below were based on this final set.

The recognition algorithm was trained using a manually produced ground truth (see Chap. 14) of the top 23 species, as shown in Fig. 16.1. Here, recognition is a significantly unbalanced data analysis problem, where the most common species (*Dascyllus reticulatus*) was on the order of 1000 times more numerous than the $23rd$ species. Altogether, these top 23 species represented 99.7 % of fish observed in the ground truth dataset.

Of the 27.4 million analyzed trajectories, the most commonly recognized species were: *Dascyllus reticulatus* (47 % of the dataset), *Scolopsis bilineata* (7 %), *Plectroglyphidodon dickii* (11 %), and *Amphiprion clarkii* (9 %). In the full dataset the ratio of *D. reticulatus* (most common) to *Pempheris vanicolensis* (least common) was 5585:1.

The recognition algorithm performance on the ground truth, when considered over all fish in the top 23 species, averaged 97 % correct. However, given the imbalance in the species, we also calculated the average correct recognition rate of each of the 23 species and then averaged these together. In this case, the average recognition rate was 75 %, which gives a measure of performance for all species, not simply the most numerous. However, since the real dataset is unbalanced, the 97 % correct recognition rate suggests that the performance on the whole video database is also good. See Chap. 11 for more details.

Fig. 16.1 *Top* 35 observed species with the number of observations in the ground truth (and the number of corresponding trajectories in parentheses). The ground truth was based on fish similarity agreement by at least 5 markers, and the species labels were given by the partner marine ecologists. Chapter 14 has more details

16.3 Statistics of the Dataset

Answering queries about the detected fish using the 400 Gb SQL database can be quite slow. As a consequence, we developed summary tables recording the individual fish detections, trackings and species classifications, which were used in the analyses presented below. The two summary tables are:

1. **VideoSummary**—the number of 10 min video clips recorded as a function of 1000 days (from 1 Oct 2010–26 June 2013) × 12 one hour time slots (from 6:00

to 17:59) $\times$ the number of cameras (9) $\times$ the different video quality categories (6). The video data array is `VideoSummary(1000,12,9,6)`.

2. **FishSummary**—the number of recognized fish (trajectories), indexed as for `VideoSummary`, with an additional index of 23 known +1 unknown species. A fish was classified by its most likely species (with the likelihood distribution ignored here). The fish data array is `FishSummary(1000, 12, 9, 6, 24)`.

One interesting factor was the nearby passing of 5 typhoons (sustained wind speed of at least 118 km/h) during the observation period: Megi (Oct 22 2010, west of Taiwan, day=22/1000), Songda (May 26 2011, east of Taiwan, day=238/1000), Nanmadol (Aug 28 2011, over Taiwan, day=332/1000), Saola (Aug 1 2012, over Taiwan, day=671/1000), and Tembin (Aug 22 2012, over Taiwan, day=692/1000). These dates are plotted in the figures below with a vertical red dashed line on the date-based plots. The results below show that the typhoons had only short term effects on the number of fish observed.

These two summary arrays now allow investigations of a number of questions. Some background questions about cameras and videos are considered first:

1. **What are the number of active cameras (top plot) and camera-hours (bottom plot) recorded per day?** A camera is active if it recorded a video on that day. A camera-hour is counted if any one of the six 10 min clips was recorded. The data is aggregated over all times and quality levels. The plots show: (1) the HoBiHu cameras were active only until day 231, and the LanYu cameras only through day 365, but the NPP cameras were active almost every day (top plot). (2) Typhoons caused only short term loss of data (both plots). (3) If the cameras were active, we achieved almost a full day of acquisition (bottom plot).

Active cameras *versus* day

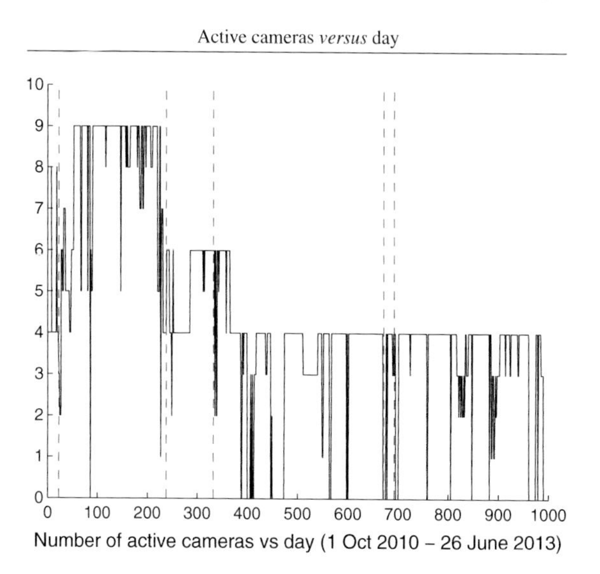

Camera hours acquired *versus* day

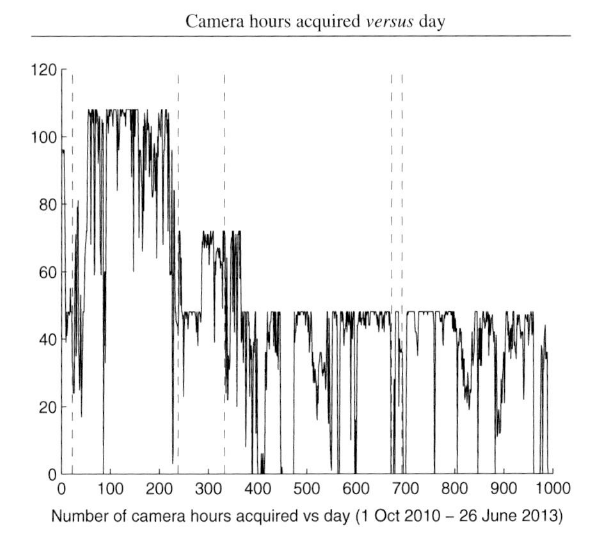

2. **What are the number of videos recorded per day (top plot) and per hour (bottom plot)?** These are aggregated over all sites and quality levels. The plots show (top) for the most part 6 videos/hour times 12 h and 9 cameras for a maximum of 648 videos/day. There was a system error near day 205. After day 400 there were only 4 working cameras. (bottom) Approximately the same number of videos were recorded at each hour, with a slight reduction at dawn and dusk.

Videos recorded *versus* day

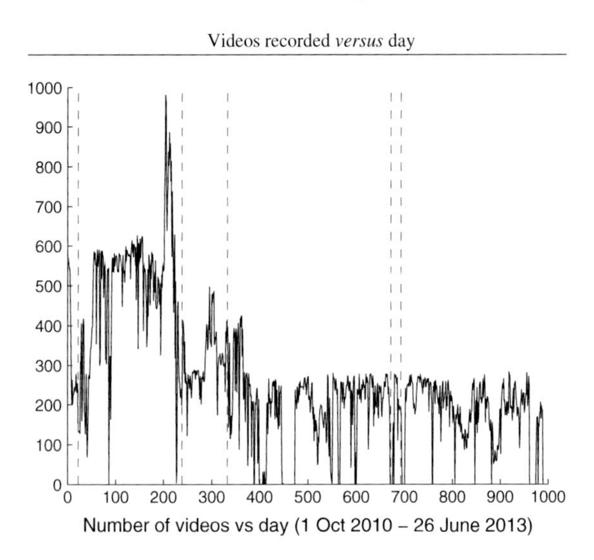

Videos recorded *versus* hour

3. **What are the number of videos per camera (top plot) and per site (bottom plot)?** The plots are aggregated over all days, times and quality levels. The bottom plot aggregates (1,2,3,4) the four cameras at the Nuclear Power Plant, (5,6,7) the three at HoBiHu harbor and (8,9) the two at LanYu island. The plots are as expected, given the successive loss of cameras over the 1000 days. The maximum number of videos for site NPP was approximately 4 cameras * 1000 days * 12 h * 6 videos = 288,000, for HoBiHu, it is about 3 * 230 * 12 * 6 = 49,680 and LanYu, it was about 2 * 360 * 12 * 6 = 51,840, which is close to the recorded number. Further, the top plot shows that all cameras in each area were approximately equally operational.

Videos recorded *versus* camera

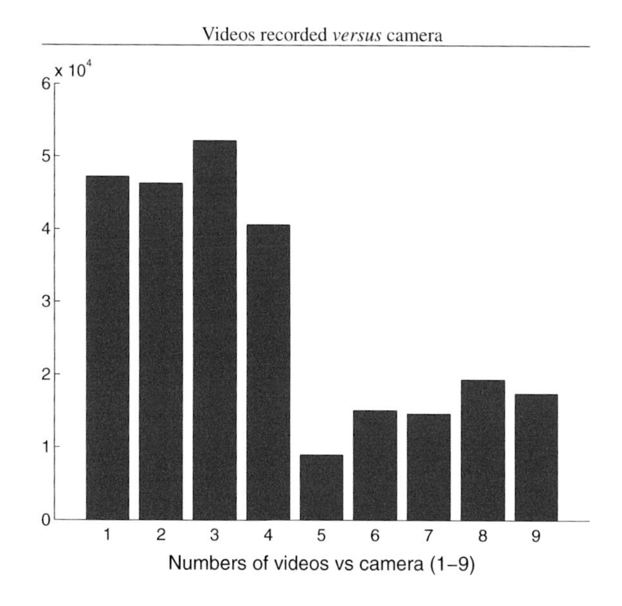

Videos recorded *versus* site

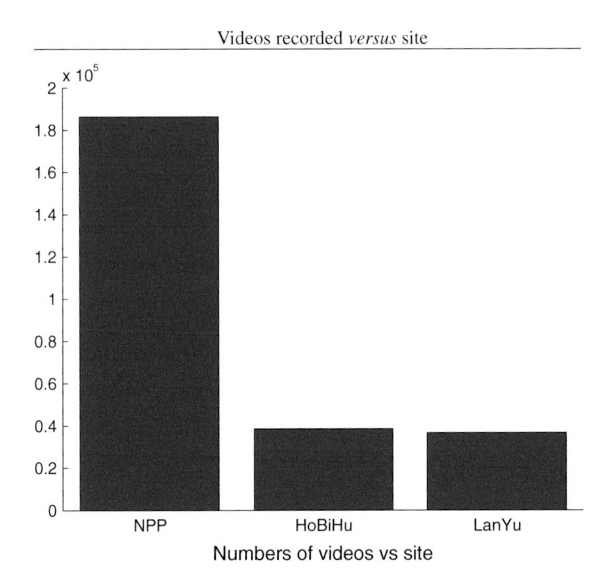

4. **How many videos were analyzed (top plot) and how many fish were detected in each video (bottom plot) for each quality classification?** The plots were aggregated over all days, times, sites and species. These are a subset (261,751) of those recorded in Table 16.1, based on those that were processed, had duplicate resolutions removed, 06:00–18:00 and no encoding errors. The quality measures (horizontal axis) were 1: algae on the lens, 2: slightly blurred water, 3: complex changing backgrounds, 4: highly blurred water, 5: clear/normal, and 6: unknown. The top plot shows that many of the videos were recorded with slightly blurred water (column 2). The bottom plot shows the median number of fish (with $\pm 1\sigma$ error bars) detected in a video for each of the six quality settings, again analyzed over all sites, days, times, and species. It shows that detection works best on the clear/normal (column 5) and slightly blurred (column 2) videos, as one can expect. This assumes that the same distribution of environmental effects are experienced for each camera condition. This is probably not the case with the highly blurred water, which is likely to occur during storms, and thereby also affects the fish. Note that the lower plot is truncated at 250 (red ticks are outliers).

Videos recorded *versus* quality

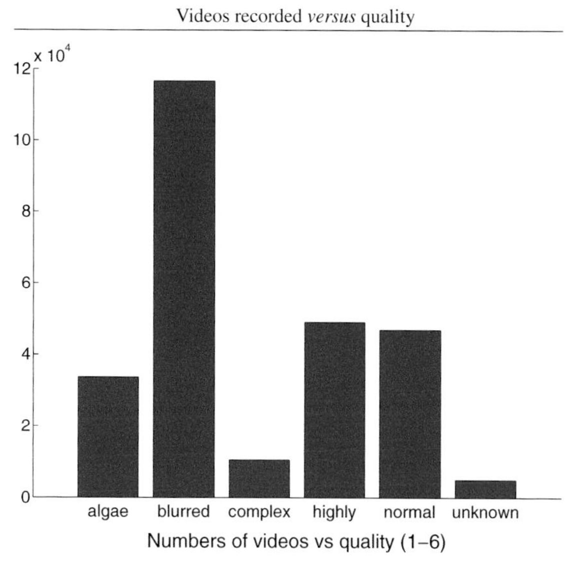

Median fish observed per video *versus* quality of video

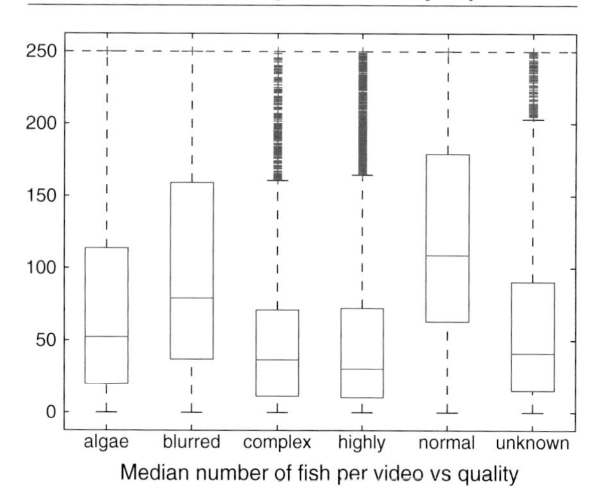

Some questions about the fish abundance over the whole observation period are now considered:

5. **What are the numbers of fish observed per day (top plot) and the number of fish observed per video per day (bottom plot)?** The plots were aggregated over all times, sites, qualities, and species. The bottommost plot shows the number of fish from the topmost plot divided by the number of videos captured on that day. The plots show that there is a lot of variability in the observations, and there are some noisy measurements (probably due to undetected compression artifact failures) at days 547, 694 and 859.

Fish observed *versus* day

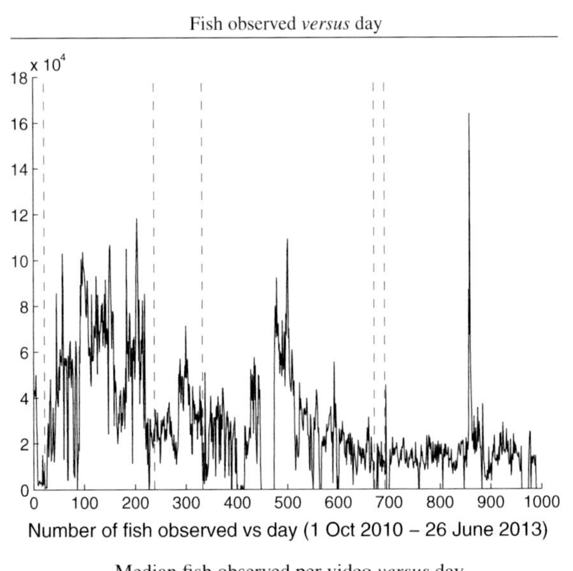

Median fish observed per video *versus* day

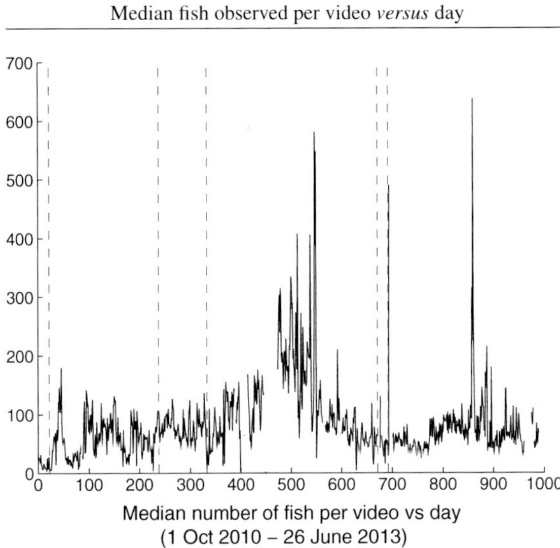

The bulges in the data near days 503 and 866 (February 15, 2012 and 2013) in the plot above led us to wonder if there was a seasonal effect. Focusing just on the NPP site, the median number of fish per video aggregated over the 4 cameras is shown in the next figure. It appears that the day 138 (15 February 2011) observations are not strong enough to confirm the hypothesis of a repeating seasonal increase.

Median fish observed at NPP sites per video *versus* day

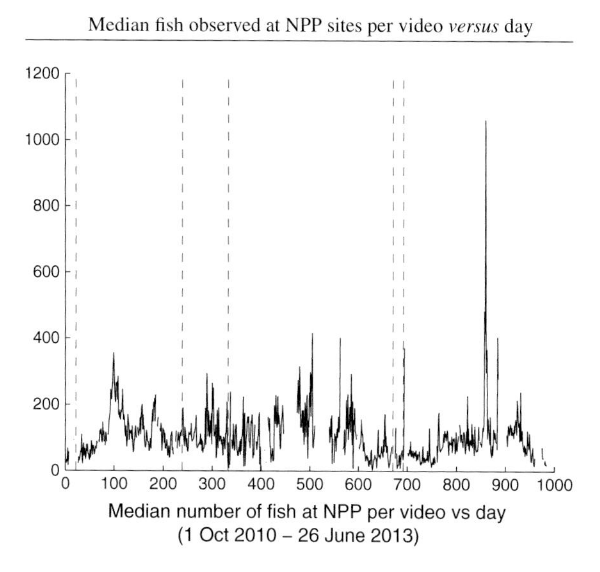

6. **Does the number of fish per video vary according to the time of day?** This plot is aggregated over all days, cameras, qualities and species and is the median number of fish per hour, with a robust estimation of the ± 1 standard deviation error bars. The black curve is over all species, cyan: *D. reticulatus*, red: *S. bilineata*, green: *P. dickii*, and blue: *A. clarkii*. This plot shows a slight increase in the median value of the total count at dawn and dusk, but the variances are quite large. It's hard to make any conclusions for the individual species.

Median fish per video *versus* time of day from data subset

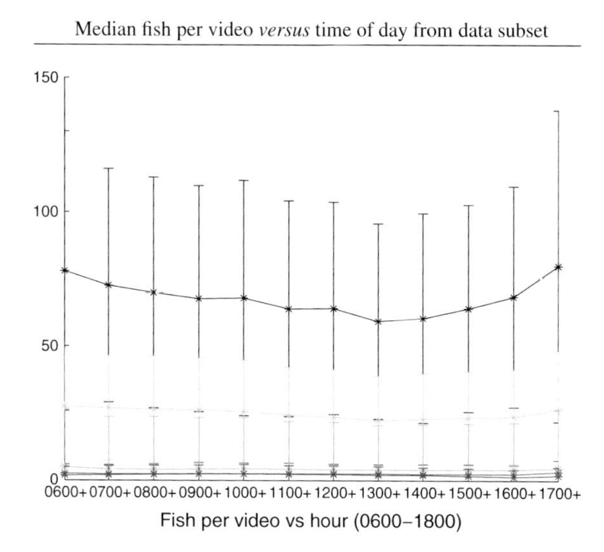

7. **Does the number of fish per video vary according to the time of day?—analysis using the User Interface**
   A similar result can be seen in these plots. The top plot shows the number of fish per video from the whole dataset (which also includes the unreliable 18:00–19:00 data), but displayed using the stacked feature of the user interface.[1] In this figure, the color coding is light blue: *D. reticulatus*, green: *S. bilineata*, orange: *P. dickii*, and light tan: *A. clarkii*. The bottom figure shows the detection results over all species, but only for week 16 of 2011.[2]

Average fish (top 4 species) per video *versus* time of day from full dataset

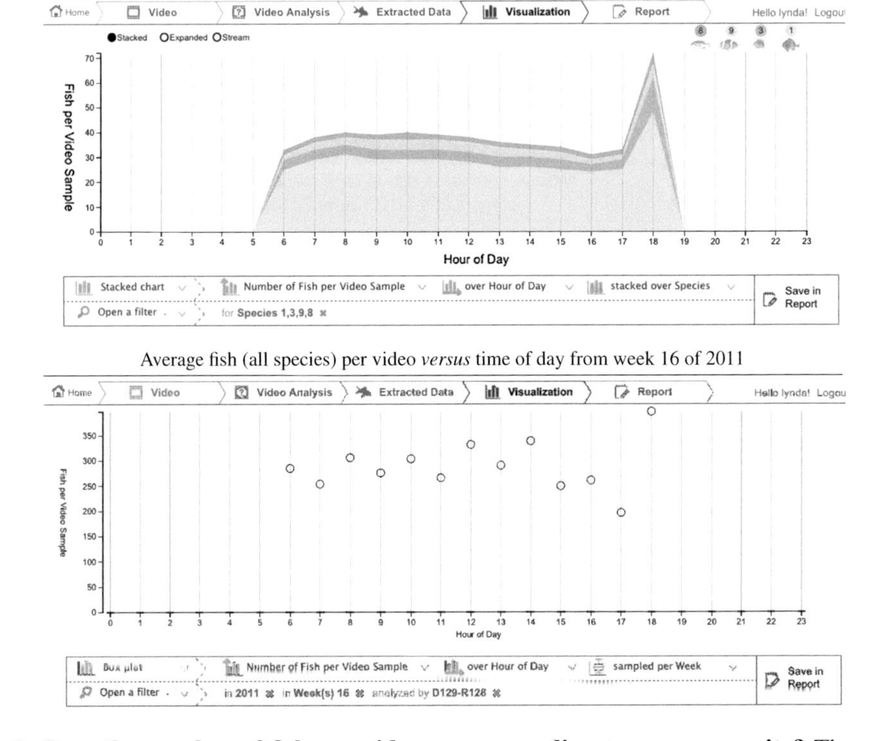

Average fish (all species) per video *versus* time of day from week 16 of 2011

8. **Does the number of fish per video vary according to camera or site?** The plots were aggregated over day, time, quality and species, and show (top plot) the median number of recognized fish per camera, with a robustly estimated 1 standard deviation error bar. Cameras 1, 2, 3, and 4 were at site NPP; cameras 5, 6, and 7 were at site HoBiHu and cameras 8 and 9 were at site LanYu. (bottom)

[1]Plot (see Appendix A): http://f4k.project.cwi.nl/demo/ui/visualization/?f_s=all&f_cs=0,1&f_vt=all&f=f_c,f_s,f_so,f_vt,f_y&f_c=all&f_h=all&f_w=all&t=B&f_so=all&y=NFC&x=H&z=W&f_y=all.

[2]Plot: http://f4k.project.cwi.nl/demo/ui/visualization/?f_s=all&f_cs=0,1&f_vt=all&f=f_c,f_s,f_so,f_vt,f_y,f_w&f_c=all&f_h=all&f_w=16&t=B&f_so=D129-R128&y=NFC&x=H&z=W&f_y=2011.

The data is aggregated by site. The plots show greater fish abundance and variability at the NPP site (which is a function of the local environment and field of view). Note that the data is truncated at 250 (red ticks are outliers).

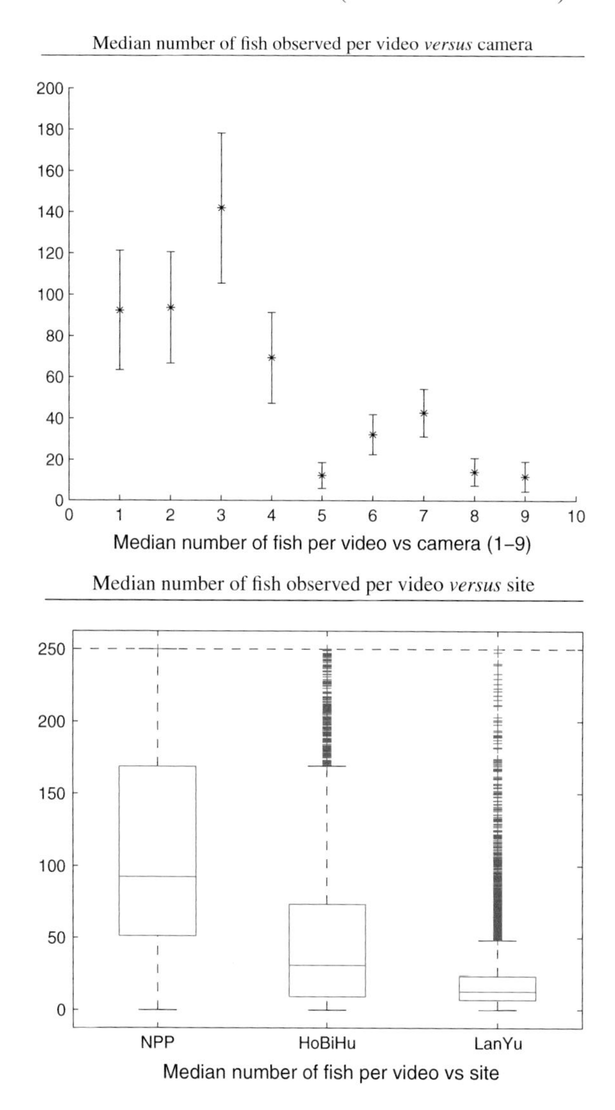

9. **Does the percentage of the fish belonging to a given species vary per day?** (top) The count of fish is aggregated over times, cameras, and qualities for the top four species {black/species 1: *D. reticulatus*, green/species 3: *P. dickii*, red/species 8: *S. bilineata*, blue/species 9: *A. clarkii*}. The four species are plotted as a percentage of the total abundance on that date. The top plot shows that *D. reticulatus* dominates the population, with the exception of days 513–611, where the recognition algorithm reported a considerably higher number of

other species, in particular *A. clarkii*. The bottom plot shows a related analysis for 2011, showing the relative percentages per week for the same four species aggregated in the User Interface.[3] In this figure, the color coding is light blue: *D. reticulatus*, green: *S. bilineata*, orange: *P. dickii*, and light tan: *A. clarkii*.

Percentage of total for top 4 species *versus* day (data subset)

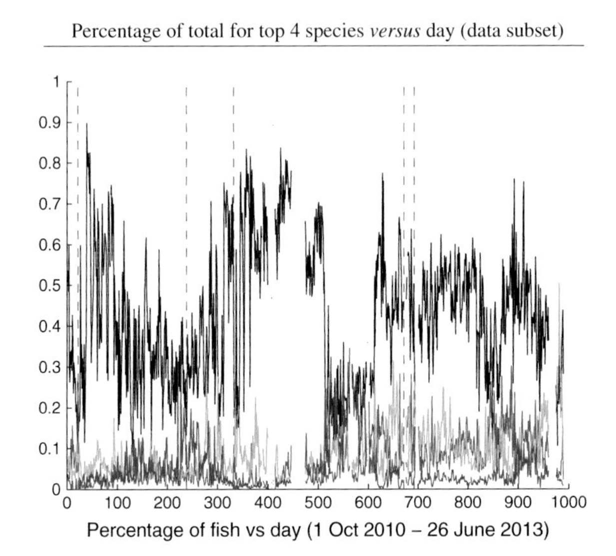

Percentage of total for top 4 species *versus* day (User Interface, full dataset)

10. **What is the relative abundance of the fish belonging to different species?** (top) The total number of recognized fish for each of the 23 species is plotted, again showing species 1, 3, 8, 9 dominate (as listed above). Species 24 is the aggregated unknown species quantity. A similar report can be produced for the full dataset directly from the User Interface, as seen in the bottom plot (but with an uncompacted species numbering scheme).[4]

---

[3]Plot: http://f4k.project.cwi.nl/demo/ui/visualization/?f_s=1,3,9,8&f_cs=0,1&f_vt=all&f=f_c,f_s,f_so,f_vt,f_y&f_c=all&f_h=all&f_w=all&t=T&f_so=all&y=NFC&x=W&z=S&f_y=2011.

[4]Plot: http://f4k.project.cwi.nl/demo/ui/visualization/?f_s=all&f_cs=0,1&f_vt=all&f=f_c,f_s,f_so,f_vt,f_y&f_c=all&f_h=all&f_w=all&t=S&f_so=all&y=FC&x=SPEC&z=&f_y=all.

Total fish observed over all 23+1 species from the dataset subset

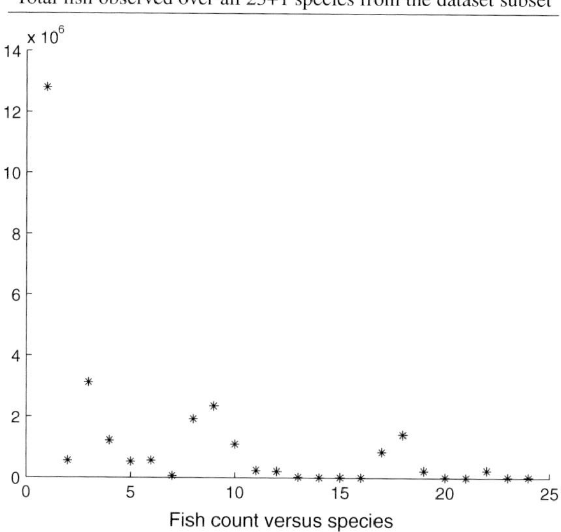

Total fish observed over all 23 species from full dataset (User Interface)

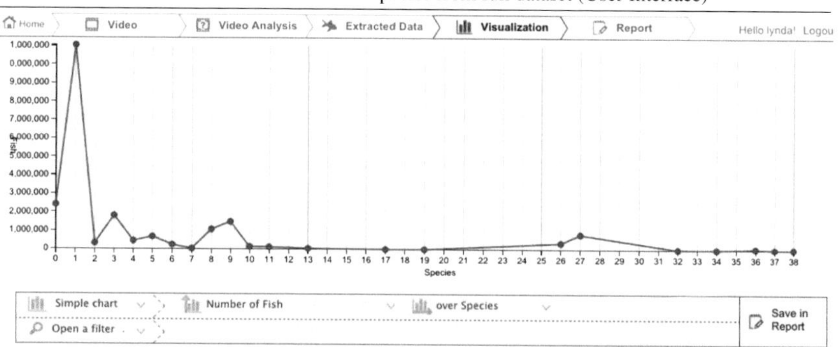

Lastly, some plots from specialized data subsets are shown. On the Fish4Knowledge project website http://groups.inf.ed.ac.uk/f4k/ there is a facility that allows people to download a subset of the videos and the analyzed results associated with those video clips. Here are some results (like those seen above), but for the subsets. There are 2 subsets: (1) all video clips for the time 08:00–08:10 for all cameras, days, and quality levels and (2) all video clips for the full day April 22, 2011. These two subsets allow an exploration of the daily and seasonal patterns of fish abundances.

There are no surprising results here, but they show researchers what they could expect if they downloaded the data subsets.

1. **What are the numbers of 08:00 videos recorded per day (top plot) and the number of fish observed per 08:00 video per day (bottom plot)?** The plots were aggregated over all cameras, qualities, and species. The bottom plot shows the number of fish from the day divided by the number of videos captured on that day (top plot). The plots show patterns similar to those seen above and thus we can conclude that the 08:00-08:10 dataset is representative of the rest of the day.

Number of 08:00-08:10 videos recorded *versus* day

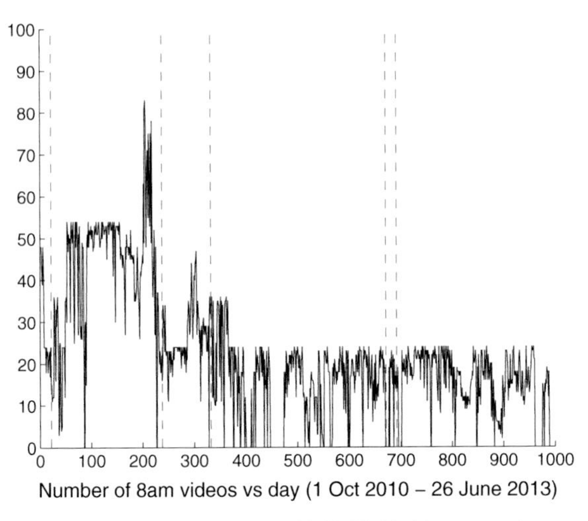

Number of Fish observed per 08:00-08:10 video *versus* day

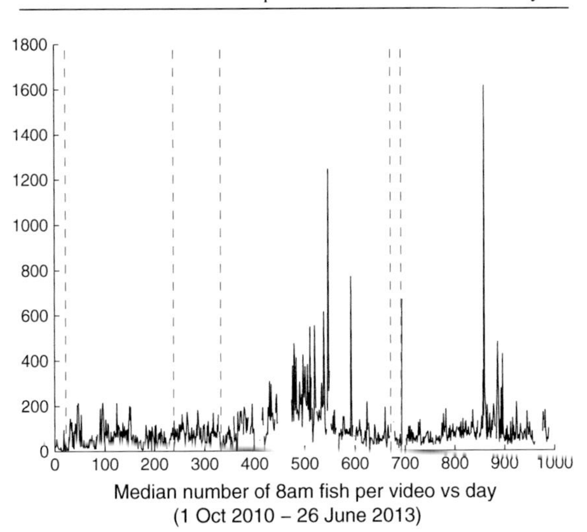

2. **What are the numbers of videos recorded and fish observed per video per hour for the dataset from April 22, 2011 (06:00–18:00)?** The plots were aggregated over all cameras, qualities, and species and show that there was approximately similar numbers of videos recorded per hour and fish observed (top), but with great variability in the number of fish observed (bottom).

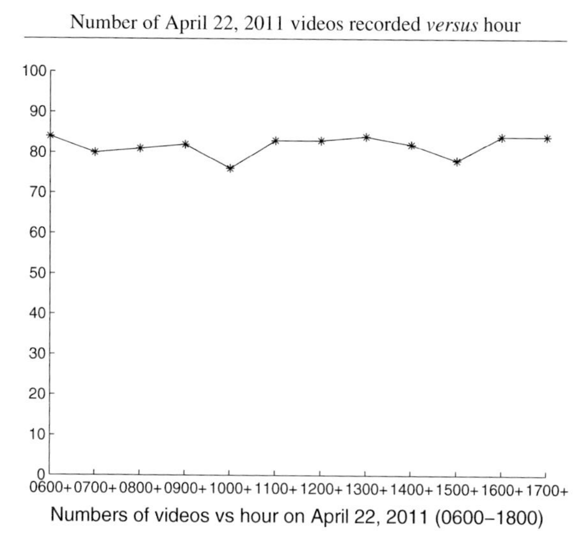

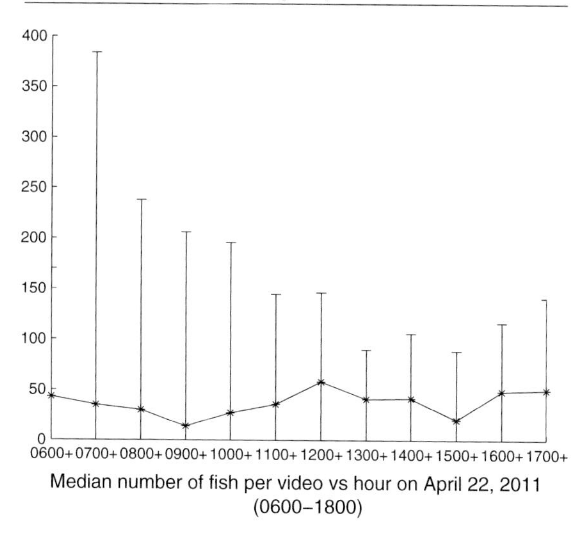

16.4 Discussion

The collection of the data over such a long time period has exposed many issues arising in such a challenging project, the most important of which is of the quality of the data. The outdoor ocean environment near the surface is visually difficult, with the most problematic effect being the dramatically varying illumination on the background because of the caustics, i.e. the light rays refracted by the changing ocean surface. Varying quality of the water media and plant life in the background also causes problems. A further issue is the occasionally defective video arising from the many technical challenges in acquiring off-shore video, cabled communication to shore, heavy compression and then use of telephone line based upload to the supercomputer center. Cameras and facilities degrade and algae grows on the lens. The result of these difficulties is a somewhat incomplete dataset, and which seems to have about a 68 % false detection rate, many of which (74 % in one experiment) are removed by the recognition algorithms.

Another factor is the time period over which the data was collected and analyzed: the enormous amount of computation required meant that it was not possible to analyze all videos with the final versions of the algorithms. Thus, results from videos processed in the autumn of 2012 were not as accurate as those processed in the summer and autumn of 2013. Recoding the recognition algorithms from Matlab to C++ would have helped greatly as potentially we could have rerun the final recognition algorithms on all of the data, but there were not enough project resources for this.

An interesting 'marine ecology' aspect of the dataset is the large percentage (67 %) of the total fish observations represented by the top three species mentioned previously, which are known to be resident, i.e. non- migratory: *D. reticulatus*, *P. dickii*, and *A. clarkii*. The implication of being resident is that multiple observations of the same individual are very likely because the fish may swim out of the field of view and then return. Because of the low camera resolution it is difficult to determine if the individual is the same as previously observed. Interestingly, Liu (2013) demonstrated that one could cluster individual *A. clarkii* (clownfish) based on their distinctive stripe patterns. From this small experiment using 785 images, we estimate that we might be over-counting resident *A. clarkii* individuals by a factor of 100.

As a consequence of the data analysis presented here, we now have a much better idea of the variability in the data, over days, weeks and years. We can also see that there is a lot of 'noise' in the data, which causes difficulties for the marine ecologists, who are uncertain about how to extract meaningful information from the noisy data. Moreover, to a considerable extent, the types of analyses reported here and implicit in the user interface are too generic, and not focused on specific ecological questions. There is a lot of information in the database, as illustrated above, but that then requires the ecologists to have both SQL programming skills, and also mathematical techniques for extracting meaningful results from the noise and variability. Future developments could give the marine ecologists better tools for analyzing the database, allowing them to address more specific research questions.

16.5 Conclusions

The Fish4Knowledge project collected one of the largest video and specialized image databases in the world, based on advances in target detection, tracking and class-based recognition. The volume of data that was collected allows analysis of fish abundances over time periods and in different situations not previously possible by diver-based observations. The magnitude of the data also allows observation of trends not visible in somewhat noisy data. The quantity of the data has helped improve the understanding of tropical coral reef fish near Taiwan, with one of the most interesting aspects being the recovery of fish populations after typhoons.

Reference

Liu, X. 2013. Identifying individual clown fish. Master's thesis, School of Informatics, University of Edinburgh.

Chapter 17
The Fish4Knowledge Virtual World Gallery

Yun-Heh Chen-Burger and Austin Tate

Abstract In this chapter, we describe our project dissemination efforts via a programmable, configurable, 3D Virtual World environment in Second Life and OpenSimulator.

17.1 Introduction

The technologies for 3D interactive environments for multiple simultaneous users are quite advanced and virtual environments are widely used in many areas, such as gaming, movies, animation, design, engineering, health and safety testing, informational, educational and multi-media applications. As the Fish4Knowledge project has an important visual aspect to show marine life observations, it will be a natural and useful step to be able to use such media to communicate the Fish4Knowledge project results, in addition to traditional academic outlets, such as web sites, scientific conferences and journal publications.

Among several 3D virtual world environments, we have chosen to build an exhibition of our project as a virtual gallery in Second Life (SL) (secondlife.com) for several reasons. One of the project partners, the University of Edinburgh, already owned virtual land in SL. On this land, there is the well-established Virtual University of Edinburgh (Vue), sponsored and presented by several schools and institutes within the university, including the School of Informatics, Information Services, e-Learning, Business School, Veterinary Medicine, Social and Political Sciences and Alumni Services. In addition, on a part of this virtual land, some of the long-distance

Y.-H. Chen-Burger (✉)
School of Mathematical & Computer Sciences, Computer Science
Heriot-Watt University, Edinburgh EH14 4AS, Edinburgh, UK
e-mail: y.j.chenburger@hw.ac.uk

A. Tate
School of Informatics, University of Edinburgh,
10 Crichton St, Edinburgh EH8 9AB, Edinburgh, UK
e-mail: bat@inf.ed.ac.uk

R.B. Fisher et al. (eds.), *Fish4Knowledge: Collecting and Analyzing Massive Coral Reef Fish Video Data*, Intelligent Systems Reference Library 104,
DOI 10.1007/978-3-319-30208-9_17

Fig. 17.1 Front of the F4K Virtual Gallery

learning courses are supported directly through the Vue facilities. Interested readers are directed to +http://vue.ed.ac.uk/ for more details.

The University of Edinburgh, at the time of consideration, already had a long history of SL deployment and its virtual land is well populated and used. It would therefore be ideal, if F4K could build its virtual gallery as a part of Vue. More importantly, SL allows its users to easily develop and program its environment. That is essential for us, as we plan to provide a tailored 3D environment to suit our needs.

As a result, we have selected Second Life as an experimental platform to host our F4K 3D Virtual Gallery. We were able to secure a piece of virtual land within Vue to build our gallery. What distinguishes this 3D virtual project demonstration area from our standard project web site (http://groups.inf.ed.ac.uk/f4k/) is that the project demonstration is intended to be fun, interactive and educational. It is not just for academics, but also for everyone who has an interest in marine life and ocean conservation. We intended to use this virtual platform to attract younger people and their educators who have an interest in using computing technologies for educational purposes to get curious about our work and marine research in general. Figure 17.1 shows the front of the F4K virtual gallery.

17.2 The Fish4Knowledge Second Life Gallery—Ground Level

The Fish4Knowledge virtual exhibition gallery is situated at a beautiful spot by a lagoon at the heart of Vue. In this gallery, visitors are able to "walk" leisurely around our virtual building, via their avatars, to read about our project work and watch our

Fig. 17.2 Ground level of the exhibition hall (East Facing)

Fig. 17.3 Ground level of the exhibition hall (South-West Facing)

underwater fish monitoring movies. They are able to learn and be entertained in a surreal environment where the sunset is reflected by the nearby lagoon shining through the large floor-to-ceiling glass window walls. Alternatively, visitors can choose to visit our gallery on starry nights, or at any other times of the (virtual) day to enjoy the shimmery sea waves. Figures 17.2, 17.3, 17.4 and 17.5 show our ground level project exhibit area.

Upon arriving at the ground level exhibition hall, the visitors can sit on our comfy virtual sofas to enjoy the surrounding or walk around the posters to view them. They can interact with or meet other visitors there or arrange to meet project representatives to talk about the project and its results. Once a poster is selected for viewing, the visitor

Fig. 17.4 Ground level of the exhibition hall (West Facing)

Fig. 17.5 Ground level of the exhibition hall (West-North Facing)

can use a combination of [left-click] for locating and arrow keys to navigating their view. Currently, there are about a dozen project posters on display, with topics ranging from high performance computing, image processing, human-computer interaction, marine biology to virtual workflow machines.

On the left hand side of Fig. 17.3, there is a large screen displaying a looped marine life video that was captured in the coastal sea off South Taiwan. This video is an example of the videos that the Fish4Knowledge team process.

17.3 The Fish4Knowledge Second Life Gallery—Underwater Level

On the north side of this gallery, there is a passageway that leads visitors down to an underwater virtual aquarium. We call this the "Virtual Fish Lab". Here we exhibit example fish that we observe in real life. Some of our virtual fish are interactive and will react in different ways when stumbled upon or interacted with. Our virtual

Fig. 17.6 Tunnel entrance to the underwater level (external view)

Fig. 17.7 Sample view to the underwater level 1

Fig. 17.8 Sample view to the underwater level 2

fish will talk to visitors about their lives, via some simple conversational skills. Figure 17.6 shows the tunnel entrance to the underwater level. Figures 17.7 and 17.8 show sample screen shots of the lower level. On the left hand side of Fig. 17.8, there is a large screen. This screen continuously displays a looped marine life video that was captured in the coastal sea off South Taiwan that is a part of videos that have been processed by the Fish4Knowledge Team.

17.4 The Fish4Knowledge Virtual World Gallery in OpenSimulator

The OpenSimulator-based "Openvue" (Open Virtual University of Edinburgh) grid and the free-to-use public OpenSimulator-based "OSGrid" have replicas of the Second Life F4K gallery. The official F4K project page at www.aiai.ed.ac.uk/project/f4k provides URLs for these replicas, including the F4K pavilion and its underwater gallery on OpenSim. An "OpenSim Archive" (OAR) file has also been created to support the replication of such facilities.

17.5 Conclusion

The central design idea of the underwater virtual aquarium is to provide a fun, interactive and educational space that gives its visitors a "surreal" experience—in that visitors can "walk around", "touch" things, interact with objects or talk with virtual fish. When there is more than one visitor in this space, people can choose to share their experiences through Second Life's and OpenSimulator's live voice and text-chat facilities. When appropriate, the Fish4Knowledge team can hold exhibition events where project works can be presented. This user experience is intended to be different from those provided by conventional publications, web sites and 2D media.

Web Resources

- Access to the Fish4Knowledge Virtual World Gallery and Aquarium can be obtained via `http://www.aiai.ed.ac.uk/project/f4k/`.
- A replica of the above Gallery is also available via The OpenSimulator-based "Openvue" (Open Virtual University of Edinburgh) grid and the free-to-use public OpenSimulator-based "OSGrid":
  `http://blog.inf.ed.ac.uk/atate/2013/05/14/...`
  `fish4knowledge-pavilion-and-underwater-observatory...`
  `-in-second-life-and-opensim/`
- Fish4Knowledge 3D Second Life Virtual Gallery Web Resources are listed at `http://groups.inf.ed.ac.uk/f4k/secondlife.htm`.

Chapter 18
Conclusions

Robert B. Fisher

Abstract This chapter gives an overview of the Fish4Knowledge project achievements, some critical assessment of what worked well and what could be improved, future directions, and a list of project publications classified by topic.

18.1 Summary of Achievements

The Fish4Knowledge project was a great success, in that it was the first project to acquire and analyze such a large amount of video data of the natural environment. To be more precise, it acquired and stored 87,000 hours of video from 9 underwater cameras (see Chaps. 1 and 4). These videos were analyzed to detect and track new objects that appeared (see Chaps. 9 and 10). This task is quite similar to other video surveillance applications, especially since the cameras were fixed. However, here, the major challenge was the lighting. Because the cameras were very close to the ocean surface (2–4 m), the undulating ocean surface refracts the sunlight to create changing illumination patterns on the background seabed and coral. Other outdoor observation applications have to cope with changing illumination, but usually over the course of minutes as the sun or clouds move, in contrast to nearly every frame being different as is the case here. Much effort was put into developing detection and tracking techniques that were effective in this context, and which resulted in a detection algorithm that had an F_1 score (harmonic mean of precision and recall) rate of 0.81 and a frame-to-frame correct tracking decision rate of 82 %. This was remarkable because, of the more than 524,000 10 min video clips analyzed, only about 14 % had water quality that the algorithms estimated was clear. The other clarity categories had 35 % of videos with slightly blurred images (silt in water), 12 % were highly blurred (after a storm), 21 % had some form of encoding error, and 9 % had algae growing on the lens. Yet, in spite of these challenges, we were still

R.B. Fisher (✉)
School of Informatics, University of Edinburgh,
10 Crichton St, Edinburgh EH8 9AB, UK
e-mail: rbf@inf.ed.ac.uk

R.B. Fisher et al. (eds.), *Fish4Knowledge: Collecting and Analyzing Massive Coral Reef Fish Video Data*, Intelligent Systems Reference Library 104,
DOI 10.1007/978-3-319-30208-9_18

able to detect on the order of 1.4 billion fish instances spread across all categories of video, which were collected into 145 million trajectories.

The detected and tracked fish were then recognized by species (see Chap. 11). As the project started, we thought that the key challenge would be the potentially large number of species of fish in the Taiwan sea (up to 3,000). In fact, because of the problem constraints (near ocean surface, coral reef, daylight hours, minimum detectable size), we only observed about 50 or so species, although there may be many less uncommon species stored in the detection database. What turned out to be the main challenges were: (1) the greatly imbalanced distribution of the observations of the different species and (2) the low quality of the individual detections. The imbalanced class problem, where the most commonly observed species (*Dascyllus reticulatus*) was observed approximately 600 times more frequently than the 23rd most commonly observed species (*Siganus fuscescens*), arose because some of the species were what the marine ecologists call 'resident'. This meant that any resident individual that was observed once was likely to be observed many times. For example, Liu (2013) analyzed a subset of 785 images of the clownfish *Amphiprion clarkii* and hypothesized that there were only 6–7 individuals present in that subset.

One consequence of this imbalance was a difficulty in developing species recognition algorithms that worked well on the minority species as well as the majority species. An example of this difficulty can be seen with the simplest of classifiers: declare all detected fish to be *D. reticulatus*, and the classifier would immediately have an accuracy of 44 %.

Image quality was not great, partly due to the video quality problems mentioned above, and partly due to the 320 × 240 format for most of the videos, which meant that the detected fish were often quite small and blurry. This led to difficulties in extracting discriminating features, such as stripes, textures and color distributions. Nonetheless, in the end the final algorithms were suitable for the 23 most common species which accounted for about 99 % of the observed fish, with an accuracy of about 97 % over the individually detected fish, when assessed against the manually derived ground-truth (see Chap. 14). (Note that even our marine ecology experts did not always agree on the fish species.) The large number of tracked and recognized fish allowed some of the first large-scale in situ analysis of fish behavior (see Chap. 12).

Although the analysis of the fish data is the most 'public' result of the project, there were many other interesting aspects, the most notable was the 'big-data' system aspects of the project. To acquire the data was a substantial logistical challenge (see Chap. 4), where the cameras needed to be securely mounted underwater, data and power cables were needed from onshore collection stations to the cameras, upload of the video from the collection stations to the NCHC supercomputer resources required careful engineering to optimize the video quality and quantity against the communication network capabilities. Easily accessible storage of 92 Tb of video data required careful design, in that there were over 500 thousand individual video clips that needed to be accessed unpredictably, by location, camera, date and time. We ultimately chose one of the SQL variants (see Chaps. 5 and 7) as the format for storing the 400 gigabytes of processed results. However, in spite of the efficiency of modern SQL implementations, we still needed to develop summary tables to allow a response in several seconds to user queries (as contrasted with minutes). Processing the video data also required a substantial effort—we wanted to prioritize the higher quality videos while exploiting all available computing resources. At times this meant up to 1,000 cores were used for the processing, which meant developing a task scheduler that could keep track of which videos were at each stage of analysis, and only initiate the next stage when the previous stage was finished (see Chaps. 6 and 8). Given the number of processing cores involved and the total amount of computation (approximately 400 core-years—see Chap. 3), there were a variety of failures that needed to be detected, such as: dead cores, crashed programs, programs stuck permanently awaiting some resource, network data corruption, etc. Processing all videos under these conditions required both human and scheduler sophistication. In the end, all videos except those assessed to have encoding errors were processed.

The final technical achievement was the user interface, which attempted to address the question of how to extract and present useful information from the large database. Many of the questions that the marine ecologists posed to us (see Chaps. 2 and 16) were related to the quantity of observations and the quality of the results. In other words, how was the abundance of fish or species related to each other, or to time, or to season, etc. And how much can we trust the data, since it is impossible for a human to visually check all 1+ billion detections and classifications. The user interface team developed a facet-based display (see Chap. 13), that allowed users to apply different filters to the data and select different presentations of the data. New visual methods for presenting the amount of uncertainty in the data were also developed (see Chap. 15) as well as an attractive SecondLife interactive environment (see Chap. 17), which was built to help promote the project results in a format that might appeal to the public imagination (although attractive, the Fish4Knowledge pavilion had very few visitors).

Being one of the first 'big-data' computer vision projects devoted to answering research questions for domain experts, we learned the importance of user-accessible ground-truth data and its presentation, and the importance of interpreting the computer vision results in terms of the domain expert's statistical needs. Insights gained from the project allowed us to explore the boundary between computer-vision centric measures such as F_1 and statistical methods used in the marine ecology domain.

The final achievement was on the human level: the project helped develop the research skills of 13 young researchers, as well as 7 more experienced researchers, through about 36 person-years of effort. Certainly, each person made their own special contribution to the project, but we tried to ensure that all were engaged at some level in all phases and activities of the project. This gave everyone new skills in data capture and analysis, parallel processing and system control, data presentation and understanding user requirements.

18.2 Critical Assessment

What worked well in the project:

1. **Video-based 'big data'**: Big data is clearly a hot topic at the moment, but most big data systems are based on analyzing text streams (e.g., from Twitter or Google) or large sensor data streams (e.g., from the Large Hadron Collider). Probably the most similar large databases are from Flickr and astronomical catalogs (e.g., Galaxy Zoo). These services are based on large image collections, whereas the Fish4Knowledge project, at least at the initial data analysis stages, is based on video. It is quite possibly the largest analyzed non-classified video dataset. (The YouTube video dataset is larger, but it is not known to be analyzed.)
2. **Good detections even with hard to analyze video**: Much surveillance research is based on detecting changes in scenes observed by fixed cameras, as in this project. However, here the constantly changing illumination patterns create a considerable challenge for the detection algorithms, which, nonetheless, were effective.
3. **Recognition even with very imbalanced classes and blurry images**: The recognition algorithms were successful, with up to 97 % of observed fish correctly recognized, which is remarkable considering the greatly imbalanced nature of the dataset and low quality images.
4. **New methods for more efficient and fun ground-truthing**: Generating 'ground-truth' datasets for training and evaluating algorithm performance is always necessary, but can be tedious, and often requires many person-months of repetitive image labeling. The Fish4Knowledge project developed a number of new approaches to ground-truthing that were either 4–5 times more efficient (through a clustering approach) or more fun (by being implemented within a computer game).
5. **New ways to explore ecological data**: Even if a marine ecologist had access to theSQL result table from the analysis of the video data, it would still be a challenge to analyze, because it would still require substantial programming skills. This is probably still the case for the more detailed explorations of the database, but thefacet-based user interface we designed allows at least preliminary exploration of the full database in a compact, yet option-rich manner. The use of summary tables that accumulate counts of fish by species, date, time, camera, image quality, and site allowed the user interface to respond quickly.
6. **Research Impact**: It is not long since the project finished, but over 2,500 researchers have looked at the video and image datasets collected here, potentially for future experiments. Moreover, both code and the user interface to the fish detection database are publicly available.

What worked less well were in these areas:

1. **Consistency of the Results**: The detection and recognition data in the database was the result of processing performed over most of a full year. In part this was due to the large amount of computation required and the amount of computing resources available. The spread was also the result of continuing development

of the detection, tracking and recognition algorithms. The consequence of this spread in development and analysis is that the earlier results are not of the same quality as the later results: more false and missed detections, trackings that paired fish incorrectly, and species not recognized because the earliest production algorithm only recognized the top 10 species (although this was more than 90 % of the observations). We tried to maintain a record of the data provenance by associating each result with the algorithms used to compute the result. There was, however, insufficient project and computing time available to make a final pass through the data with the best algorithms. It is confusing for the analysts to know how to use the results obtained from different algorithms (each with their own imperfections). New methods for combining data with different stages of provenance are needed. While domain experts (and computer scientists) may find mixtures of results from different versions algorithms undesirable, it is likely to become a common situation in digital science projects, where data quantities overwhelm the capability to process the data. Improved analytic tools are likely to be developed as scientific and algorithmic developments occur, but they will never be applied to all old data. The handling of mixed provenance data will be an important research question for many future 'big data' projects, whether environmental or otherwise.

2. **Quality of Results**: While we attempted to assess the accuracy of the algorithms as given above, this assessment was based on a set of manually analyzed ground-truth data. The amount of ground-truth data was so small compared to the full dataset (e.g., 10^5 detections out of 10^9 were used for the recognition ground truth). This leads one to worry if there was enough variation included in the data used for training the different algorithms. Based on the rejection statistics from the final recognition algorithm, it is possible that something like 30 % of the detections in the full, uncleaned, database are false detections. But we really do not know at this stage. Clearly, more research is needed into methods for helping clean these sorts of huge datasets (as well as how to put cleaner data into the database in the first place).
3. **Researcher Needs**: We developed a user interface that allows easy exploration of the full population data in a number of ways. This overcame the problem of how to do marine science with only a few hours of data samples obtained by human divers. But the marine ecologists we were working with, in the end, did not seem too excited by what we had developed. This disinterest seemed to arise from two perspectives: (1) the sorts of questions that the marine ecologists wanted to ask were much more focused on specific issues than the broad population statistics that we were obtaining. (2) There seemed to be a continuing concern about whether our data and analyses were accurate. We believe that there are many errors, but a population and statistical perspective can still allow meaningful conclusions (and there must also be many errors in human data collected using divers). Perhaps one could view the approach developed here as a new way of doing marine ecology, and it will take a while before it becomes an accepted approach.

18.3 What Lies in the Future

While much was achieved in the course of the project, each achievement raised the prospect of further developments. We outline a few of these here:

- **Image Capture**: Much of the data was 320 × 240 and 5 frames per second. Higher quality would improve detection rates, allow capture of smaller and more distant fish, and improve the consistency of tracking, as fish would not move as much between frames.
- **Image Analysis**: Higher resolution and thus also larger fish images would allow better analysis of the fish boundary, colors and textures, which could lead to higher species recognition accuracy.
- **Image Analysis**: Exploiting foreground knowledge of the different species for better detection and tracking. All fish were considered equally, but clearly different species have different behaviors, especially in terms of speeds and preferences of positioning with respect to the coral. The benefit of this extra information could be investigated.
- **Image Analysis**: Dataset 'cleaning'. Although the detection algorithms performed well considering the difficult illumination and other effects, it seems likely that there are a substantial number of false fish detections and recognitions. Developing additional filters to remove or relabel these would enhance the value of the dataset for the marine ecologists, and increase their confidence in this approach to data collection and analysis. Developing techniques for marking data as potentially suspect, or computing data confidence could be useful, as this would allow future analysts to be aware of the statistics of the quality of the results, or to develop algorithms that focus on the more reliable data.
- **Image Analysis**: Speed. Many of the algorithms were coded in Matlab. Recoding in e.g., C/C++ could potentially reduce computation time by a factor of 10–50. Recoding for a GPGPU could potentially speed up some algorithms by a further factor of 10, but a full exploitation of this speedup would require the 100 CPU cores typically used to each have its own GPGPU, which is clearly expensive. This recoding would require substantial human labor potentially 1–2 person years of effort.
- **Workflow**: Self-monitoring and self-repair. With so many cores in operation, processes would periodically crash. Investigation into more sophisticated methods for massive computation self-monitoring and self-repairing workflow would reduce the occasional manual monitoring, restart and repair of processing queues and machines.
- **Computational System**: Investigation into methods for communication and resource control across heterogeneous architectures. Currently, all data is shared via the SQL database, which adds read/write and network latencies, and bottlenecks when up to 1,000 cores are in action. The current data processing routines are sufficiently slow that the extra latency is relatively tiny, but one might implement direct sharing of results along pipes or messages between consecutive processes. Another option would be to migrate processes to where the data is stored. Both

would require more sophisticated process control, as currently consecutive process stages could execute days apart.

- **Domain Expert Acceptance**: Developing methods for conveying the correctness of results from massive data sets. Given that it is impossible to completely clean a massive dataset, methods are needed for helping users form valid conclusions and assess the scale of error in those conclusions.
- **System Evaluation**: Developing methods for more efficiently acquiring ground-truth. Both algorithm development (and training) and evaluation need ground truth data, but acquiring this so far has required considerable human effort. Further development of efficient methods would be beneficial, and some preliminary benefits of clustering similar examples have been seen.
- **System Evaluation**: Developing methods for estimating the correctness of results when only a tiny proportion of the data can be manually evaluated. Traditional image algorithm analysis has evaluated performance by comparison of a substantial portion of the dataset to some 'ground-truth' measure. Here, even our substantial ground-truth (starting from 10^5 samples) is only a tiny 0.0001 of the full dataset. New methods of analysis are needed, as well as a better understanding by domain experts of how much ground truth is needed to give them confidence in the analysis.
- **System Reusability**: We have been asked by other projects about reuse of our code and databases. In theory, this is a good idea, but there are many problems: the image capture and communication hardware was specialized for the particular Taiwan configuration, the video processing was configured for the specific supercomputer architecture, the processing cores were dedicated, the workflow scheduler was implemented for the available processor pool, the recognition algorithms were trained for a specific hand-acquired ground truth dataset, etc. The detection and tracking software seems more portable, but still depends on fixed cameras and largely stationary background.
- **User Needs**: The focus on computing abundance did not excite the marine ecologists as much as we had hoped, as they were often focused on more specific questions. In many cases, the system probably recorded video containing the data that they wanted. How to make this data easily accessible without requiring substantial programming skills is an open question. In theory, researchers proficient in SQL could analyze the database, but this is still a burden that we would like to be able to avoid.

18.4 Project Publications

This is a mostly complete list (at the time of the book printing) of the publications that were produced as either part of the Fish4Knowledge project, or afterwards from the research team using the data acquired during the project. Many of the papers are online either from the project website: http://groups.inf.ed.ac.uk/f4k/publication.htm or the individual authors' websites.

18.4.1 Fish Detection and Tracking

1. M. Aldinucci, C. Spampinato, M Drocco, M. Torquati, S. Palazzo, D. Giordano. "A Parallel Edge Preserving Algorithm for Salt and Pepper Image Denoising", Proc. 3rd IEEE Int. Conf. on Image Processing Theory, Tools & Applications, Istanbul, Turkey, 15–18, 2012.
2. S. Palazzo, I. Kavasidis, C. Spampinato, "Covariance based modeling of underwater scenes for fish detection", Proc. Int. Conf. on Image Processing, Melbourne, 2013.
3. S. Palazzo, C. Spampinato, C. Beyan. "Event Detection in Underwater Domain by Exploiting Fish Trajectory Clustering", Proc. ACM International Workshop on Multimedia Analysis for Ecological Data, MAED'12, Nara, Japan, 2012.
4. S. Roccella, "Video Enhancement by Superresolution", MS thesis (in Italian), Univ of Catania, 2013.
5. C. Spampinato, E. Beauxis-Aussalet, S. Palazzo, C. Beyan, J. van Ossenbruggen, J. He, B. Boom, X. Huang, X. "A rule-based event detection system for real-life underwater domain", Machine Vision and Applications, 2013, pp 1–19.
6. C. Spampinato, S. Palazzo, D. Giordano, I. Kavasidis, F-P Lin, Y-T Lin, "Covariance Based Fish Tracking In Real-Life Underwater Environment", Proc. Int. Conference on Computer Vision Theory and Applications—VISAPP 2012, Rome, Italy, 2012.
7. C. Spampinato, S. Palazzo, D. Giordano, "Evaluation of Tracking Algorithm Performance without Ground-Truth Data", Proc. Int. Conf. on Image Proc. (ICIP), 2012.
8. C. Spampinato, B. Boom, J. He (Eds), VIGTA'12: Proc. 1st Int. Workshop on Visual Interfaces for Ground Truth Collection in Computer Vision Applications, ACM AVI 2012, Capri, Italy, May 21–21, 2012.
9. C. Spampinato, S. Palazzo. "Enhancing Object Detection Performance by Integrating Motion Objectness and Perceptual Organization", Proc. 21st Int. Conf. on Pattern Recognition (ICPR), 2012, Tsukuba Science City, Japan.
10. K. Vougioukas, B. J. Boom, R. B. Fisher, "Adaptive Deblurring of Surveillance Video Sequences that Deteriorate over Time", Proc. Int. Conf. on Image Processing, Melbourne, 2013.
11. Konstantinos Vougioukas, "Adaptive filters to remove blurring effects over time for underwater surveillance", MSc Dissertation, School of Informatics, Univ. of Edinburgh, 2012.

18.4.2 Fish Species Classification and Behavior Analysis

12. C. Beyan, R. B. Fisher, "A Filtering Mechanism for Normal Fish Trajectories", Proc. 21st Int. Conf. on Pattern Recognition (ICPR), 2012.

13. C. Beyan, R. B. Fisher, "Detection of Abnormal Fish Trajectories Using a Clustering Based Hierarchical Classifier", Proc. British Machine Vision Conference, 2013.
14. C. Beyan, R. B. Fisher, "Detecting Abnormal Fish Trajectories using Clustered and Labeled Data", Proc. Int. Conf. on Image Processing, Melbourne, 2013.
15. C. Beyan, "Classifying Imbalanced Data Sets Using Similarity Based Hierarchical Decomposition", Pattern Recognition, Pattern Recognition, 2014.
16. C. Beyan, B. J. Boom, J. M. P. Liefhebber, K.-T. Shao, R. B. Fisher, "Natural Swimming Speed of *Dascyllus reticulatus* Increases with Water Temperature", Marine Science, to appear.
17. C. Beyan, "Detection of Unusual Fish Trajectories from Underwater Videos", PhD thesis, Univ. of Edinburgh, 2015.
18. P. X. Huang, B. J. Boom, R. B. Fisher, "Underwater Live Fish Recognition using a Balance-Guaranteed Optimized Tree", Proc. Asian Conf. on Computer Vision, Daejeon, Korea, 2012.
19. P. X. Huang, B. J. Boom, R. B. Fisher, "Hierarchical Classification for Live Fish Recognition", Proc. BMVC student workshop, September 2012.
20. P. X. Huang, B. J. Boom, R. B. Fisher; "Hierarchical classification with reject option for live fish recognition", Machine Vision and Applications, 2014.
21. X. Huang, "'Balance-Guaranteed Optimized Tree with Reject option for live fish recognition", PhD thesis, Univ. of Edinburgh, 2014.
22. Yan Li, "Fish Component Recognition", MSc Dissertation, School of Informatics, Univ. of Edinburgh, 2012.
23. Xiao Liu, "Identifying individual clown fish", MSc Dissertation, School of Informatics, Univ. of Edinburgh, 2013.
24. Y.-H. Shiau, F.-P. Lin, C.-C. Chen, "Using Sparse Representation for Fish Recognition and Verification in Real World Observation", Proc. Int. Workshop on Visual observation and Analysis of Animal and Insect Behavior (VAIB), in conjunction with ICPR 2012, Tsukuba, Japan, 2012.
25. C. Spampinato, S. Palazzo, B. Boom, J. van Ossenbruggen, I. Kavasidis, R. Di Salvo, F.-P. Lin, D. Giordano, L. Hardman, R. B. Fisher, "Understanding Fish Behavior during Typhoon Events in Real-Life Underwater Environments", Multimedia Tools and Applications, 2012.
26. C. Spampinato, S. Palazzo "Hidden Markov Models For Detecting Anomalous Fish Trajectories In Underwater Footage", Proc. IEEE Machine Learning for Signal Processing Workshop, September 23–26, 2012.
27. C. Spampinato, A. Faro, S. Palazzo, "Event Detection in Crowds of People by Integrating Chaos and Lagrangian Particle Dynamics", Proc. 3rd Int. Conf. on Information and Multimedia Technology (ICIMT 2011), Dubai, UAE, December 28–30, 2011.

18.4.3 User Needs and Information Presentation

28. E. Beauxis-Aussalet, E. Arslanova, L. Hardman, and J. van Ossenbruggen. "A case study of trust issues in scientific video collections". Proc. 2nd ACM international workshop on Multimedia Analysis for Ecological Data. ACM, 2013.
29. E. Beauxis-Aussalet, S. Palazzo, G. Nadarajan, E. Arslanova, C. Spampinato, L. Hardman. "A Video Processing and Data Retrieval Framework for Fish Population Monitoring". Proc. 2nd ACM international workshop on Multimedia Analysis Ecological Data. ACM, 2013.
30. E. Beauxis-Aussalet, L. Hardman. "Simplifying the Visualization of Confusion Matrix". Proc. 26th Benelux Conference on Artificial Intelligence (BNAIC). 2014.
31. E. Beauxis-Aussalet, L. Hardman. "Multifactorial Uncertainty Assessment for Monitoring Population Abundance using Computer Vision". Proc. International Conference on Data Science and Advanced Analytics (DSAA). IEEE, 2015.
32. E. Beauxis-Aussalet, L. Hardman. "Multi-Purpose Exploration of Uncertain Data for the Video Monitoring of Ecosystems". Proc. Eurovis international workshop on visualization in environmental sciences (EnvirVis). Eurographics, 2015.
33. E. Beauxis-Aussalet, E. Arslanova, L. Hardman. "Supporting Non-Experts' Awareness of Uncertainty: Negative Effects of Simple Visualizations in Multiple Views". Proc. 33rd European Conference on Cognitive Ergonomics (ECCE). ACM, 2015.
34. B. J. Boom, E. Beauxis-Aussalet, L. Hardman, R. B. Fisher. "Uncertainty-Aware Estimation of Population Abundance using Machine Learning". ACM Multimedia Systems Journal, 2015, pp1–13.
35. J. He, V. Hollink, A. de Vries. "Combining implicit and explicit topic representations for result diversification", Proc. 35th Int. ACM SIGIR Conf. on Research and development in information retrieval (SIGIR'12), pp 851–860, 2012.
36. Jiyin He, Maarten de Rijke, Merlijn Sevenster, R. van Ommering, and Yuchen Qian. "Generating Links to Background Knowledge: A Case Study Using Narrative Radiology Reports", Proc. 20th ACM Conference on Information and Knowledge Management (CIKM 2011), pp. 1867–1876, Glasgow, ACM, October, 2011.
37. Jiyin He, Marc Bron and Maarten de Rijke. "A Query Performance Analysis for Result Diversification", Proc. ICTIR '11: 3rd International Conference on the Theory of Information Retrieval, pp. 351–355, Sep. 2011.
38. Vera Hollink, Jiyin He, and Arjen P. de Vries. "Explaining query modifications: an alternative interpretation of term addition and removal". Proc. 34th European Conference on Information Retrieval (ECIR'12), Barcelona, Spain, 2012.
39. S. Theis, "Comparing efficiency and cognitive load of uncertain data visualizations", Thesis Master Information Studies, Universiteit van Amsterdam, July 2011.

18.4.4 System Architecture and Overview

40. B. J. Boom, P. X. Huang, C. Spampinato, S. Palazzo, J. He, C. Beyan, E. Beauxis-Aussalet, J. van Ossenbruggen, G. Nadarajan, J. Y. Chen-Burger, D. Giordano, L. Hardman, F.-P. Lin, R. B. Fisher, "Long-term underwater camera surveillance for monitoring and analysis of fish populations", Proc. Int. Workshop on Visual observation and Analysis of Animal and Insect Behavior (VAIB), in conjunction with ICPR 2012, Tsukuba, Japan, 2012.
41. B. J. Boom, J. He, S. Palazzo, P. X. Huang, H.-M. Chou, F.-P. Lin, C. Spampinato, R. B. Fisher; "A research tool for long-term and continuous analysis of fish assemblage in coral-reefs using underwater camera footage", Ecological Informatics, 2014.
42. H.M Chou, S.I Lin, J.H Chang, F.P Lin. "A Practical Use Case of Cloud Computing Supports Research Collaboration—Data Management and Analysis for Long Term Video Monitoring of Coral Reef Fishes", Proc. Cloud for Research Collaboration Workshop, 2013
43. S.W. Lo and F.P. Lin, "Video Query Using Temporal Signature and Similarity Matching", Proc. 2nd Int. Conf. on Engineering and Technology Innovation, Kaohsiung, Taiwan, 2012.
44. S.W. Lo and F.P. Lin, "A Shot Detection Approach to Synchronize Stereoscopic Video", Proc Int. Symp. on Computer, Communication, Control, 2012. Also J. Applied Mathematics & Information Sciences, pp 43–49, June 2013.
45. S. McDonagh, C. Beyan, P. X. Huang, R. B. Fisher, "Applying semi-synchronised task farming to large-scale computer vision problems", Int. Journal of High Performance Computing, 2014, online doi:10.1177/1094342014532965.
46. G. Nadarajan, Y.-H. Chen-Burger, R. B. Fisher. "Semantics and Planning Based Workflow Composition for Video Processing". Journal of Grid Computing, Special Issue on Scientific Workflows, 2013, pp 523–551.
47. G. Nadarajan, C.-L. Yang, Y.-J. Cheng, S.-I. Lin, Y.-H. Chen-Burger, F.-P. Lin. "Real-time Data Streaming Architecture and Intelligent Workflow Management for Processing Massive Ecological Videos". Proc. Applications and Theories for Scientific Big Data Workshop (ATSBD), in conjunction with ASE/IEEE BigData2013, Sept 2013.
48. G. Nadarajan, C.-L. Yang, Y.-H. Chen-Burger. "Multiple Ontologies Enhanced with Performance Capabilities to Define Interacting Domains within a Workflow Framework for Analyzing Large Undersea Videos", Proc. 5th Int. Conf. on Knowledge Engineering and Ontology Development (KEOD 2013), Sept 2013.
49. G. Nadarajan, Y.-H. Chen-Burger. "Goal, Video Description and Capability Ontologies for Fish4Knowledge Domain", Proc. Special Session on Intelligent Workflow, Cloud Computing and Systems, QUES-AMSTA 12, Croatia, June 2012.
50. G. Nadarajan, Y.-H. Chen-Burger, R. B. Fisher, "SWAV: Semantics-based Workflows for Automatic Video Analysis", Proc. Special Session on Intelligent Work-

flow, Cloud Computing and Systems, QUES-AMSTA 11, Manchester, U.K., July 2011.
51. G. Nadarajan, Y.-H. Chen-Burger, R. B. Fisher, C. Spampinato, "A Flexible System for Automated Composition of Intelligent Video Analysis", Proc. Int. Symp. on Image and Signal Processing and Analysis, Dubrovnik, Sept. 4–6, 2011.
52. G. Santoro, C. Pino, D. Giordano. "A Semantic Based Retrieval System of Arctic Animal Images", Proc. ACM Int. Workshop on Multimedia Analysis for Ecological Data, MAED'12, Nara, Japan, 2012.
53. Y.H. Shiau, F.P. Lin, C.C. Chen, "Real-Time Fish Observation and Fish Category Database Construction", Proc. 21st Int. Conf. on Pattern Recognition, November 11–15, 2012, Tsukuba, Japan.
54. Y.H. Shiau, S.I. Lin, Yi-Hsuan Chen, S.-W. Lo, C.C. Chen, "Fish Observation, Detection, Recognition and Verification in The Real World", Proc. Int. Conf. on Image Processing, Computer Vision, and Pattern Recognition, Las Vegas , USA, 2012.
55. C. Spampinato, V. Mezaris, J. van Ossenbruggen, "Multimedia Analysis for Ecological Data", Proc. ACM Multimedia, Nara, Japan, November 2012.
56. Ihunanya Martina Ugwuh, "Exploiting Parallelism in Video and Image Processing Tasks: A-state-of-the-art survey", MSc Dissertation, School of Informatics, Univ. of Edinburgh, 2012.

18.4.5 System Evaluation and Data Ground Truthing

57. Bastiaan J. Boom, Phoenix X Huang, Robert B. Fisher. "Approximate Nearest Neighbor Search to Support Manual Image Annotation of Large Domain-specific Datasets", Proc. Int. Workshop on Video and Image Ground Truth in Computer Vision Applications (VIGTA'13), VIGTA 2013, St Peterburgh
58. B. J. Boom, P. X. Huang, J. He, R. B. Fisher, "Supporting Ground-Truth annotation of image datasets using clustering", Proc. 21st Int. Conf. on Pattern Recognition (ICPR), 2012.
59. R. Di Salvo, D. Giordano, I. Kavasidis, "A Crowdsourcing approach to support video annotation", Proc. Int. Workshop on Video and Image Ground Truth in Computer Vision Applications (VIGTA'13), St. Petersburg, 2013.
60. R. B. Fisher, B. J. Boom, P. X. Huang; "Preliminary Experiments with the Fish4Knowledge Dataset", Proc. Int. Workshop on Visual observation and Analysis of Vertebrate and Insect Behavior (VAIB), in conjunction with ICPR 2014, Stockholm, 2014.
61. J. He, J. van Ossenbruggen, and A. P. de Vries. "Fish4label: Accomplishing an Expert Task without Expert Knowledge". Proc. Open Research Areas in Information Retrieval (OAIR-13), 10th International Conference in the RIAO series, 2013.

62. J. He, J. van Ossenbruggen, and A. P. de Vries. "Do you need experts in the crowd? A case study in image annotation for marine biology", Proc. Open Research Areas in Information Retrieval (OAIR-13), 10th International Conference in the RIAO series, 2013.
63. I. Kavasidis, S. Palazzo, R. Di Salvo, D. Giordano, C. Spampinato, "An innovative web-based collaborative platform for video annotations", Multimedia Tools and Applications, 2013, pp 1-20.
64. I. Kavasidis, C. Spampinato, D. Giordano, "Generation of Ground Truth for Object Detection While Playing an Online Game: Productive Gaming or Recreational Working?", Proc. 2013 IEEE on Computer Vision and Pattern Recognition Workshops (CVPR-13), Portland, USA, June 2013.
65. I. Kavasidis, S. Palazzo, R. Di Salvo, D. Giordano, C. Spampinato, "A Semi-automatic Tool for Detection and Tracking Ground Truth Generation in Videos", Proc. First Int. Workshop on Visual Interfaces for Ground Truth Collection in Computer Vision Applications, Capri, May 2012.
66. I. Kavasidis, S. Palazzo. "Quantitative Performance Analysis of Object Detection Algorithms on Underwater Video Footage", Proc. ACM Int. Workshop on Multimedia Analysis for Ecological Data, MAED'12, Nara, Japan, 2012.

Reference

Liu, X. 2013. Identifying Individual Clown Fish. Master's thesis, School of Informatics, University of Edinburgh.

Glossary

Note: although some of these terms have wider usages, we gloss the terms as used in the context of the Fish4Knowledge project.

Abundance a measure of the number of biological individuals in an area
Accuracy the percentage of a set of detections or classifications that is correct
ALPS Advanced Large-scale Parallel Supercluster at Taiwan's National Center for High Performance Computing
Affinity Propagation a technique for clustering samples by using their similarity
Average Precision the classification Precision of each class averaged over all classes
Average Recall the classification Recall of each class averaged over all classes
Background model an approach to describing the scene background that lies behind the moving foreground objects
Batch job a computer processing task that has been set to run without a person waiting, e.g. at a display screen, for the results
BGOTR Balance-Guaranteed Optimized Tree with Reject option—the hierarchical classification method used in this project for identifying the fish species
Big-data a set of problems involving large datasets, e.g. many millions of data items
CART Classification And Regression Tree—a method for hierarchical classification
Classifier an algorithm for deciding which class an object (or some data) belongs to
Clustering a method for grouping objects into different subsets
Confusion Matrix a method for displaying how many members of one class are identified or misidentified as members of each possible class
Crowdsourcing solving a problem by using work (often paid) from a large number of people
Detection discovering an object in an image or video
European Union a political grouping of European countries that funded much of the research described in this book
F1 Score the harmonic mean combining precision and recall
F4K an abbreviation for the EU Framework 7 funded Fish4Knowledge project
Facet-based user interface an approach to building a User-Interface that allows users to see different views of their data
False Negative (FN) a Positive data sample that is classified as Negative
False Positive (FP) a Negative data sample that is classified as Positive

R.B. Fisher et al. (eds.), *Fish4Knowledge: Collecting and Analyzing Massive Coral Reef Fish Video Data*, Intelligent Systems Reference Library 104,
DOI 10.1007/978-3-319-30208-9

Field of View the portion of a scene observed by a camera

Fish4Knowledge the name of the European Union funded project that led to the work described in this book

Filtering a set of methods for removing unwanted data, e.g. noise, bad data, irrelevant data, etc

Foreground model an approach to describing the objects that move in front of a background

Gaussian mixture model/GMM a method for modeling a probability distribution that has more than one mode

Ground truth a specification of the 'true answer' of some calculation, usually for comparison to the calculated answer for error analysis, or for use in training an algorithm

Hierarchical classification deciding what type an object is by first deciding which group of classes it belongs to, and then which subgroup of classes, and so on until a single class is identified

Hierarchical decomposition breaking up a set of data into smalled subsets, and then splitting the subsets into smaller subsets, etc

Imbalanced classes when there is a considerable difference in the likelihood of seeing a sample from one class as compared to a sample from another class

Kernel density estimation a way of modeling and estimating the, possibly complicated, probability distribution of a variable based on samples from the distribution

OpenSimulator An open source virtual world server. Also called OpenSim.

Ontology a way of specifying the relationships between different concepts (e.g. the fact that a cat is an animal). In this book, we have taken a semi-formal approach, where standard formal notations have been used, accompanied by informal explanatory descriptions.

PCA Principal Component Analysis—a data analysis method that identifies relationships in the data

Precision the percentage of objects extracted in a database that are of the correct type

Query locating something in a database using some index information

RDF Resource Description Framework—a specification for web-based data interchange

Recall the percentage of a set of the total possible objects of a given type that are succesfully located in a database

Second Life™ a large online virtual world platform by Linden Labs, programmable and tailorable by users, which has been used for creative, social, business and educational purposes.

Similarity Score a measure of how similar two items are

SKOS Simple Knowledge Organization System—a schema for storing data in the web

Species—the bottom level classification of a biological organism (followed by genus, family, etc)

SQL Structured Query Language—a special purpose language for extracting results from a relational database

SVM Support Vector Machine—an algorithm for deciding which class an object belongs to

Texton a small patch of texture in an image

Tracking connecting the same object detected in consecutive video frames together

Trajectory the path that a detected and tracked object makes as it moves

True Negative (TN) a negative sample which is correctly classified as Negative

True Positive (TP) a positive sample which is correctly classified as Positive

Type I Error a False Positive error where a False sample is classified as True

Type II Error a False Negative error where a True sample is classified as False

VIBE—a technique for detecting moving objects against a background

Video clip a short segment of video

VIP task a projessing job based on one of a set of Visual Image Processing modules

Virtual Machine when a computer acts as if it were a different machine (by simulation)

Vue/Virtual University of Edinburgh a virtual space in Second Life and OpenSimulator containing a Fish4Knowledge building

Windrider a supercomputer at the Taiwan National Center for High Performance Computing

Workflow composition linking together a combination of sequential and parallel smaller tasks to form a larger task

Virtual workflow machine a workflow machine whose execution procedure is not hard-coded, but which will dynamically select and execute processes based on the process instructions (process model) at run time.

Visualization presenting information by way of an image or video

Appendix A
User Interface and Usage Scenario

Emma Beauxis-Aussalet and Lynda Hardman

Abstract This appendix describes a user interface for visualizing the fish monitoring data extracted from Fish4Knowledge collection of video footage (Sect. A.1). It provides functionalities for exploring the computer vision results, as well as their uncertainties. Its interaction design offers intuitive and flexible navigation through the complex, multidimensional dataset and metadata about provenance and software features. It addresses the user needs discussed in Chap. 2, and implements the means to deal with uncertainty as introduced in Chap. 13. Information is organized in a series of tabs reflecting the levels of information processing: *Video*, *Video Analysis*, *Raw Data*, *Visualization* and *Report*. The tabs give access to explanation and visualization of uncertainty introduced at each information processing level. The user interface is described in Sects. A.1 and A.2 illustrates its usage for exploring fish counts, investigating potential uncertainties, and reporting ecological findings. The interface is available at this URL http://f4k.project.cwi.nl/demo/ui/.[1]

A.1 User Interface

The user interface deals with two key challenges while remaining intuitive and user-friendly. First, it deals with the complexity of the multiple dimensions of the extracted information, and with the description of the underlying processes that provided the information. Second, it deals with the diversity of user goals, as ecologists may perform a variety of data analysis tasks focusing on specific species, locations or time periods. The user interface design uses tabs to deliver manageable units of information that reflect the information processing sequence: data collection (Video tab), data

[1]This research prototype was tested from Chrome only. For backup purposes, and faster access from Asia, the interface is also available at http://gleoncentral.nchc.org.tw.

R.B. Fisher et al. (eds.), *Fish4Knowledge: Collecting and Analyzing Massive Coral Reef Fish Video Data*, Intelligent Systems Reference Library 104,
DOI 10.1007/978-3-319-30208-9

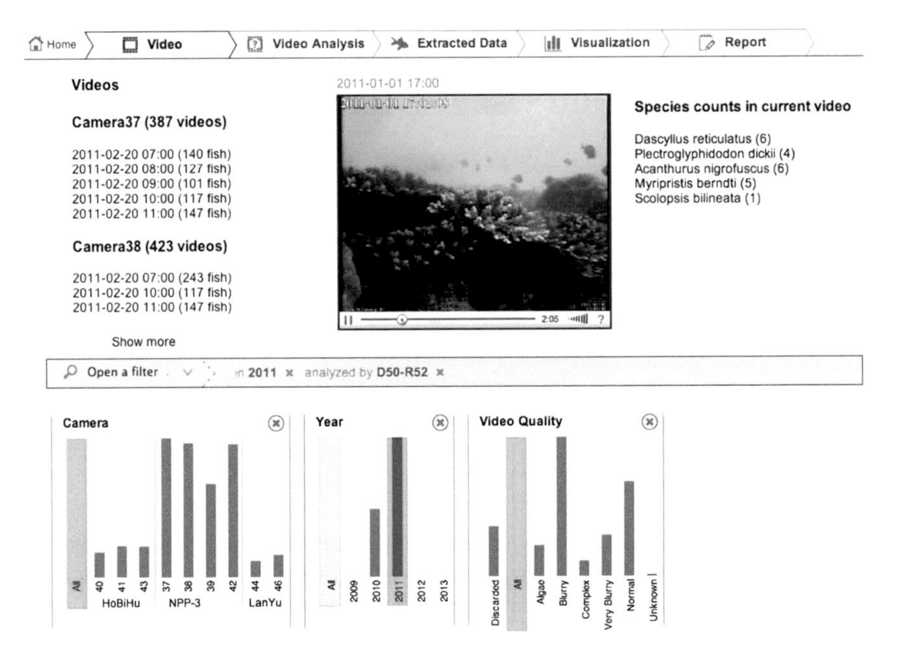

Fig. A.1 The Video Tab. In this example, the user is browsing videos from weeks 8–12 of 2011, analyzed by the software version *D50-R52*

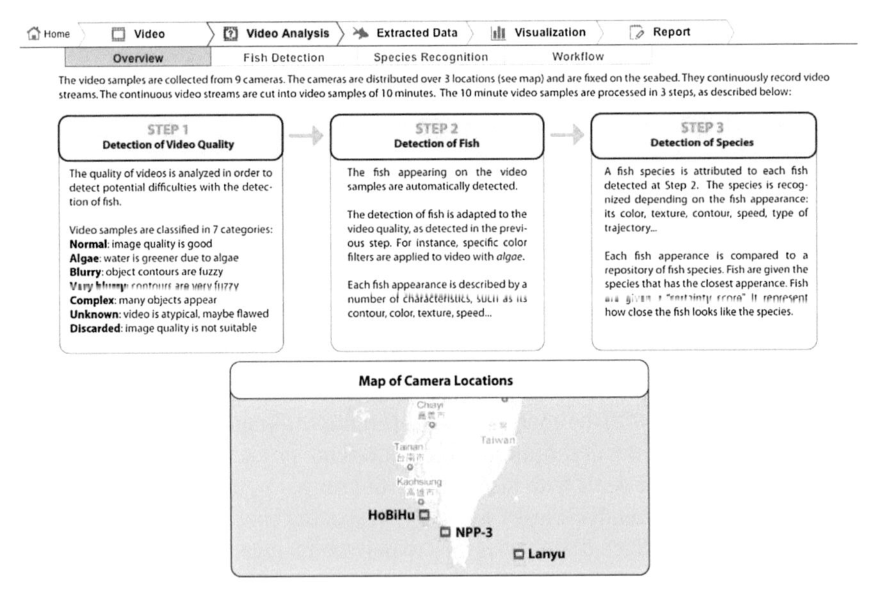

Fig. A.2 The video analysis tab—overview sub-tab provides explanations of the video processing steps

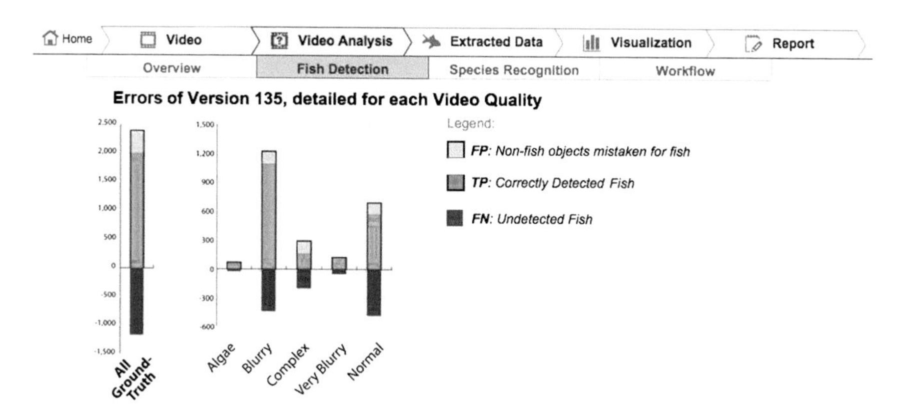

Fig. A.3 The video analysis tab—fish detection sub-tab provides simplified visualizations of ground-truth evaluations of the fish detection component. Evaluations are provided for each video quality

processing (Video Analysis, Extracted Data tabs) and data interpretation (Visualization, Report tabs). Information about fish populations, the data collection method, and the potential uncertainty are gradually exposed in dedicated views. Within the tabs, users can access on-demand information, depending on their specific needs. This approach supports the exploration of task-specific end-results, and provides access to the underlying complexity and uncertainties introduced by the processing within the system.

The Video tab supports video browsing, as shown in Fig. A.1. It contains filtering functionalities for specifying the videos of interest (e.g., at specific locations or time periods). With this tab, users can inspect the data collection conditions: which ecosystems are observed, over which time periods, and with which field of view and image quality (e.g., lens biofouling, water turbidity).

The Video Analysis tab provides explanations of the video processing steps, and visualizations of the computer vision errors. It exposes the technical concepts needed for understanding computer vision uncertainty. The Overview sub-tab provides explanations of the main video processing steps (Fig. A.2). The Fish Detection, and Species Recognition sub-tabs provide visualizations of ground-truth evaluations (Figs. A.3 and A.4). The Workflow sub-tab provides on-demand video processing (Fig. A.5). Users can request the analysis of specific videos (e.g., from time periods and cameras of interest), with specific software component versions (e.g., with the best accuracy for the species of interest). It serves either for processing videos that were not yet analyzed, or for experimenting with different versions of the video analysis components (e.g., to check robustness of observations).

The Extracted Data tab provides an overview of the available video data and their properties (i.e., their dimensions). It shows all the characteristics of the fish monitoring information extracted from the video footage. It also explains the 4 main metrics provided for describing fish populations and their uncertainty: Number of Fish, Number of Video Samples (e.g., to check for missing videos), Mean

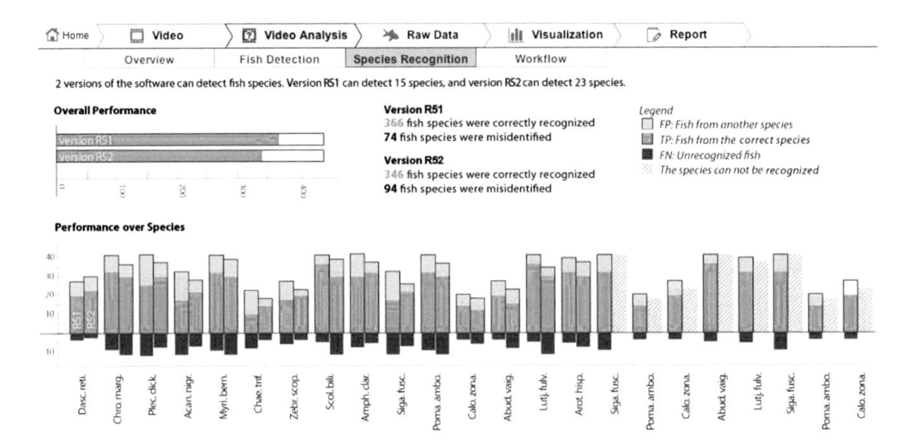

Fig. A.4 The video analysis tab—species recognition sub-tab provides simplified visualizations of ground-truth evaluations of the species recognition components. Evaluations are provided for each species

Fig. A.5 The video analysis tab—Workflow sub-tab supports user requests for specific video analyses. The ground-truth evaluation of the component versions users plan to use are shown

Abundance per 10-min (e.g., to compensate for missing videos, as discussed in Chap. 13 Sect. 13.4), and Number of Species (e.g., for studying *species richness*). This tab helps understanding how fish populations can be monitored, and identifying the information relevant for particular studies (Fig. A.6).

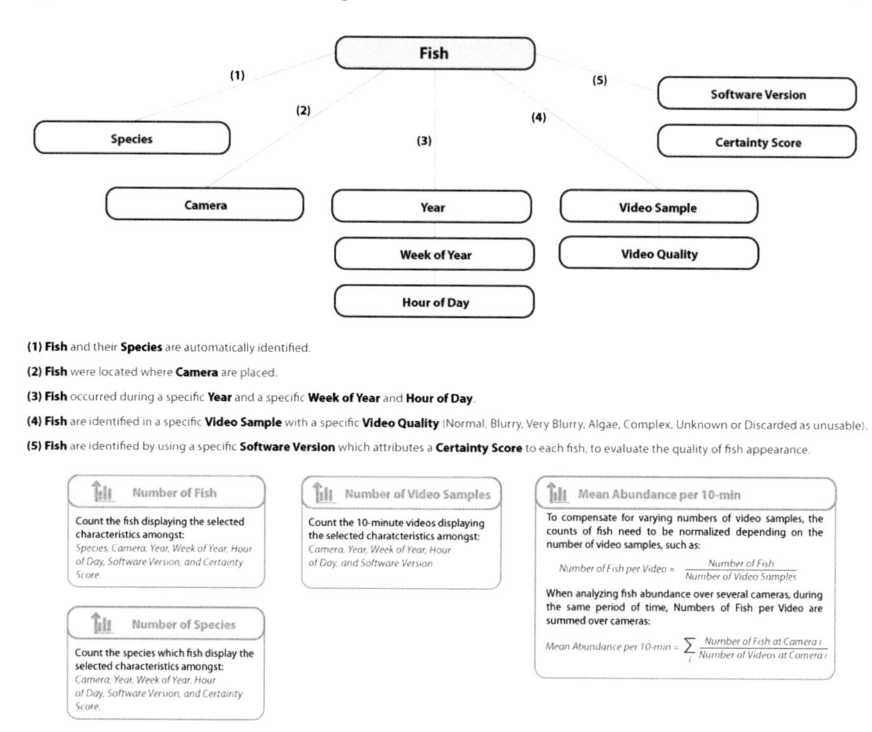

Fig. A.6 The extracted data tab provides a schema of the video analysis data, and explanations of the Y axis metrics

The Visualization tab, shown in Fig. A.7, provides means to explore fish counts and their characteristics. In Fig. A.7, *Zone A* contains the main graph, and *Zones B and C* support the adaptation of the main graph to specific user needs. In *Zone B* users can specify what the axes of the main graph represent. For instance, while the y-axis represents numbers of fish, the x-axis can represent their distribution over weeks of the year or hours of the day. Users can also select other types of graph (e.g., stacked chart or boxplot). These graph provide additional information about the visualized fish population, e.g., the proportion of each species shown in Fig. A.7. The selection of stacked charts or boxplots leads to the display of dedicated menus for adapting further the visualization. For instance fish counts can be stacked by species or by camera, as shown in Figs. A.9, A.10.

Zone C contains filter widgets for both selecting datasets of interest, and overviewing datasets over several dimensions (e.g., proportion of videos from each camera or image quality). Filter widgets are displayed on-demand. There are widgets for each dimension of the data, namely: Year, Week of Year and Hour of Day of fish occurrence, Camera, Species, Certainty Score, Video Quality and Software Version. A summary of the filters applied is provided in *Zone B*. To limit information overload, the default filters (e.g., all species, all cameras) are not mentioned in the summary. The widgets histograms display the same metric as the main graph, and applied to the

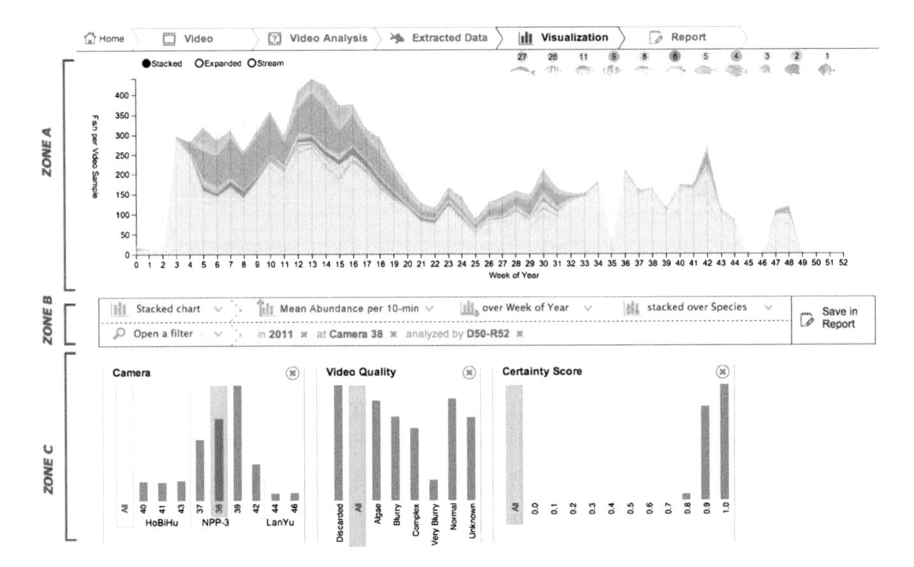

Fig. A.7 The visualization tab displaying a visualization of fish abundance over the year 2011, for the ecosystem observed at camera 38 (*Zone A*). Each species population is indicated with different colors, to show the relative population sizes. *Zone C* displays the distribution of fish over cameras, image quality, and *certainty score* (a measure of fish appearance quality indicating potential computer vision errors). Zones A and C use *Mean Abundance per 10-min* as the metric for fish abundance, to compensate for uneven numbers of video samples (Chap. 13 Sect. 13.4). *Zone B* indicates that visualizations in *Zone A* and *C* concern a dataset produced by the version *D50-R52* of computer vision software, and for the videos collected during 2011 at camera 38. Filters on all other dimensions are set to default, i.e., all species, weeks of year, hours of day, video qualities and certainty scores (Fig. A.8)

same dataset. For instance, in Fig. A.7 both the graph of *Zone A* and the histograms of *Zone C* display numbers of fish per video sample. Both use a dataset of fish detected by software version *D50-R52*, occurring in 2011 at Camera 38, and belonging to all species, certainty scores, image quality, weeks of year and hours of day. The *Camera* widget uses a dataset from all cameras, and highlights in blue which camera is selected. More filter widgets can be opened on-demand (e.g., to select species of interest, or data from a specific versions of the video analysis software). Figure A.8 shows all the filter widgets provided by the user interface.

The *Visualization tab* also supports the exploration of uncertainty due to missing videos, image quality or fish appearance quality. Videos can be missing due to camera maintenance, encoding errors, or unfinished processing queues. Figures A.9, A.10, A.11, A.12 show user interactions for exploring how variation in numbers of video impact fish counts, and how to explore fish abundance while abstracting from variations in video numbers. The quality of each fish appearance is measured using a *certainty score*, which can be displayed as a widget in *Zone C*. The score indicates how much fish look like the fish model for their species, which is used by computer vision software to determine fish species. The higher the score, the more certain is the species recognition.

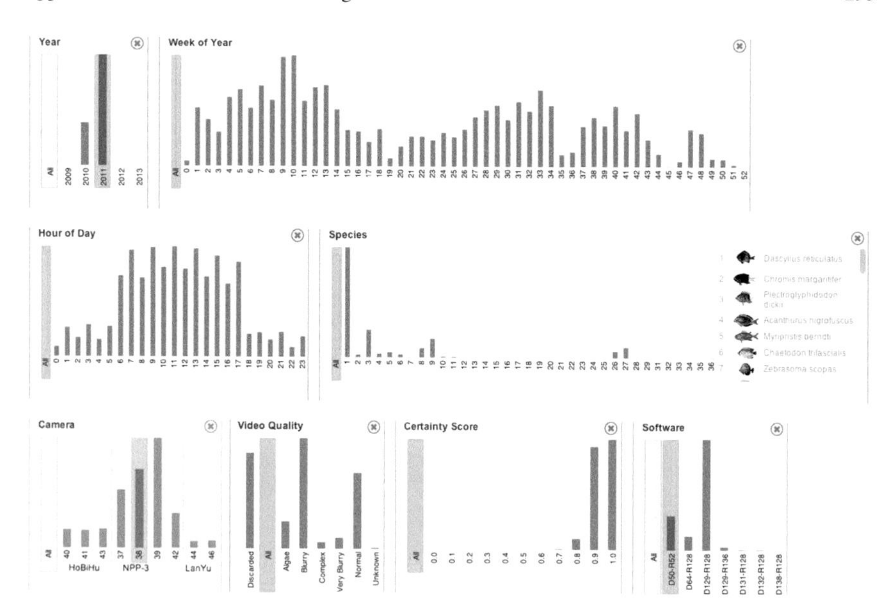

Fig. A.8 The *filter widgets* that enable the selection of the dataset of interest, i.e., the time period, location and other characteristics of the fish to study, as well as the versions of the software that produced the data. The histograms of the widgets provide an overview of various data distribution. The Y axis of the histograms represents the same metric as for the main visualization. This example shows the distribution of fish counts over several dimensions. We can see that only 4 versions of the software can provide fish counts for the periods and locations selected by users

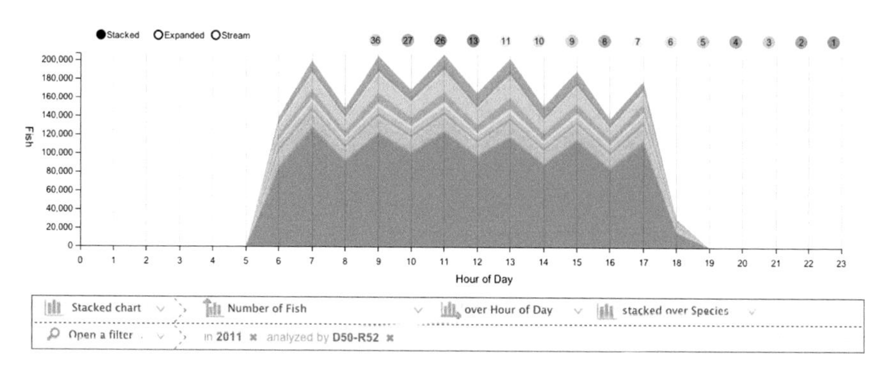

Fig. A.9 Visualizations of fish counts stacked by species, showing the total fish observed in 2011 at each hour of the day

The Report tab supports manual grouping and annotation of graphs created in the Visualization tab, as shown in Fig. A.13. Visualizations can be added to and removed from a *report*, and their interpretation can be described with a visualization title and a comment. Using the *Download* button, users can save the report they are currently working on. Downloaded reports consist of a text file containing a list of parameters. They can be stored or shared with other ecologists as any kind of text file. To visualize

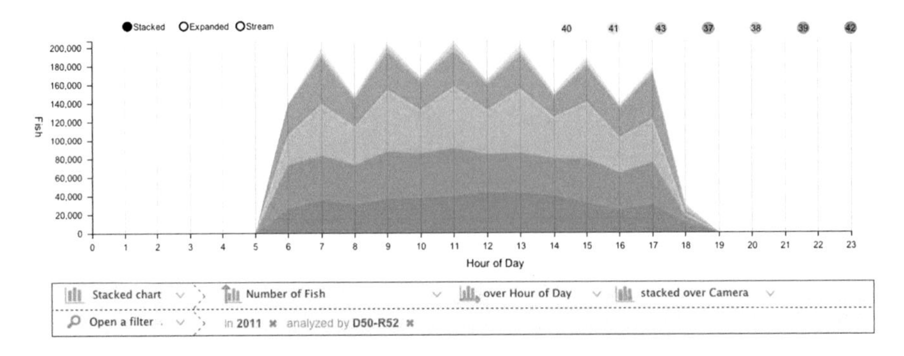

Fig. A.10 Visualizations of the same fish counts as in Fig. A.9 but stacked by camera

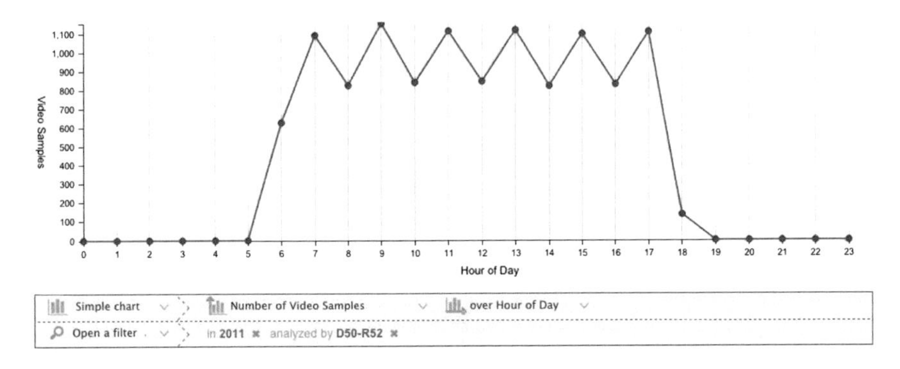

Fig. A.11 Visualization of the numbers of videos from which fish counts in Figs. A.9, A.10 were extracted

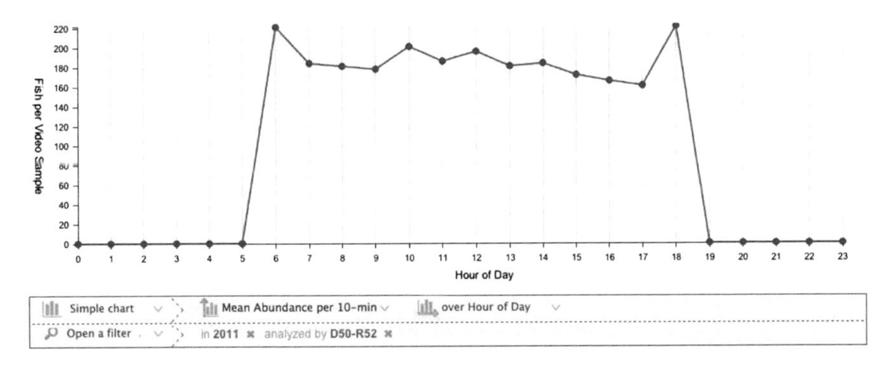

Fig. A.12 Visualization of the mean fish abundance per 10-min (Chap. 13 Sect. 13.4). It balances the variations of numbers of video and their impact on fish counts shown in Figs. A.9, A.10, A.11

a downloaded report, users can upload the parameter files with the *Upload* button of the *Report* tab. With this tab, ecologists can document their sense-making process and findings, and collaborate with each other.

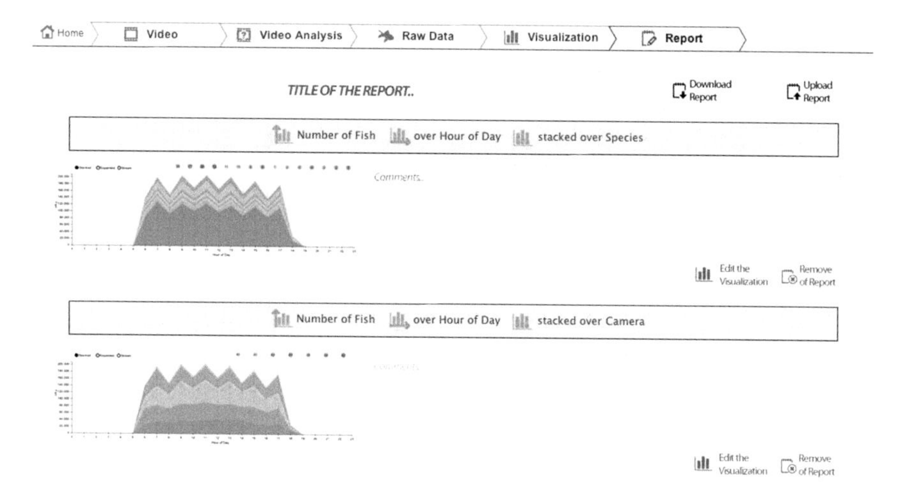

Fig. A.13 The report tab

A.2 Usage Scenario

This section describes the typical usage of the Fish4Knowledge tool, using personas and storytelling. As a sample user, we consider the profile of Erica who is an ecologist interested in the study of fish abundance in Taiwanese ecosystems. Her research focuses on cyclic evolutions of fish abundance over seasons, months and days. She wants to measure community size (i.e., the total number of fish regardless of the species) for different periods of time, and to visualize changes of fish counts over time using different time units (e.g., fish count for each day or each week). To assess her scientific findings, she needs to control and explain her data collection and data analysis process. She also needs data visualizations to illustrate her findings.

1. Exploring fish counts—Erica wants to measure the total abundance of fish, taking every species into account, for each week of the last 3 years. To do so, since numbers of video is likely to vary over weeks, she uses the built-in measurement of fish abundance expressed as average fish count per video. She indicates the time unit (i.e., week) and the period of interest (i.e., from January 2010 till December 2012), and the Fish4Knowledge interface provides her with a graphic visualization of fish abundance. Erica is surprised by the fish abundance in 2010, particularly for the month of March. She decides to explore data for that month only. She modifies the timeframe of study and the time unit to visualize a daily fish abundance for March 2010. Figures A.14 shows the initial visualization, overviewing fish counts for several years, and Fig. A.15 shows the next visualization, zooming in the fish counts for March 2010.

2. Controlling the visualization—Considering the surprising fish counts for March 2010, Erica does not trust the system for correctly analyzing the videos. She wants to check what the system has done to produce the fish counts. She opens

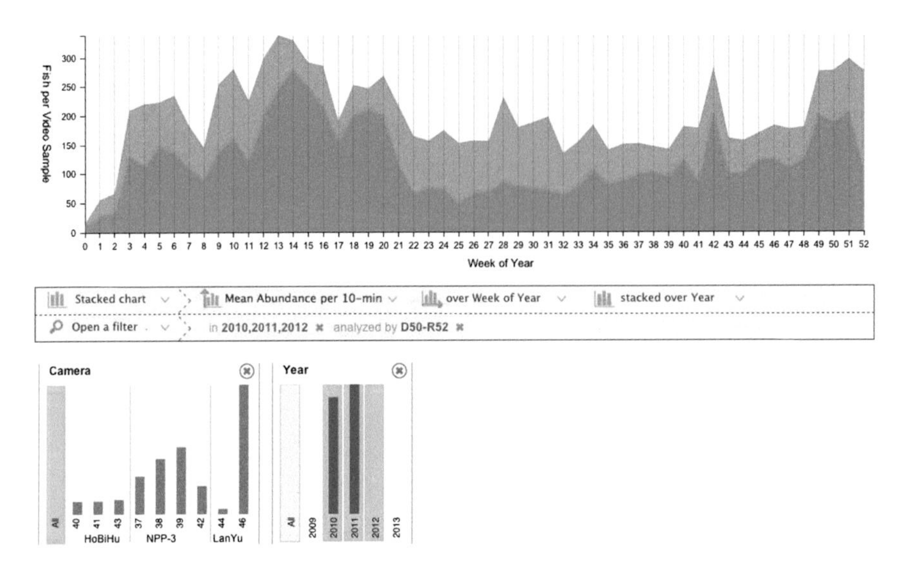

Fig. A.14 A Visualization of annual fish counts

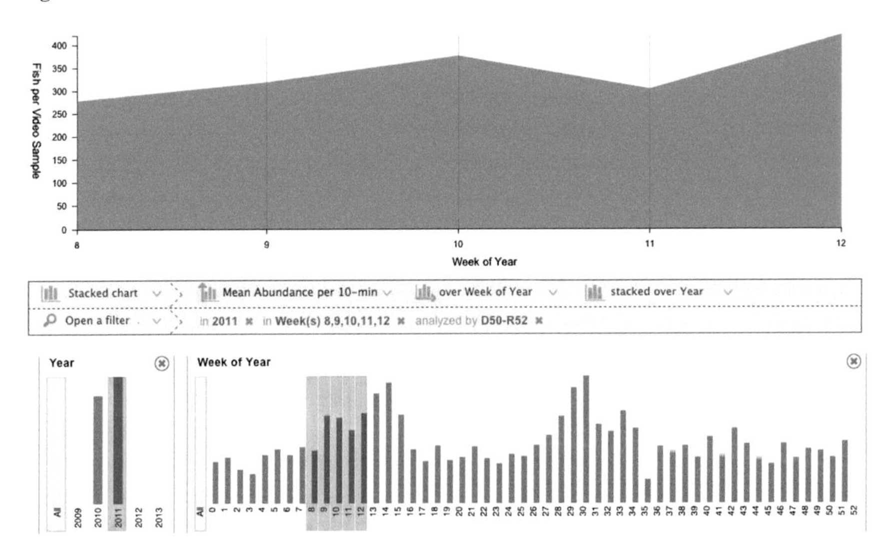

Fig. A.15 A Visualization of fish counts from March 2011

the series of tabs which guide her through the information processing steps. The *Video Analysis* tab (Fig. A.2) explains the computer vision processes that produced the fish counts she was visualizing. The *Extracted Data* tab (Fig. A.6) shows the characteristics of the extracted data. With these explanations, she understands that 4 main points can be investigated:

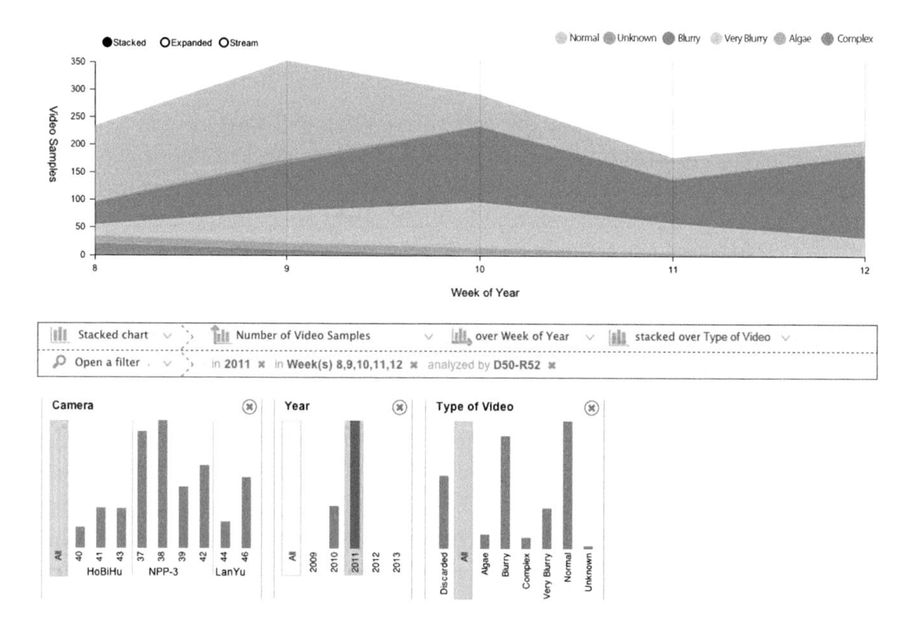

Fig. A.16 A Visualization of the numbers of video samples available for March 2011, and the proportions of videos from each image quality. The *lower* part of the interface shows the distribution of videos per camera, year, and type of video quality

- *Videos*: How many videos were taken into account, and what were their image quality?
- *Video Analysis*: What software components analyzed the videos and produced the fish counts?
- *Video Data*: What computer vision data were used for the visualization? What filters can be applied to select the most relevant information?
- *Visualization*: What other representations of the data allow to explore further the fish counts?

3. Controlling the videos—Erica wants to check if a homogeneous set of cameras and videos were available in 2010. She wants to investigate potential camera breakdown, and inconsistencies in image quality and fields of view. To do so, she opens the *Video* tab giving access to the video footage. She inspects the videos available for each camera, period of time, and image quality. She watches a few videos in 2010, and she even looks for video in 2011 to verify that the image quality and field of view is similar. From the *Visualization* tab, Erica can also control the video samples that produced the fish counts. She can visualize the variations in numbers of processed and unprocessed videos, and potential video scarcity. She can investigate the impact of numbers of video on raw numbers of fish, as shown in Figures A.9, A.10, A.11, A.12. She can also visualize the proportions of videos belonging to each type of image quality, as shown in Fig. A.16. Hence, she can control if a significant number of videos are missing or are of poor quality. For the dataset she is studying, the numbers of videos and their image quality are found consistent.

4. Controlling the video analysis software—Erica still wants to find out what makes the overall abundance different in 2010. She has checked that it is not due to inconsistencies in the set of videos, and she continues by checking the software components that analyzed the videos. She particularly looks for differences in the versions of the components. While inspecting the software components in the *Video Analysis* tab (e.g., Fig. A.3), she notices that the Fish Detection software, responsible for detecting fish among other objects, had a major version change. Also, the new software version produces much fewer computer vision errors compared to older versions. In the *Visualization* tab, she can verify that most of video analysis performed before September 2010 used older versions of the Fish Detection software. To check how this influenced the result, she launches the analysis of the videos before September 2010 with the newest software version. To do so, she uses the interface in the *Workflow* sub-tab shown in Fig. A.5. When the video analysis is redone, Erica inspects again the visualization. She sees that the differences between the fish count in 2010 and the other years remains the same.

5. Controlling the selected video data—Erica now wants to check what data was retrieved from the database. In *Raw Data* tab exposing the data schema (Fig. A.6), she notices an attribute called *Certainty Score* which contains a score representing the similarity of each fish with their species model. This score can be used as a threshold to select the fish to take into account, and filter out fish with higher chances of errors, i.e., fish with low scores. Erica decides to set a higher threshold, to discard fish with very low scores and see how it influences the results. She can set the threshold using the *Certainty Score* widget shown in Fig. A.7. When she visualizes again the fish abundance, she sees that the overall abundance has lower values, but there are no major changes in the trends, and the results for 2010 are still different from the other years. Erica also tries different *Certainty Score* thresholds, but there are still no major changes in the observed pattern of fish abundance.

6. Exploring further data visualizations—Erica did not find any technical reason to explain the unusual fish counts that appear for the year 2010. But now she knows exactly what is represented in the visualization, and she trusts the system. Thus Erica starts thinking about biological and environmental reasons that can explain the population dynamics in 2010. She wonders if the changes in 2010 occurred for the whole island of Taiwan, or if they were localized in a specific area. She investigates this hypothesis by visualizing fish abundance for various areas. Erica can visualize fish counts per location, by either using stacked charts as shown in Fig. A.9, or the *Camera* widget as in Fig. A.14, or by setting the x-axis of the main graph as representing each location. Erica can compare localized fish count by gathering visualizations for specific areas using the *Report* tab (e.g., Fig. A.13).

With such set of visualizations, she clearly finds out that the changes occurred in the southern part of Taiwan, and that differences in that area influenced the overall fish abundance she was visualizing before. She continues her investigation by selecting specific species to take into account in the visualization. Erica chooses a set of *indicative* species that are likely to react to specific pollutions and environmental events. To do so, she uses the *Species* widget, one of the filtering widget offered by the interface (Fig. A.8). Once she has studied the abundance of each species of interest, Erica concludes that the unusual fish abundance observed in March 2010 in southern part of Taiwan is possibly due to a chemical leak, or to a peak of water temperature.

7. Report and comment findings—To illustrate her findings, Erica creates a set of specifically customized visualizations. She gathers the visualizations of interest using the *Report* tab (e.g., Fig. A.13). She writes her comments and interpretations for each visualization. She downloads her report and sends it to other colleagues that can help her investigate further the kind of chemical leaks and environmental events that took place in March 2010.

A.3 Conclusion

Our interaction design let users specify which visualizations are relevant for their goal. Information of interest is displayed on-demand (e.g., open tabs, open widgets in *Zone C*, change graph axes, display details in stacked charts and boxplots), and hidden when no longer relevant (e.g., close widgets, switch back to simple graph). Our design supports a wide range of data analysis, using specific information among the many dimensions of data collected from videos, while limiting display cluttering and information overload. It addresses our context where ecologists pursue a variety of research goals, while being unfamiliar with computer vision data.

A.4 Acknowledgement

We are grateful for the valuable contributions of Elvira Arslanova; Prof. Shao, and his colleagues from Academia Sinica; Prof. Fan, Dr. Hai Ocean Chiang and their colleagues from the National Museum of Marine Science & Technology of Taiwan; Dr.Ir. Nagelkerke, Dr. Tulp and their colleagues from the IMARES Research Institute of Wageningen University; Dr. Lavaleye, Dr. Duineveld and their colleagues from the Royal Institute for Sea Research (NIOZ); Tiziano Perrucci and Martin van Harmelen.

Appendix B
Database Tables Related to F4K Workflow

Gayathri Nadarajan, Cheng-Lin Yang

See Fig. B.1.

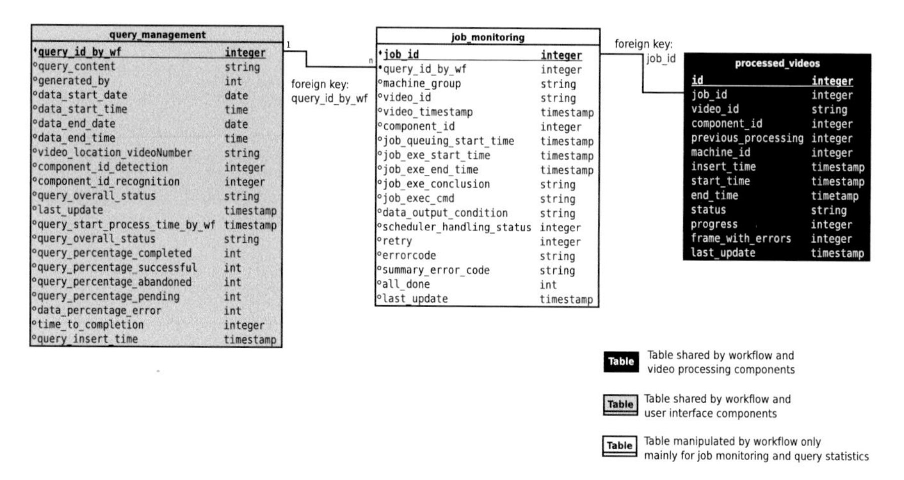

Fig. B.1 F4K Database tables manipulated and shared by the SWELL workflow system. The descriptions of *query_management* and *job_monitoring* tables are given in Figs. B.2 and B.3 respectively

R.B. Fisher et al. (eds.), *Fish4Knowledge: Collecting and Analyzing Massive Coral Reef Fish Video Data*, Intelligent Systems Reference Library 104,
DOI 10.1007/978-3-319-30208-9

Field	key	updated by	Data Type	Field Description
query _ID_by_WF	primary key	Database	integer	This is automatically generated and incremented by the database when CWI inserts a new query
query_content		CWI	string	query content is the task to be executed: inc. detect_track_fish, recognise_fish, execution_progress, abort.
generated_by		CWI/WF	integer	This indicates who generated the query, 1 indicates CWI and 0 indicates WF
data_start_date		CWI	date	Please provide a start date for the video to be processed, format is yyyy-mm-dd.
data_start_time		CWI	time	When not provided, the default is 06:00:00. Otherwise, please provide a start time for the video to be processed, e.g. 06:00:00 is 6am, and 14:00:00 is 2pm, etc. Please provide in hh:mm:ss
data_end_date		CWI	string	Please provide an end date for the video to be processed, format is yyyy-mm-dd.
data_end_time		CWI	date	When not provided, the default is 17:50:00. Otherwise, please provide a start time for the video to be processed, e.g. 06:00:00 is 6am, and 14:00:00 is 2pm, etc. Please provide in hh:mm:ss
video_location_videoNumber		CWI	time	This field indicates where the videos are shot. This is based on a combination of fields names of the 'cameras' table. These fields are called "Location" and "Video Number". Valid values are NPP-3/1, NPP-3/2, NPP-3/3, NPP-3/4, HoBiHu/1, HoBiHu/2, HoBiHu/3, LanYu/1, LanYu/2. However, the user can indicate more than one location/camera in one query. Therefore a valid query may also be, e.g. 'NPP-3/1,NPP-3/2,NPP-3/3,NPP-3/4' (with comma, no space between them).
component_id_detection	(this is an optional field)	CWI		This field allows CWI to specify which component ID to use. Currently, the user can specify the detection component ID. Please see the task mapping for task definition and useage for reference.
component_id_recognition	(this is an optional field)	CWI	integer	This field allows CWI to specify which component ID to use. Currently, the user can specify the recognition component ID. Please see the task mapping for task definition and useage for reference.
query_insert_time_by_CWI		CWI	time stamp	this field is updated when CWI made an insertion of a new query into the DB.
query_start_process_time_by_WF		WF engine	time stamp	this field is updated when WF engine picks up this query and start processing it.

Fig. B.2 Description of the main fields of the *query_management* table

Field	key	updated by	Data Type	Field Description
job_id	primary_key	WF engine	integer	job_id is obtained from the resource scheduler (LSF/SGE). Each query may spawn thousands of jobs.
query_id_by_WF	foreign_key	WF engine	integer	WF will create this ID when starting to process this query
machine group		WF engine	string	indicates the types of machines that run the job. Valid strings are 'VM' and 'WR'.
video_id		WF engine	string	this value is taken from the "video" table.
video_timestamp		WF engine	timestamp	Valid values are the time stamp which is the start time of the 10 minutes video clip.
component_id		WF engine	integer	This field denotes the component_id of the software module related to this job
Job_queuing_start_time		WF engine		the time a job is submitted to the scheduler.
job_exe_start_time		WF engine		the time a job starts to be executed.
job_exe_end_time		WF engine		the time a job finishes execution
job_exe_conclusion		WF Monitor		This field indicates the final job execution conclusion of a job. Valid values are "successful', 'pending', 'abandoned' and 'error'. Summation of these figures are used as input for the Query Mgmt Table.
job_exec_cmd		WF engine	string	This is the scheduler command used to run this job.
data_output_condition		WF Monitor	string	Field under consideration. Currently there is no direct way to retrieve this daya. To check the output stores in the fish DB, when a job is finished. If there is no appropriate entry in the fish DB. The value = missing; otherwise the value = ok.
scheduler_handling_status		WF engine/ WF Monitor	integer	This field is initially inserted by WF engine (when it is pending). It is then later on updated by the WF Monitor for dynamic job execution status. This field describes the status of a job. Example values from LSF scheduler are pending (waiting in queue), running, done (job completed normally), psusp (job suspended while in pending status), ususp (job suspended by user), ssusp (job suspended by LSF system while running), etc.
retry		WF Monitor		This field indicates the number of times this job as been retried
errorcode		WF Monitor		This field is defined in the error_code field in the "F4K WF Status Monitoring and Error Handling" definition table.
summary_errorcode		WF Monitor		This field is defined in the error_code field in the "F4K WF Status Monitoring and Error Handling" definition table.
time_to_completion		WF Monitor	integer	This indicates additional time needed for query completion. This is calculated by multipling the time needed/module/video and percentage still need to be completed. E.g. 60 minutes on average/module x/video * 50% still needs to be done = 30 minutes still needs to be done. time_to_completion = 30.
all_done		WF Monitor	integer	
last_update		WF engine	timestamp	

Fig. B.3 Description of the main fields of the *job_monitoring* table

Appendix C
F4K Database Schema

Hsiu-Mei Chou

The full SQL database is illustrated in Figs. C.1 and C.2 and given in detail in the following text.

```
-- MySQL Script generated by MySQL Workbench
-- 11/10/15 10:20:24
-- Model: New Model    Version: 1.0
-- MySQL Workbench Forward Engineering

SET @OLD_UNIQUE_CHECKS=@@UNIQUE_CHECKS, UNIQUE_CHECKS=0;
SET @OLD_FOREIGN_KEY_CHECKS=@@FOREIGN_KEY_CHECKS,
   FOREIGN_KEY_CHECKS=0;
SET @OLD_SQL_MODE=@@SQL_MODE, SQL_MODE='TRADITIONAL,
   ALLOW_INVALID_DATES';

-- -----------------------------------------------------
-- Schema f4k_db
-- -----------------------------------------------------
CREATE SCHEMA IF NOT EXISTS `f4k_db` DEFAULT CHARACTER SET latin1 ;
USE `f4k_db` ;

-- -----------------------------------------------------
-- Table `f4k_db`.`algorithm`
-- -----------------------------------------------------
CREATE TABLE IF NOT EXISTS `f4k_db`.`algorithm` (
  `algorithm_id` INT(11) NOT NULL,
  `name` VARCHAR(45) NULL DEFAULT NULL,
  `type` VARCHAR(45) NULL DEFAULT NULL,
  `description` TEXT NULL DEFAULT NULL,
  `version` VARCHAR(10) NULL DEFAULT NULL,
```


R.B. Fisher et al. (eds.), *Fish4Knowledge: Collecting and Analyzing Massive Coral Reef Fish Video Data*, Intelligent Systems Reference Library 104,
DOI 10.1007/978-3-319-30208-9

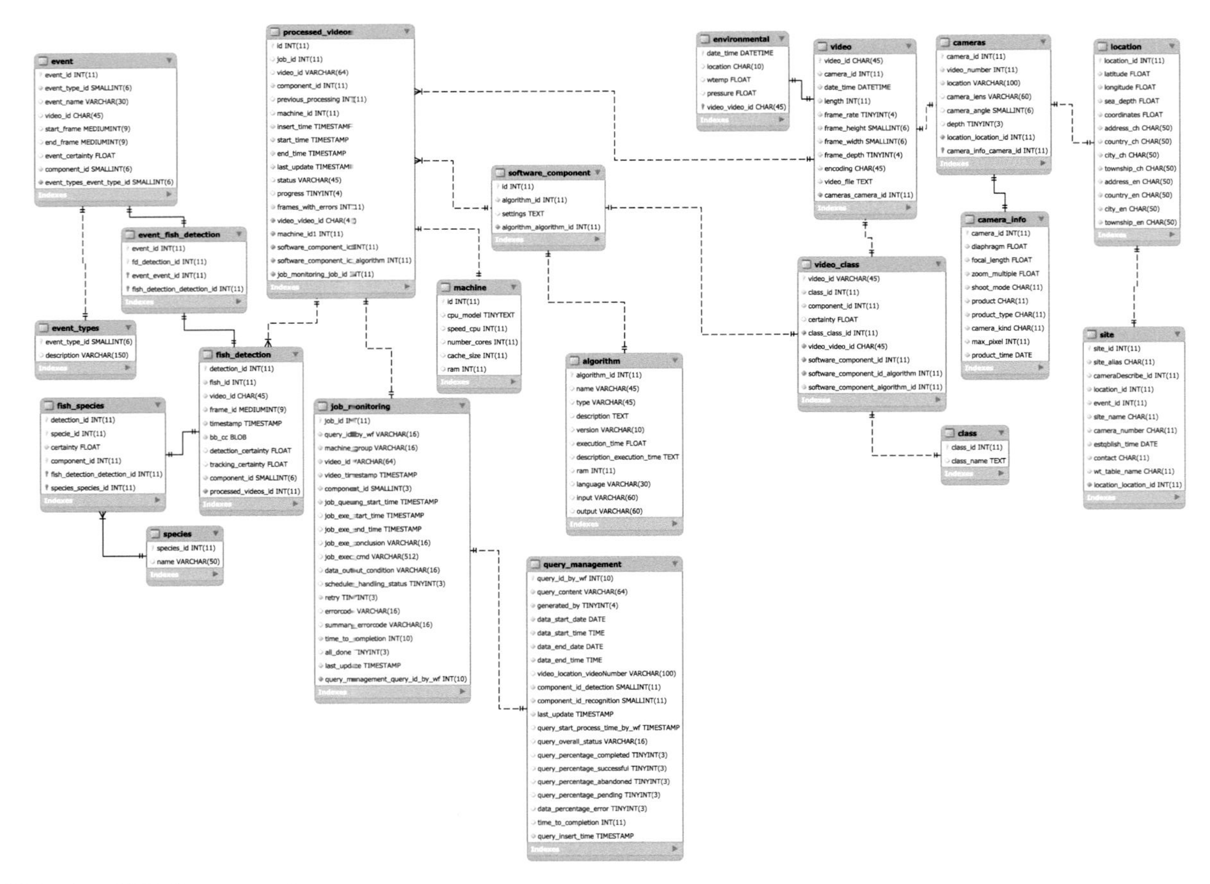

Fig. C.1 F4K database EER diagram part 1

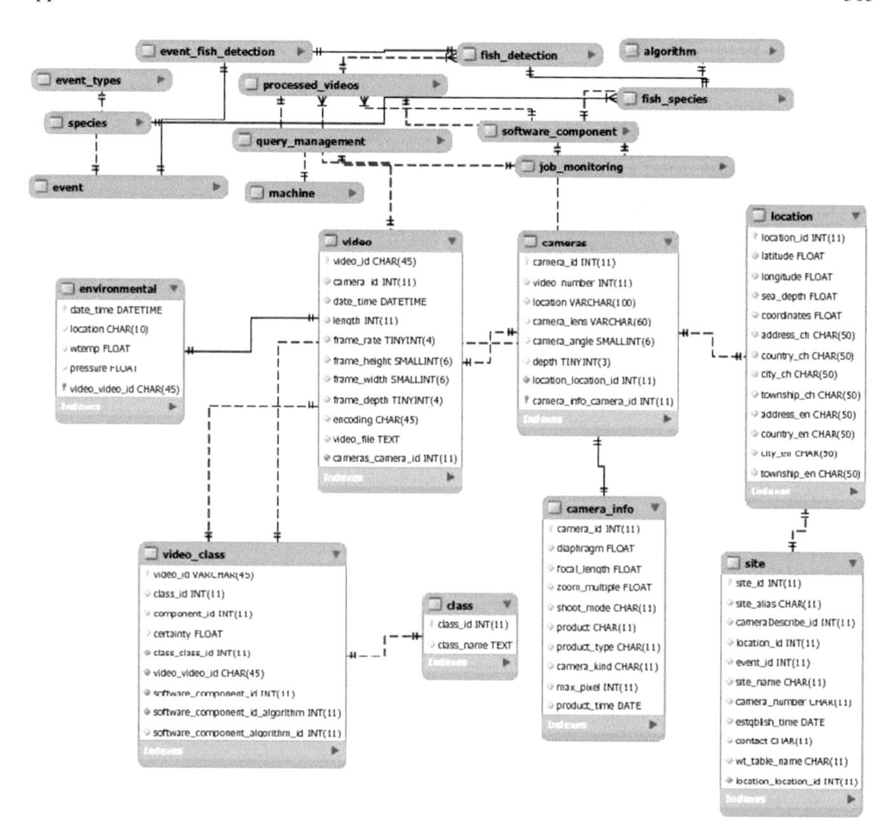

Fig. C.2 F4K database EER diagram part 2

```
  `execution_time` FLOAT NULL DEFAULT NULL,
  `description_execution_time` TEXT NULL DEFAULT NULL,
  `ram` INT(11) NULL DEFAULT NULL,
  `language` VARCHAR(30) NULL DEFAULT NULL,
  `input` VARCHAR(60) NULL DEFAULT NULL,
  `output` VARCHAR(60) NULL DEFAULT NULL,
  PRIMARY KEY (`algorithm_id`))
ENGINE = MyISAM
AUTO_INCREMENT = 58
DEFAULT CHARACTER SET = latin1;

-- -----------------------------------------------------
-- Table `f4k_db`.`camera_info`
-- -----------------------------------------------------
CREATE TABLE IF NOT EXISTS `f4k_db`.`camera_info` (
  `camera_id` INT(11) NOT NULL AUTO_INCREMENT,
  `diaphragm` FLOAT NOT NULL,
  `focal_length` FLOAT NOT NULL,
```

```
  'zoom_multiple' FLOAT NOT NULL,
  'shoot_mode' CHAR(11) NOT NULL,
  'product' CHAR(11) NOT NULL,
  'product_type' CHAR(11) NOT NULL,
  'camera_kind' CHAR(11) NOT NULL,
  'max_pixel' INT(11) NOT NULL,
  'product_time' DATE NOT NULL,
  PRIMARY KEY ('camera_id'))
ENGINE = MyISAM
DEFAULT CHARACTER SET = latin1;


-- -----------------------------------------------------
-- Table 'f4k_db'.'location'
-- -----------------------------------------------------
CREATE TABLE IF NOT EXISTS 'f4k_db'.'location' (
  'location_id' INT(11) NOT NULL AUTO_INCREMENT,
  'latitude' FLOAT NOT NULL,
  'longitude' FLOAT NOT NULL,
  'sea_depth' FLOAT NOT NULL,
  'coordinates' FLOAT NOT NULL,
  'address_ch' CHAR(50) NOT NULL,
  'country_ch' CHAR(50) NOT NULL,
  'city_ch' CHAR(50) NOT NULL,
  'township_ch' CHAR(50) NOT NULL,
  'address_en' CHAR(50) NOT NULL,
  'country_en' CHAR(50) NOT NULL,
  'city_en' CHAR(50) NOT NULL,
  'township_en' CHAR(50) NOT NULL,
  PRIMARY KEY ('location_id'))
ENGINE = MyISAM
AUTO_INCREMENT = 7
DEFAULT CHARACTER SET = latin1;


-- -----------------------------------------------------
-- Table 'f4k_db'.'cameras'
-- -----------------------------------------------------
CREATE TABLE IF NOT EXISTS 'f4k_db'.'cameras' (
  'camera_id' INT(11) NOT NULL AUTO_INCREMENT,
  'video_number' INT(11) NOT NULL,
  'location' VARCHAR(100) NOT NULL DEFAULT '',
  'camera_lens' VARCHAR(60) NULL DEFAULT NULL,
  'camera_angle' SMALLINT(6) NULL DEFAULT NULL,
  'depth' TINYINT(3) NULL DEFAULT NULL,
  'location_location_id' INT(11) NOT NULL,
  'camera_info_camera_id' INT(11) NOT NULL,
  PRIMARY KEY ('camera_id', 'camera_info_camera_id'),
  UNIQUE INDEX 'location' ('location' ASC, 'video_number' ASC),
```

```
  INDEX 'fk_cameras_location1_idx' ('location_location_id' ASC),
  INDEX 'fk_cameras_camera_info1_idx' ('camera_info_camera_id'
     ASC))
ENGINE = MyISAM
AUTO_INCREMENT = 47
DEFAULT CHARACTER SET = latin1;


-- -------------------------------------------------------
-- Table 'f4k_db'.'class'
-- -------------------------------------------------------
CREATE TABLE IF NOT EXISTS 'f4k_db'.'class' (
  'class_id' INT(11) UNSIGNED NOT NULL AUTO_INCREMENT,
  'class_name' TEXT NULL DEFAULT NULL,
  PRIMARY KEY ('class_id'))
ENGINE = InnoDB
AUTO_INCREMENT = 8
DEFAULT CHARACTER SET = latin1;

-- -------------------------------------------------------
-- Table 'f4k_db'.'video'
-- -------------------------------------------------------
CREATE TABLE IF NOT EXISTS 'f4k_db'.'video' (
  'video_id' CHAR(45) NOT NULL,
  'camera_id' INT(11) NOT NULL,
  'date_time' DATETIME NOT NULL,
  'length' INT(11) NOT NULL,
  'frame_rate' TINYINT(4) NOT NULL,
  'frame_height' SMALLINT(6) NOT NULL,
  'frame_width' SMALLINT(6) NOT NULL,
  'frame_depth' TINYINT(4) NOT NULL,
  'encoding' CHAR(45) NOT NULL,
  'video_file' TEXT NOT NULL,
  'cameras_camera_id' INT(11) NOT NULL,
  PRIMARY KEY ('video_id'),
  INDEX 'camid' ('camera_id' ASC),
  INDEX 'dt' ('date_time' ASC),
  INDEX 'fk_video_cameras_idx' ('cameras_camera_id' ASC))
ENGINE = MyISAM
DEFAULT CHARACTER SET = latin1;


-- -------------------------------------------------------
-- Table 'f4k_db'.'environmental'
-- -------------------------------------------------------
CREATE TABLE IF NOT EXISTS 'f4k_db'.'environmental' (
  'date_time' DATETIME NOT NULL DEFAULT '0000-00-00 00:00:00',
  'location' CHAR(10) NULL DEFAULT NULL,
  'wtemp' FLOAT NULL DEFAULT NULL,
```

```
  'pressure' FLOAT NULL DEFAULT NULL,
  'video_video_id' CHAR(45) NOT NULL,
  PRIMARY KEY ('date_time', 'video_video_id'),
  INDEX 'fk_environmental_video1_idx' ('video_video_id' ASC))
ENGINE = MyISAM
DEFAULT CHARACTER SET = latin1;


-- -----------------------------------------------------
-- Table 'f4k_db'.'event_types'
-- -----------------------------------------------------
CREATE TABLE IF NOT EXISTS 'f4k_db'.'event_types' (
  'event_type_id' SMALLINT(6) NOT NULL AUTO_INCREMENT,
  'description' VARCHAR(150) NULL DEFAULT NULL,
  PRIMARY KEY ('event_type_id'))
ENGINE = MyISAM
AUTO_INCREMENT = 456
DEFAULT CHARACTER SET = latin1;


-- -----------------------------------------------------
-- Table 'f4k_db'.'event'
-- -----------------------------------------------------
CREATE TABLE IF NOT EXISTS 'f4k_db'.'event' (
  'event_id' INT(11) NOT NULL AUTO_INCREMENT,
  'event_type_id' SMALLINT(6) NOT NULL,
  'event_name' VARCHAR(30) NULL DEFAULT NULL,
  'video_id' CHAR(45) CHARACTER SET 'utf8' NULL DEFAULT NULL,
  'start_frame' MEDIUMINT(9) NULL DEFAULT NULL,
  'end_frame' MEDIUMINT(9) NULL DEFAULT NULL,
  'event_certainty' FLOAT NULL DEFAULT NULL,
  'component_id' SMALLINT(6) NOT NULL,
  'event_types_event_type_id' SMALLINT(6) NOT NULL,
  PRIMARY KEY ('event_id'),
  INDEX 'fk_event_event_types1_idx' ('event_types_event_type_id'
      ASC))
ENGINE = MyISAM
AUTO_INCREMENT = 4189
DEFAULT CHARACTER SET = latin1;


-- -----------------------------------------------------
-- Table 'f4k_db'.'machine'
-- -----------------------------------------------------
CREATE TABLE IF NOT EXISTS 'f4k_db'.'machine' (
  'id' INT(11) UNSIGNED NOT NULL AUTO_INCREMENT,
  'cpu_model' TINYTEXT NULL DEFAULT NULL,
  'speed_cpu' INT(11) NULL DEFAULT NULL,
  'number_cores' INT(11) NULL DEFAULT NULL,
```

```
  'cache_size' INT(11) NULL DEFAULT NULL,
  'ram' INT(11) NULL DEFAULT NULL,
  PRIMARY KEY ('id'))
ENGINE = MyISAM
AUTO_INCREMENT = 30
DEFAULT CHARACTER SET = latin1;


-- -----------------------------------------------------
-- Table 'f4k_db'.'software_component'
-- -----------------------------------------------------
CREATE TABLE IF NOT EXISTS 'f4k_db'.'software_component' (
  'id' INT(11) UNSIGNED NOT NULL,
  'algorithm_id' INT(11) NOT NULL DEFAULT '0',
  'settings' TEXT NULL DEFAULT NULL,
  'algorithm_algorithm_id' INT(11) NOT NULL,
  PRIMARY KEY ('id'),
  INDEX 'fk_software_component_algorithm1_idx'
     ('algorithm_algorithm_id' ASC))
ENGINE = MyISAM
DEFAULT CHARACTER SET = latin1;


-- -----------------------------------------------------
-- Table 'f4k_db'.'query_management'
-- -----------------------------------------------------
CREATE TABLE IF NOT EXISTS 'f4k_db'.'query_management' (
  'query_id_by_wf' INT(10) NOT NULL AUTO_INCREMENT,
  'query_content' VARCHAR(64) NOT NULL DEFAULT '',
  'generated_by' TINYINT(4) UNSIGNED NOT NULL DEFAULT '1',
  'data_start_date' DATE NOT NULL DEFAULT '0000-00-00',
  'data_start_time' TIME NOT NULL DEFAULT '06:00:00',
  'data_end_date' DATE NOT NULL DEFAULT '0000-00-00',
  'data_end_time' TIME NOT NULL DEFAULT '17:50:00',
  'video_location_videoNumber' VARCHAR(100) NULL DEFAULT NULL,
  'component_id_detection' SMALLINT(11) NOT NULL DEFAULT '0',
  'component_id_recognition' SMALLINT(11) NOT NULL DEFAULT '0',
  'last_update' TIMESTAMP NOT NULL DEFAULT CURRENT_TIMESTAMP ON
     UPDATE CURRENT_TIMESTAMP,
  'query_start_process_time_by_wf' TIMESTAMP NULL DEFAULT
     '0000-00-00 00:00:00',
  'query_overall_status' VARCHAR(16) NULL DEFAULT 'pending',
  'query_percentage_completed' TINYINT(3) UNSIGNED NULL DEFAULT '0',
  'query_percentage_successful' TINYINT(3) UNSIGNED NULL DEFAULT '0',
  'query_percentage_abandoned' TINYINT(3) UNSIGNED NULL DEFAULT '0',
  'query_percentage_pending' TINYINT(3) UNSIGNED NULL DEFAULT '0',
  'data_percentage_error' TINYINT(3) UNSIGNED NULL DEFAULT '0',
  'time_to_completion' INT(11) UNSIGNED NULL DEFAULT '0',
  'query_insert_time' TIMESTAMP NOT NULL DEFAULT '0000-00-00 00:00:00',
```

```
  PRIMARY KEY ('query_id_by_wf'))
ENGINE = InnoDB
AUTO_INCREMENT = 155
DEFAULT CHARACTER SET = utf8;


-- -----------------------------------------------------
-- Table 'f4k_db'.'job_monitoring'
-- -----------------------------------------------------
CREATE TABLE IF NOT EXISTS 'f4k_db'.'job_monitoring' (
  'job_id' INT(11) UNSIGNED NOT NULL,
  'query_id_by_wf' VARCHAR(16) NOT NULL DEFAULT '',
  'machine_group' VARCHAR(16) NOT NULL DEFAULT 'WR',
  'video_id' VARCHAR(64) NOT NULL,
  'video_timestamp' TIMESTAMP NOT NULL DEFAULT '0000-00-00 00:00:00',
  'component_id' SMALLINT(3) UNSIGNED NOT NULL,
  'job_queuing_start_time' TIMESTAMP NOT NULL DEFAULT '0000-00-00
      00:00:00',
  'job_exe_start_time' TIMESTAMP NULL DEFAULT '0000-00-00 00:00:00',
  'job_exe_end_time' TIMESTAMP NULL DEFAULT '0000-00-00 00:00:00',
  'job_exe_conclusion' VARCHAR(16) NULL DEFAULT '',
  'job_exec_cmd' VARCHAR(512) NULL DEFAULT NULL,
  'data_output_condition' VARCHAR(16) NULL DEFAULT '',
  'scheduler_handling_status' TINYINT(3) UNSIGNED NULL DEFAULT '0',
  'retry' TINYINT(3) UNSIGNED NOT NULL DEFAULT '0',
  'errorcode' VARCHAR(16) NULL DEFAULT NULL,
  'summary_errorcode' VARCHAR(16) NULL DEFAULT NULL,
  'time_to_completion' INT(10) UNSIGNED NOT NULL DEFAULT '0',
  'all_done' TINYINT(3) UNSIGNED NULL DEFAULT '0',
  'last_update' TIMESTAMP NOT NULL DEFAULT CURRENT_TIMESTAMP ON
      UPDATE CURRENT_TIMESTAMP,
  'query_management_query_id_by_wf' INT(10) NOT NULL,
  PRIMARY KEY ('job_id'),
  INDEX 'fk_job_monitoring_query_management_idx'
      ('query_management_query_id_by_wf' ASC),
  CONSTRAINT 'fk_job_monitoring_query_management'
    FOREIGN KEY ('query_management_query_id_by_wf')
    REFERENCES 'f4k_db'.'query_management' ('query_id_by_wf')
    ON DELETE NO ACTION
    ON UPDATE NO ACTION)
ENGINE = InnoDB
DEFAULT CHARACTER SET = utf8;


-- -----------------------------------------------------
-- Table 'f4k_db'.'processed_videos'
-- -----------------------------------------------------
CREATE TABLE IF NOT EXISTS 'f4k_db'.'processed_videos' (
  'id' INT(11) NOT NULL AUTO_INCREMENT,
```

```
  'job_id' INT(11) NULL DEFAULT '-1',
  'video_id' VARCHAR(64) NULL DEFAULT NULL,
  'component_id' INT(11) NOT NULL,
  'previous_processing' INT(11) NULL DEFAULT '0',
  'machine_id' INT(11) NULL DEFAULT NULL,
  'insert_time' TIMESTAMP NOT NULL DEFAULT '0000-00-00 00:00:00',
  'start_time' TIMESTAMP NOT NULL DEFAULT '0000-00-00 00:00:00',
  'end_time' TIMESTAMP NOT NULL DEFAULT '0000-00-00 00:00:00',
  'last_update' TIMESTAMP NOT NULL DEFAULT CURRENT_TIMESTAMP ON
      UPDATE CURRENT_TIMESTAMP,
  'status' VARCHAR(45) NULL DEFAULT NULL,
  'progress' TINYINT(4) NULL DEFAULT NULL,
  'frames_with_errors' INT(11) NOT NULL DEFAULT '0',
  'video_video_id' CHAR(45) NOT NULL,
  'machine_id1' INT(11) UNSIGNED NOT NULL,
  'software_component_id' INT(11) UNSIGNED NOT NULL,
  'software_component_id_algorithm' INT(11) NOT NULL,
  'job_monitoring_job_id' INT(11) UNSIGNED NOT NULL,
  PRIMARY KEY ('id'),
  INDEX 'vid' ('video_id' ASC),
  INDEX 'compntid' ('component_id' ASC),
  INDEX 'pvstatus' ('status' ASC),
  INDEX 'fk_processed_videos_video1_idx' ('video_video_id' ASC),
  INDEX 'fk_processed_videos_machine1_idx' ('machine_id1' ASC),
  INDEX 'fk_processed_videos_software_component1_idx'
      ('software_component_id' ASC, 'software_component_id_algorithm'
       ASC),
  INDEX 'fk_processed_videos_job_monitoring1_idx'
      ('job_monitoring_job_id' ASC))
ENGINE = MyISAM
AUTO_INCREMENT = 1181716
DEFAULT CHARACTER SET = latin1;


-- -----------------------------------------------------
-- Table 'f4k_db'.'fish_detection'
-- -----------------------------------------------------
CREATE TABLE IF NOT EXISTS 'f4k_db'.'fish_detection' (
  'detection_id' INT(11) NOT NULL AUTO_INCREMENT,
  'fish_id' INT(11) NOT NULL,
  'video_id' CHAR(45) CHARACTER SET 'utf8' NOT NULL DEFAULT '',
  'frame_id' MEDIUMINT(9) NOT NULL DEFAULT '0',
  'timestamp' TIMESTAMP NOT NULL DEFAULT CURRENT_TIMESTAMP ON
      UPDATE CURRENT_TIMESTAMP,
  'bb_cc' BLOB NOT NULL,
  'detection_certainty' FLOAT NULL DEFAULT NULL,
  'tracking_certainty' FLOAT NULL DEFAULT NULL,
  'component_id' SMALLINT(6) NOT NULL,
  'processed_videos_id' INT(11) NOT NULL,
```

```
  PRIMARY KEY (`detection_id`),
  INDEX `fish_id` (`fish_id` ASC),
  INDEX `frame_id` (`frame_id` ASC),
  INDEX `video_id` (`video_id` ASC),
  INDEX `fk_fish_detection_processed_videos1_idx`
      (`processed_videos_id` ASC))
ENGINE = MyISAM
AUTO_INCREMENT = 1446725357
DEFAULT CHARACTER SET = latin1;


-- -----------------------------------------------------
-- Table `f4k_db`.`event_fish_detection`
-- -----------------------------------------------------
CREATE TABLE IF NOT EXISTS `f4k_db`.`event_fish_detection` (
  `event_id` INT(11) NOT NULL,
  `fd_detection_id` INT(11) NOT NULL,
  `event_event_id` INT(11) NOT NULL,
  `fish_detection_detection_id` INT(11) NOT NULL,
  PRIMARY KEY (`event_id`, `fd_detection_id`, `event_event_id`,
      `fish_detection_detection_id`),
  INDEX `fk_event_fish_detection_event1_idx` (`event_event_id` ASC),
  INDEX `fk_event_fish_detection_fish_detection1_idx`
      (`fish_detection_detection_id` ASC))
ENGINE = MyISAM
DEFAULT CHARACTER SET = utf8;


-- -----------------------------------------------------
-- Table `f4k_db`.`species`
-- -----------------------------------------------------
CREATE TABLE IF NOT EXISTS `f4k_db`.`species` (
  `species_id` INT(11) NOT NULL AUTO_INCREMENT,
  `name` VARCHAR(50) NULL DEFAULT NULL,
  PRIMARY KEY (`species_id`),
  UNIQUE INDEX `name` (`name` ASC))
ENGINE = MyISAM
AUTO_INCREMENT = 39
DEFAULT CHARACTER SET = latin1;


-- -----------------------------------------------------
-- Table `f4k_db`.`fish_species`
-- -----------------------------------------------------
CREATE TABLE IF NOT EXISTS `f4k_db`.`fish_species` (
  `detection_id` INT(11) NOT NULL,
  `specie_id` INT(11) NOT NULL,
  `certainty` FLOAT NOT NULL,
  `component_id` INT(11) NOT NULL,
```

```
  'fish_detection_detection_id' INT(11) NOT NULL,
  'species_species_id' INT(11) NOT NULL,
  PRIMARY KEY ('detection_id', 'specie_id', 'component_id',
      'fish_detection_detection_id', 'species_species_id'),
  INDEX 'fk_fish_species_fish_detection1_idx'
      ('fish_detection_detection_id' ASC),
  INDEX 'fk_fish_species_species1_idx' ('species_species_id' ASC))
ENGINE = MyISAM
DEFAULT CHARACTER SET = latin1;


-- -----------------------------------------------------
-- Table 'f4k_db'.'site'
-- -----------------------------------------------------
CREATE TABLE IF NOT EXISTS 'f4k_db'.'site' (
  'site_id' INT(11) NOT NULL AUTO_INCREMENT,
  'site_alias' CHAR(11) NOT NULL,
  'cameraDescribe_id' INT(11) NOT NULL,
  'location_id' INT(11) NOT NULL,
  'event_id' INT(11) NOT NULL,
  'site_name' CHAR(11) NOT NULL,
  'camera_number' CHAR(11) NOT NULL,
  'estqblish_time' DATE NOT NULL,
  'contact' CHAR(11) NOT NULL,
  'wt_table_name' CHAR(11) NOT NULL,
  'location_location_id' INT(11) NOT NULL,
  PRIMARY KEY ('site_id'),
  INDEX 'fk_site_location1_idx' ('location_location_id' ASC))
ENGINE = MyISAM
DEFAULT CHARACTER SET = latin1;

-- -----------------------------------------------------
-- Table 'f4k_db'.'video_class'
-- -----------------------------------------------------
CREATE TABLE IF NOT EXISTS 'f4k_db'.'video_class' (
  'video_id' VARCHAR(45) NOT NULL,
  'class_id' INT(11) NOT NULL,
  'component_id' INT(11) NOT NULL,
  'certainty' FLOAT NULL DEFAULT NULL,
  'class_class_id' INT(11) UNSIGNED NOT NULL,
  'video_video_id' CHAR(45) NOT NULL,
  'software_component_id' INT(11) UNSIGNED NOT NULL,
  'software_component_id_algorithm' INT(11) NOT NULL,
  'software_component_algorithm_id' INT(11) NOT NULL,
  PRIMARY KEY ('video_id'),
  INDEX 'fk_video_class_class1_idx' ('class_class_id' ASC),
  INDEX 'fk_video_class_video1_idx' ('video_video_id' ASC),
  INDEX 'fk_video_class_software_component1_idx'
      ('software_component_id'
```

```
      ASC, 'software_component_id_algorithm'
      ASC, 'software_component_algorithm_id' ASC),
  CONSTRAINT 'fk_video_class_class1'
    FOREIGN KEY ('class_class_id')
    REFERENCES 'f4k_db'.'class' ('class_id')
    ON DELETE NO ACTION
    ON UPDATE NO ACTION,
  CONSTRAINT 'fk_video_class_video1'
    FOREIGN KEY ('video_video_id')
    REFERENCES 'f4k_db'.'video' ('video_id')
    ON DELETE NO ACTION
    ON UPDATE NO ACTION,
  CONSTRAINT 'fk_video_class_software_component1'
    FOREIGN KEY ('software_component_id' ,
      'software_component_id_algorithm')
    REFERENCES 'f4k_db'.'software_component' ('id' ,
      'algorithm_id')
    ON DELETE NO ACTION
    ON UPDATE NO ACTION)
ENGINE = InnoDB
DEFAULT CHARACTER SET = latin1;


SET SQL_MODE=@OLD_SQL_MODE;
SET FOREIGN_KEY_CHECKS=@OLD_FOREIGN_KEY_CHECKS;
SET UNIQUE_CHECKS=@OLD_UNIQUE_CHECKS;
```

Index

R.B. Fisher et al. (eds.), *Fish4Knowledge: Collecting and Analyzing Massive Coral Reef Fish Video Data*, Intelligent Systems Reference Library 104,
DOI 10.1007/978-3-319-30208-9

W

Printed in the United States
By Bookmasters